INFORMATION, CODING AND MATHEMATICS

THE KLUWER INTERNATIONAL SERIES IN ENGINEERING AND COMPUTER SCIENCE

COMMUNICATIONS AND INFORMATION THEORY

Consulting Editor

Robert Gallager

Other books in the series:

CODES, GRAPHS, AND SYSTEMS, edited by Richard E. Blahut and Ralf Koetter, ISBN: 0-7923-7686-2

CODES, CURVES AND SIGNALS: ***Common Threads in Communications***, edited by Alexander Vardy; ISBN: 0-7923-8374-5

PERSPECTIVES IN SPREAD SPECTRUM, Amer A. Hassan, John E. Hershey, and Gary J. Saulnier; ISBN: 0-7923-8265-X

WIRELESS PERSONAL COMMUNICATIONS: ***Advances in Coverage and Capacity***, Jeffrey H. Reed, Theodore S. Rappaport, Brian D. Woerner; ISBN: 0-7923-9788-6

ASYMPTOTIC COMBINATORIAL CODING THEORY, Volodia Blinovsky; ISBN: 0-7923-9988-9

PERSONAL AND WIRELESS COMMUNICATIONS: ***Digital Technology and Standards,*** Kun Il Park; ISBN: 0-7923-9727-4

WIRELESS INFORMATION NETWORKS: ***Architecture, Resource Managment, and Mobile Data,*** Jack M. Holtzman; ISBN: 0-7923-9694-4

DIGITAL IMAGE COMPRESSION: ***Algorithms and Standards***, Weidong Kou; ISBN: 0-7923-9626-X

CONTROL AND PERFORMANCE IN PACKET, CIRCUIT, AND ATM NETWORKS, XueDao Gu, Kazem Sohraby and Dhadesugoor R. Vaman; ISBN: 0-7923-9625-1

DISCRETE STOCHASTIC PROCESSES, Robert G. Gallager; ISBN: 0-7923-9583-2

WIRELESS PERSONAL COMMUNICATIONS: ***Research Developments***, Brian D. Woerner, Theodore S. Rappaport and Jeffrey H. Reed; ISBN: 0-7923-9555-7

PLANNING AND ARCHITECTURAL DESIGN OF INTEGRATED SERVICES DIGITAL NETWORKS, A. Nejat Ince, Dag Wilhelmsen and Bülent Sankur; ISBN: 0-7923-9554-9

WIRELESS INFRARED COMMUNICATIONS, John R. Barry; ISBN: 0-7923-9476-3

COMMUNICATIONS AND CRYPTOGRAPHY: ***Two sides of One Tapestry***, Richard E. Blahut, Daniel J. Costello, Jr., Ueli Maurer and Thomas Mittelholzer; ISBN: 0-7923-9469-0

WIRELESS AND MOBILE COMMUNICATIONS, Jack M. Holtzman and David J. Goodman; ISBN: 0-7923-9464-X

INTRODUCTION TO CONVOLUTIONAL CODES WITH APPLICATIONS, Ajay Dholakia; ISBN: 0-7923-9467-4

CODED-MODULATION TECHNIQUES FOR FADING CHANNELS, S. Hamidreza Jamali, and Tho Le-Ngoc; ISBN: 0-7923-9421-6

WIRELESS PERSONAL COMMUNICATIONS: ***Trends and Challenges***, Theodore S. Rappaport, Brian D. Woerner, Jeffrey H. Reed; ISBN: 0-7923-9430-5

ELLIPTIC CURVE PUBLIC KEY CRYPTOSYSTEMS, Alfred Menezes; ISBN: 0-7923-9368-6

SATELLITE COMMUNICATIONS: Mobile and Fixed Services, Michael Miller, Branka Vucetic and Les Berry; ISBN: 0-7923-9333-3

WIRELESS COMMUNICATIONS: Future Directions, Jack M. Holtzman and David J. Goodman; ISBN: 0-7923-9316-3

DISCRETE-TIME MODELS FOR COMMUNICATION SYSTEMS INCLUDING ATM, Herwig Bruneel and Byung G. Kim; ISBN: 0-7923-9292-2

APPLICATIONS OF FINITE FIELDS, Alfred J. Menezes, Ian F. Blake, XuHong Gao, Ronald C. Mullin, Scott A. Vanstone, Tomik Yaghoobian; ISBN: 0-7923-9282-5

WIRELESS PERSONAL COMMUNICATIONS, Martin J. Feuerstein, Theodore S. Rappaport; ISBN: 0-7923-9280-9

SEQUENCE DETECTION FOR HIGH-DENSITY STORAGE CHANNEL, Jaekyun Moon, L. Richard Carley; ISBN: 0-7923-9264-7

DIGITAL SATELLITE COMMUNICATIONS SYSTEMS AND TECHNOLOGIES: Military and Civil Applications, A. Nejat Ince; ISBN: 0-7923-9254-X

IMAGE AND TEXT COMPRESSION, James A. Storer; ISBN: 0-7923-9243-4

VECTOR QUANTIZATION AND SIGNAL COMPRESSION, Allen Gersho, Robert M. Gray;

INFORMATION, CODING and MATHEMATICS

Proceedings of Workshop honoring Prof. Bob McEliece on his 60th birthday

edited by

Mario Blaum
IBM Research Division

Patrick G. Farrell
Lancaster University

Henk C. A. van Tilborg
Eindhoven University of Technology

Springer Science+Business Media, LLC

Electronic Services <http://www.wkap.nl>

Library of Congress Cataloging-in-Publication Data

A C.I.P. Catalogue record for this book is available
from the Library of Congress.

ISBN 978-1-4419-5289-9 ISBN 978-1-4757-3585-7 (eBook)
DOI 10.1007/978-1-4757-3585-7

Printed on acid-free paper.

Robert J. McEliece

TABLE OF CONTENTS

Preface

These Proceedings contain 21 papers that were presented at the Workshop celebrating Prof. Bob McEliece's 60th birthday at Caltech on May 24th and 25th, 2002.

Bob has had a tremendous influence in different fields of information theory. He has worked in all aspects of coding, block codes, bounds, convolutional codes, turbo codes. He has obtained important results in cryptography and Shannon Theory. As a tribute to his achievements, I am proud to present the 21 papers that integrate this volume, which are related to different aspects of Bob's work. I will not summarize to the reader the contents of each of these papers. The authors themselves make an excellent case by showing Bob's influence in their results. Just let me say that when I was editing the book, I felt like reading immediately some of the papers. I did not have the time to go into them in detail (otherwise, you would not be reading this Preface), but eventually I will. I am sure that the reader will share my enthusiasm for those papers on his or her favorite topic.

Describing Bob's professional achievements could take forever, and those are well known. He is being honored by much more than technical excellence, though. He is one of the most beloved members of our Information Theory community. He is known for selflessly sharing his knowledge with everybody. Bob is always moved by a sense of curiosity. At 60, he is somehow slowing down as a runner (he is still an avid one), but not as a researcher. If anything, his production seems to increase as the years go by. The number of graduate students that have graduated under his direction now number in the dozens (I don't know the exact number). Most of them are now prestigious professionals on their own. Bob is also an excellent teacher and lecturer. His skill at transmitting ideas is remarkable. In professional conferences, his talks are usually the most eagerly followed.

Personally, I first met Bob in the fall of 1982. I was then a second-year graduate student in Mathematics at Caltech. Bob had just arrived to Caltech from Illinois. I was looking for an adviser, and I soon found one! My interaction with him has influenced me for the rest of my life. We would meet once a week, if at all, and those were quality meetings! Bob had an uncanny ability to make a remark that would trigger an idea. It is difficult to explain, he never told me what to do exactly, but he knew how to inspire a solution to a problem. I also found out that Bob's curiosity is not limited to technical results. He is well informed about current events, and I always enjoyed discussing with him any topic under the sun (I've never shared his enthusiasm for McDonald's food and Star Wars movies, though). He has had a great influence over my life. I have with him a debt of gratitude, that I am sure is shared by all of his ex-students. I present these Proceedings to him as a token of our appreciation.

These Proceedings were a collective effort. I want to thank my co-editors and great friends, Paddy Farrell and Henk van Tilborg. This is not the first time we worked together on a project, and having them at my side is always reassuring. I want to thank the chairmen of the different sessions of the Workshop: Sol Golomb, Bahram Honary, Mike Pursley, Oliver Collins, Ilya Dumer and Will Lin. They were the ones that really sought the great contributions that you find in this volume. I thank the authors, not only for their great results, but also for having to put up with my inexperience as an editor. I have burdened them with repeated and sometimes contradictory requests, and they always took them with a sense of humor. Last but not least, I want to thank Kluwer's editors Jennifer Evans and Anne Murray, their patience with me was outstanding. I promise them that next time I will be much more efficient!

The Workshop organization involved many people and probably I will not be able to thank them all. I hope nobody feels postponed. David MacKay created and maintained the Web site of the Workshop, http://www.inference.phy.cam.ac.uk/bob/. Meina Xu and Shirley Beatty were outstanding with the local organization at Caltech. Laif Swanson was always there to make the right suggestion. And I want to thank my friend Shuki Bruck for more reasons than I can enumerate.

Finally, I would like to acknowledge the generous financial support provided by the Lee Center for Advanced Networking and the Electrical Engineering department at Caltech, this is a nice gift and a wonderful recognition from Bob's intellectual home.

And this is enough for a preface, so, dear reader, enjoy this book.

Mario Blaum, San José, California

A Computational Theory of Surprise

Pierre Baldi
Department of Information and Computer Science
California Institute for Telecommunications and Information Technology
University of California, Irvine
Irvine, CA 92697-3425
pfbaldi@ics.uci.edu

Abstract

While eminently successful for the *transmission* of data, Shannon's theory of information does not address semantic and subjective dimensions of data, such as relevance and surprise. We propose an observer-dependent computational theory of surprise where surprise is defined by the relative entropy between the prior and the posterior distributions of an observer. Surprise requires integration over the space of models in contrast with Shannon's entropy, which requires integration over the space of data. We show how surprise can be computed exactly in a number of discrete and continuous cases using distributions from the exponential family with conjugate priors. We show that during sequential Bayesian learning, surprise decreases like $1/N$ and study how surprise differs and complements Shannon's definition of information.

Keywords: Information, Surprise, Relevance, Bayesian Probabilities, Entropy, Relative Entropy.

1 Introduction

The notion of information is central to science, technology, and many other human endeavors. While several approaches for quantifying information have been proposed, the most successful one so far has been Claude Shannon's definition introduced over half a century ago [20, 18, 4, 8]. According to Shannon, the information contained in a data set D is given by $-\log P(D)$, and the average information over all possible data sets $\mathcal{D}$ is the entropy $H(P(D)) = -\int_{\mathcal{D}} P(D) \log P(D) dD$.

Although it has been eminently successful for the development of modern telecommunication and computer technologies, Shannon's definition does not capture all aspects of information and comes with a number of shortcomings that may in part explain why the theory has not been as successful as one would have hoped in other areas of science such as biology, psychology, or economics.

A first concern is that it fails to account how data can have different significance for different observers. This is rooted in the origin of the probabilities used in the definition of information. These probabilities are defined according to an observer or a model M which Shannon does not describe explicitly so that the information in a data set is rather the negative log-likelihood

$$I(D, M) = -\log P(D|M) \tag{1}$$

and the corresponding entropy is the average over all data sets

$$I(\mathcal{D}, M) = H(P(D|M)) = -\int_{\mathcal{D}} P(D|M) \log P(D|M) dD \tag{2}$$

As pointed out by Edward Jaynes ([14]), this observer is essentially the communication engineer designing the communication system and, as such, M is fixed. However, not only information ought to be a property of the data, it should also be highly dependent on the observer, because the same data may carry completely different meanings for different observers. Consider for instance the genomic DNA sequence of the AIDS virus. It is a string of about 10,000 letters over the 4-letter DNA alphabet, of great significance to researchers in the biological or medical sciences, but utterly uninspiring to a layman. Within Shannon's framework, one could consider two observers $\mathcal{O}_1$ and $\mathcal{O}_2$ (or two model classes $\mathcal{M}_1$ and $\mathcal{M}_2$) with models M_1 and M_2 and assign information $-\log P(D|M_1)$ and $-\log P(D|M_2)$ to the data relative to each model. This however remains unsatisfactory. In particular, even if the two likelihoods were the same, the data D could carry different amounts of information for $\mathcal{O}_1$ and $\mathcal{O}_2$ depending on their expectations. Thus information ought to depend on the observer and also on his expectations.

Indeed Shannon's theory of information explicitly ignores any notions of relevance or semantics in the data. As pointed out in the title of Shannon's seminal article, it is a theory of *communication*, in the sense of transmission rather than information. It concentrates on the problem of "reproducing at one point either exactly or approximately a message selected at another point" regardless of the relevance of the message. But there is clearly more to information than data reproducibility and somehow information ought to depend also on the model or hypothesis M, or rather on the class $\mathcal{M}$ of such models.

Shannon's theory also produces a well-known paradoxical effect that is often puzzling to new students in information theory. How is it that "white snow", the most boring of all television programs, carries the most Shannon information? On one hand, it is clear that the uniform distribution has the highest entropy and reproducing a snow pattern *exactly* requires a very large number of bits. On the other hand, producing "snow-like" patterns is very easy. How can we reconcile the two viewpoints in a rigorous way? Notice that this paradox has nothing to do with the complexity of the generative model being used. A high order Markov model of the television images would still make snow highly improbable and therefore highly informative from Shannon's standpoint.

In short, there seems to be room for developing concepts of information that complement or extend Shannon's definition. The main purpose here is to develop a computational theory of subjective information surprise, or surprise. Surprise, no matter how one defines it, is obviously related to Shannon's information: a rare event is in general surprising and ought to carry a great deal of Shannon information due to its low probability. But beyond this obvious relationship, a theory of surprise should to be able to measure information surprise that is contained in data (1) in an observer-dependent way; (2) related to his changes in expectation; (3) through a definition that clearly establishes a connection with the foundations of probability theory; and (4) clarifies the "white snow" paradox and related concerns.

If such a definition exists, it must first of all be related to the foundations of the notion of probability, which can be approached from a frequentist or subjectivist, also called Bayesian, point of view[3, 6]. Here we follow the Bayesian approach which has been prominent in recent years and has led to important developments in many fields [12, 10]. The definition we propose stems directly from the Bayesian foundation of probability theory, and the relation given by Bayes theorem between the prior and posterior probabilities of an observer (see also [23]). The amount of surprise in the data for a given observer can be measured by looking at the change that has taken place in going from the prior to the posterior probabilities.

2 Information and Surprise

In the subjectivist framework, degrees of belief or confidence are associated with hypotheses or models. It can be shown that under a small set of reasonable axioms, these degrees of belief can be represented by real numbers and that when rescaled to the [0,1] interval these degrees of confidence must obey the rules of probability and in particular Bayes theorem [9, 19, 15]. Specifically, if an observer has a model M for the data, associated with a prior probability $P(M)$, the arrival of a data set D leads to a reevaluation of the probability in terms of the posterior distribution

$$P(M|D) = \frac{P(D|M)P(M)}{P(D)} \tag{3}$$

The effect of the information contained in D is clearly to change the belief of the observer from $P(M)$ to $P(M|D)$. Thus, a complementary way of measuring information carried by the data D is to measure the distance between the prior and the posterior. To distinguish it from Shannon's communication information, we call this notion of information the surprise information or *surprise*

$$S(D, \mathcal{M}) = d[P(M), P(M|D)] \tag{4}$$

where d is a distance or similarity measure. There are different ways of measuring a distance between probability distributions. In what follows, for standard well known theoretical reasons (including invariance with respect to reparameterizations), we use the relative entropy or Kullback-Liebler [17] divergence K which is not symmetric and hence not a distance. This lack of symmetry, however, does not matter in most cases and in principle can easily be fixed by symmetrization of the divergence. The surprise then is

$$\begin{aligned} S(D,\mathcal{M}) &= K(P(M),P(M|D)) = \int_{\mathcal{M}} P(M)\log\frac{P(M)}{P(M|D)}dM \\ &= -H(P(M)) - \int P(M)\log P(M|D)dM \\ &= \log P(D) - \int_{\mathcal{M}} P(M)\log P(D|M)dM \end{aligned} \tag{5}$$

Alternatively, we can define the single model surprise by the log-odd ratio

$$S(D,M) = \log\frac{P(M)}{P(M|D)} \tag{6}$$

and the surprise by its average

$$S(D,\mathcal{M}) = \int_{\mathcal{M}} S(D,M)P(M)dM \tag{7}$$

taken with respect to the prior distribution over the model class. In statistical mechanics terminology, the surprise can also be viewed as the free energy of the negative log-posterior at a temperature $t = 1$, with respect to the prior distribution over the space of models [2].

Note that this definition addresses the "white snow" paradox. At the time of snow onset, the image distribution we expect and the image we perceive are very different and therefore the snow carries a great deal of both surprise and Shannon's information. Indeed snow may be a sign of storm, earthquake, toddler's curiosity, or military putsch. But after a few seconds, once our model of the image shifts towards a snow model of random pixels, television snow perfectly fits the prior and hence becomes boring. Since the prior and the posterior are virtually identical, snow frames carry 0 surprise although megabytes of Shannon's information.

The similarities and differences of surprise with Shannon's information should now be clear—in particular, surprise is a dual notion that requires integration over the space of models rather than the space of data. In the next sections, we show how this integration can be carried analytically in simple cases. As is the case for Bayesian inference, however, integration over the space of models in general is likely to require Monte Carlo methods. Information is also bound to vary when data is received sequentially, i.e. during learning processes. Therefore we also show how the proposed definition is related to the behavior of learning curves. Details of calculations are provided in the Appendix.

3 Computation of Surprise

Here we consider a data set $D = \{x_1, \ldots, x_N\}$ containing N points. Surprise can be calculated exactly in a number of interesting cases. For simplicity, although this does not correspond to any restriction of the general theory, we consider only the case of conjugate priors, where the prior and the posterior have the same functional form. In this case, in order to compute the surprise defined by Equation 5, we need only to compute general terms of the form

$$F(P_1, P_2) = \int P_1 \log P_2 dx \tag{8}$$

where P_1 and P_2 have the same functional form. The surprise is then given by

$$S = F(P_1, P_1) - F(P_1, P_2) \tag{9}$$

where P_1 is the prior and P_2 is the posterior. Note also that in this case the symmetric divergence can easily be computed using $F(P_1, P_1) - F(P_1, P_2) + F(P_2, P_2) - F(P_2, P_1)$. Details for the calculation of $F(P_1, P_2)$ in the examples below are given in the Appendix. It should also be clear that in simple cases, for instance for certain members of the exponential family [7] of distributions, the posterior depends entirely on the sufficient statistics and therefore we can expect surprise also to depend only on sufficient statistics in these cases.

3.1 Discrete Data and Dirichlet Model

Consider the case where x_i is binary. The simplest class of models for D is then $M(p)$, the first order Markov models with a single parameter p representing the probability of emitting a 1. The conjugate prior on p is the Dirichlet prior (or beta distribution in the 2-D case)

$$D_1(a_1, b_1) = \frac{\Gamma(a_1 + b_1)}{\Gamma(a_1)\Gamma(b_1)} x^{a_1 - 1}(1-x)^{b_1-1} = C_1 x^{a_1-1}(1-x)^{b_1-1} \quad (10)$$

with $a_1 \geq 0$, $b_1 \geq 0$, and $a_1 + b_1 > 0$. The expectation is $a_1/(a_1 + b_1), b_1/(a_1 + b_1)$. With n sucesses in the sequence D, the posterior is a Dirichlet distribution $D_2(a_2, b_2)$ with [2]

$$a_2 = a_1 + n \quad \text{and} \quad b_2 = b_1 + (N - n) \quad (11)$$

The surprise can be computed exactly

$$\begin{aligned} S(D, \mathcal{M}) = K((D_1, D_2)) &= \log \frac{C_1}{C_2} + n[\Psi(a_1 + b_1) - \Psi(a_1)] \\ &\quad + (N - n)[\Psi(a_1 + b_1) - \Psi(b_1)] \quad (12) \end{aligned}$$

where Ψ is the derivative of the logarithm of the Gamma function (see Appendix). When $N \to \infty$, and $n = pN$ with $0 < p < 1$ we have

$$S(D, \mathcal{M}) \approx N\mathrm{K}(p, a_1) \quad (13)$$

where $K(p, a_1)$ represents the Kullback-Liebler divergence distance between the empirical distribution $(p, 1-p)$ and the expectation of the prior $(a_1/(a_1 + b_1), b_1/(a_1 + b_1))$. Thus asymptotically surprise information grows linearly with the number of data points with a proportionality coefficient that depends on the discrepancy between the expectation of the prior and the observed distribution. The same relationship can be expected to be true in the case of a multinomial model. In the case of a symmetric prior ($a_1 = b_1$), a slightly more precise approximation is provided by:

$$S(D_1, D_2) \approx N[\sum_{k=a_1}^{2a_1-1} \frac{1}{k} - H(p)] \quad (14)$$

For instance, when $a_1 = 1$ then $R(D_1, D_2) \approx N(1 - H(p))$, and when $a_1 = 5$ then $R(D_1, D_2) \approx N[0.746 - H(p)]$.

These results provide a clear explanation for the television "snow" effect. With a uniform symmetric prior, the empirical distribution with maximal entropy brings the least information. If we expect snow, the Kullback-Liebler divergence between the prior and the posterior is 0 and therefore there is essentially no surprise in the signal. As pointed out, this is not the case, however, at the time of onset of the snow where the divergence may even be large.

3.2 Continuous Data: Unknown Mean/Known Variance

When the x_i are real, we can consider first the case of unknown mean with known variance. We have a family $M(\mu)$ of models, with a Gaussian prior $G_1(\mu_1, \sigma_1^2)$. If the data has known variance σ^2, then the posterior distribution is Gaussian $G_2(\mu_2, \sigma_2^2)$ with parameters given by [10]

$$\mu_2 = \frac{\frac{\mu_1}{\sigma_1^2} + \frac{N\bar{m}}{\sigma^2}}{\frac{1}{\sigma_1^2} + \frac{N}{\sigma^2}} \quad \text{and} \quad \frac{1}{\sigma_2^2} = \frac{1}{\sigma_1^2} + \frac{N}{\sigma^2} \tag{15}$$

where $\bar{m}$ is the observed mean. In the general case

$$\begin{aligned} S(D, \mathcal{M}) = KG_1, G_2) &= \log \frac{\sigma}{\sqrt{\sigma^2 + N\sigma_1^2}} + N\frac{\sigma_1^2}{2\sigma^2} + \frac{N^2\sigma_1^2(\mu_1 - \bar{m})^2}{2\sigma^2(\sigma^2 + N\sigma_1^2} \\ &\approx \frac{N}{2\sigma^2}[\sigma_1^2 + (\mu_1 - \bar{m})^2] \end{aligned} \tag{16}$$

the approximation being valid for large N. In the special case where the prior has the same variance has the data $\sigma_1 = \sigma$ then the formula simplify a little and yield

$$S = K(G_1, G_2) = \frac{N}{2} - \frac{1}{2}\log(N+1) + \frac{N^2(\mu_1 - \bar{m})^2}{2(N+1)\sigma^2} \approx \frac{N}{2\sigma^2}[\sigma^2 + (\mu_1 - \bar{m})^2] \tag{17}$$

when N is large. In any case, surprise grows linearly with N with a coefficient that is the sum of the prior variance and the square difference between the expected mean and the empirical mean scaled by the variance of the data.

3.3 Continuous Data: Unknown Variance/Known Mean

When the x_i are real, we can then consider the case of unknown variance with known mean. We have a family $M(\sigma^2)$ of models, with a conjugate scaled inverse gamma prior

$$\Gamma_1(\nu_1, s_1) = \frac{(\frac{\nu_1}{2})^{\nu_1/2} s_1^{\nu_1}}{\Gamma(\frac{\nu_1}{2})} (\sigma^2)^{-(\frac{\nu_1}{2}+1)} e^{-\frac{\nu_1 s_1^2}{2\sigma^2}} d\sigma^2 = C_1 (\sigma^2)^{-(\frac{\nu_1}{2}+1)} e^{-\frac{\nu_1 s_1^2}{2\sigma^2}} d\sigma^2 \tag{18}$$

The posterior is then a scaled inverse gamma distribution [10] with

$$\nu_2 = \nu_1 + N \quad \text{and} \quad s_2^2 = \frac{\nu_1 s_1^2 + N\bar{\sigma}^2}{\nu_1 + N} \tag{19}$$

Here $\bar{\sigma}^2 = \sum (x_i - m)^2/N$ is the observed variance, based on the known mean m. The surprise

$$S(D, \mathcal{M}) = K(\Gamma_1, \Gamma_2) = \log \frac{C_1}{C_2} - \frac{N}{2}[\Psi(\frac{\nu}{2}) + \log \frac{2}{\nu_1 s_1^2}] + \frac{N\bar{\sigma}^2}{2s_1^2} \tag{20}$$

For large values of N,

$$S = K(\Gamma_1, \Gamma_2) \approx \frac{N}{2}[\frac{\bar{\sigma}^2}{s_1^2} + \log \frac{\nu_1 s_1^2}{2\bar{\sigma}^2} - \Psi(\frac{\nu_1}{2})] \tag{21}$$

Thus surprise information scales linearly with N, with a coefficient of proportionality that typically depends mostly on the ratio of the empirical variance to the scale parameters s_1^2, which is roughly the expectation of the prior [the expectation of the prior is $\nu_1 s_1^2/(\nu_1 - 2)$ provided $\nu_1 > 2$]. The effects of very large of very small values of $\bar{\sigma}$, or ν_1 can also be seen in the formula above. In particular, surprise is largest when the empirical variance $\bar{\sigma}^2$ goes to 0 or infinity, i.e. is very different from the prior expectation.

3.4 Continuous Data: Unknown Mean/Unknown Variance

When the x_i are real, we can finally consider the case of unknown mean with unknown variance. We have a family $M(\mu, \sigma^2)$ of models, with a conjugate prior $G_1\Gamma_1 = P(\mu|\sigma^2)P(\sigma^2) = G_1(\mu_1, \sigma^2/\kappa_1)\Gamma_1(\nu_1, s_1)$, product of a normal with a scaled inverse gamnma distribution. Thus the prior has four parameters $(\mu_1, \kappa_1, \nu_1, s_1)$, with $\kappa_1 > 0$, $\nu_1 > 0$, and $s_1 > 0$. The conjugate posterior has the same form, with similar parameters $(\mu_2, \kappa_2, \nu_2, s_2)$ satisfying (see for instance [10])

$$\mu_2 = \frac{\kappa_1}{\kappa_1 + N}\mu_1 + \frac{N}{\kappa_1 + N}\bar{m} \tag{22}$$

$$\kappa_2 = \kappa_1 + N \tag{23}$$

$$\nu_2 = \nu_1 + N \tag{24}$$

$$\nu_2 s_2^2 = \nu_1 s_1^2 + (N-1)\bar{\sigma}^2 + \frac{\kappa_1 N}{\kappa_1 + N}(\bar{m} - \mu_1)^2 \tag{25}$$

with $\bar{m} = \sum x_i/N$ and $\bar{\sigma}^2 = \sum(x_i - \bar{m})^2/(N-1)$. The surprise is

$$\begin{aligned} S(D, \mathcal{M}) &= K(G_1\Gamma_1, G_2\Gamma_2) = \frac{1}{2}\log\frac{\kappa_1}{\kappa_1 + N} + \frac{N}{2\kappa_1} + \\ &\frac{\kappa_1 + N}{2}\left[\frac{N(\bar{m} - \mu_1)}{(\kappa_1 + N)s_1}\right]^2 + \log\frac{C_1}{C_2} - \frac{N}{2}[\Psi(\frac{\nu_1}{2}) + \log\frac{2}{\nu_1 s_1^2}] + \\ &\frac{(N-1)\bar{\sigma}^2 + \frac{\kappa_1 N}{\kappa_1 + N}(\bar{m} - \mu_1)^2}{2s_1^2} \end{aligned} \tag{26}$$

For large values of N,

$$R(G_1\Gamma_1, G_2\Gamma_2) \approx \frac{N}{2}[\frac{1}{\kappa_1} + \frac{\bar{\sigma}^2}{s_1^2} + \log\frac{\nu_1 s_1^2}{2\bar{\sigma}^2} - \Psi(\frac{\nu_1}{2}) + \frac{(\bar{m} - \mu_1)^2}{s_1^2}] \tag{27}$$

Surprise information is linear in N with a coefficient that is essentially the sum of the coefficients derived in the unknown mean and unknown variance partial cases.

4 Learning and Surprise

There is an immediate connection between surprise and computational learning theory. If we imagine that data points from a training set are presented sequentially, we can consider that the posterior distribution after the N-th point becomes the prior for the next iteration (sequential Bayesian learning). In this case we can expect on average surprise to decrease after each iteration, since as a system learns what is relevant in a data set, new data points become less and less surprising. This can be quantified precisely, at least in simple cases.

4.1 Learning Curves: Discrete Data

Consider first a sequence of 0-1 examples $D = (d_N)$. The learner starts with a Dirichlet prior $D_0(a_0, b_0)$. With each example d_N, the learner updates its Dirichlet prior $D_N(a_N, b_N)$ into a Dirichlet posterior $D_{N+1}(a_{N+1}, b_{N+1})$ with $(a_{N+1}, b_{N+1}) = (a_N + 1, b_N)$ if $d_{N+1} = 1$, and $(a_{N+1}, b_{N+1}) = (a_N, b_N + 1)$ otherwise. When $d_{N+1} = 1$, the corresponding surprise is easily computed using Equations 45 and 48. For simplicity, and without much loss of generality, let us assume that a_0 and b_0 are integers, so that a_N and b_N are also integers for any N. Then if $d_{N+1} = 1$ the relative surprise is

$$S(D_N, D_{N+1}) = \log \frac{a_N}{a_N + b_N} + \sum_{k=0}^{b_N - 1} \frac{1}{a_N + k} \tag{28}$$

and similarly in the case $d_{N+1} = 0$ by interchanging the role of a_N and b_N. Thus, in this case,

$$0 \leq S(D_N, D_{N+1}) \leq \frac{1}{a_N} + \log(1 - \frac{1}{a_N + b_N}) \tag{29}$$

Asymptotically we have $a_N \approx a_0 + pN$ and therefore

$$0 \leq S(D_N, D_{N+1}) \leq \frac{1-p}{pN} \tag{30}$$

Thus surprise decreases in time with the number of examples as $1/N$.

4.2 Learning Curves: Continuous Data

In the case of continuous Gaussian data with, for instance, known variance σ^2, the learner starts with a Gaussian prior $G_0(\mu_0, \sigma_0^2)$ on the mean. With each example d_N, the learner updates its Gaussian prior $G_N(\mu_N, \sigma_N^2)$ into a Gaussian posterior $G_{N+1}(\mu_{N+1}, \sigma_{N+1}^2)$ with

$$\mu_{N+1} = \frac{\frac{\mu_N}{\sigma_N^2} + \frac{d_{N+1}}{\sigma^2}}{\frac{1}{\sigma_N^2} + \frac{1}{\sigma^2}} \quad \text{and} \quad \frac{1}{\sigma_{N+1}^2} = \frac{1}{\sigma_N^2} + \frac{1}{\sigma^2} \tag{31}$$

From Equation 16, the relative surprise is

$$S(G_N, G_{N+1}) = \log \frac{\sigma}{\sqrt{\sigma^2 + \sigma_n^2}} + \frac{\sigma_N^2}{2\sigma^2}(1 + \frac{(\mu_N - d_{N+1})^2}{\sigma^2 + \sigma_N^2}) \tag{32}$$

Asymptotically

$$S(G_N, G_{N+1}) \leq \frac{\sigma_N^2}{2\sigma^2} \tag{33}$$

From Equation 15, we have $\frac{1}{\sigma_{N+1}^2} = \frac{1}{\sigma_0^2} + \frac{(N+1)}{\sigma^2}$, or $\sigma_{N+1}^2 = \frac{\sigma_0^2\sigma^2}{\sigma^2+(N+1)\sigma_0^2}$. Thus

$$0 \leq S(G_N, G_{N+1}) \leq \frac{1}{2(N+1)} \tag{34}$$

Thus in this case surprise decreases in time with the number of examples also as $1/N$.

5 Surprise, Evidence, and Mutual Information

To measure the effect of the data on the prior and the posterior, one could have envisioned using the difference between the entropy of the prior and the entropy of the posterior. However, unlike surprise which is always positive, the difference between these two entropies can be either positive or negative and therefore is not a suitable measure.

In the formula given above for the surprise (Equation 5), we have introduced the *evidence* $P(D) = P(D|\mathcal{M}) = \int_{\mathcal{M}} P(M, D)dM$. The evidence plays a key role in Bayesian analyis and is the hinge that leads to the next cycle of Bayesian analysis beyond the class of models $\mathcal{M}$. Shannon's information could be defined with respect to the evidence in the form

$$I(D, \mathcal{M}) = -\log P(D|\mathcal{M}) \tag{35}$$

with the associated evidence entropy

$$I(\mathcal{D}, \mathcal{M}) = -\int_{\mathcal{D}} P(D|\mathcal{M}) \log P(D|\mathcal{M}) dD \tag{36}$$

For a fixed data set D, the surprise is

$$S(D, \mathcal{M}) = -I(D, \mathcal{M}) + \int_{\mathcal{M}} P(M) I(D, M) dM \tag{37}$$

is therefore the difference between the average Shannon information per model, taken with respect to the prior, and the Shannon information based on the evidence.

If we integrate the surprise with respect to the evidence

$$\int_{\mathcal{D}} P(D) S(D, \mathcal{M}) dD = \int_{\mathcal{D},\mathcal{M}} P(D) P(M) \log \frac{P(D)P(M)}{P(D, M)} dDdM \tag{38}$$

we get the Kullback-Liebler divergence $K(P(D)P(M), P(D, M))$ which is the symmetric inverse of the mutual information MI between $\mathcal{D}$ and $\mathcal{M}$ $MI(\mathcal{D}, \mathcal{M}) = K(P(D, M), P(D)P(M))$.

6 Discussion and Extensions

Surprise is different from other definitions of information that have been proposed [1] as alternatives to Shannon's entropy. Most alternative definitions, such as Rényi's entropies, are actually algebraic variations on Shannon's definition rather than conceptually different approaches. While Shannon's definition fixes the model and varies the data, surprise fixes the data and varies the model. Surprise is a measure of dissimilarity between the prior and posterior distributions and as such it lies close to the axiomatic foundation of Bayesian probability.

In a number of cases, surprise can be computed analytically both in terms of exact and asymptotic formula. The analytical results presented here could be extended in several directions including non-conjugate and other prior distributions, more complex multidimensional distributions (e.g. multinomial, inverse Wishart), and more general families of distributions (e.g. exponential family [7]). In general, however, the computation of surprise can be expected to require Monte Carlo methods to approximate integrals over model spaces. In this respect, the computation of surprise should benefit from progress in this active area of research as well as increase in computing power.

While applications remain to be developed, a theory of surprise could be used in areas as diverse as game theory, machine learning, Internet commerce, and the design of sensory systems. Consider, for instance, the design of artificial sensory systems or the reverse engineering of natural ones. Clearly, attention mechanisms play a fundamental role allowing perceptual systems to shift their resources and bring them to bear on the most surprising region of the input space. In both natural systems and some of their artificial cousins, expectations could be generated by top down connections and compared in real time with input streams generated by bottom up connections [13]. Mismatches between input and expectations could be computed using surprise theory and lead to saliency maps. These maps in turn could guide attentional mechanisms, whereby additional processing resources are dynamically allocated to the regions of the input field that are the most surprising, i.e. which carry the highest amount of information with respect to the expectations.

Likewise, we have only touched upon the connection between surprise and machine learning [22] by showing that surprise decreases as $1/N$ during sequential learning in simple cases. This analysis could be extended to more complex settings, such as artificial neural networks.

But the notion of surprise has its own limitations. In particular, it does not capture all the semantics/relevance aspects of data. When the degree of surprise of the data with respect to the model class becomes low, the data is no longer informative for the given model class. This, however, does not necessarily imply that one has a good model since the model class itself could be unsatisfactory and in need of a complete overhaul. The process by which we decide a model is unsatisfactory in an alternative free setting, the open-ended aspect of inference, remains elusive to modeling.

Conversely, highly surprising data could be a sign that learning is required or that the data is irrelevant. If while surfing the web in search of a car one stumbles on a picture of Marilyn Monroe, the picture may carry a low degree of relevance, a high degree of surprise, and a low-to-high amount of Shannon information depending on the pixel structure. Thus, relevance, surprise, and Shannon's entropy are three different facets of information that can be present in different combinations. The notion of *relevance* in particular seems to be the least understood although there have been several attempts [16, 21]. A possible direction is to consider, in addition to the space of data and models, a third space $\mathcal{A}$ of actions or interpretations and define relevance as the relative entropy between the prior $P(A)$ and the posterior $P(A|D)$ distributions over $\mathcal{A}$. Whether this approach simply shifts the problem into the definition of the set $\mathcal{A}$ remains to be seen. In any event, the quest to understand the nature of information is unlikely to be over.

Appendix A: Discrete Case

In the two-dimensional case, consider two Dirichlet distributions $D_1 = D(a_1, b_1)(x) = C_1 x^{a_1-1}(1-x)^{b_1-1}$ and $D_2 = D(a_2, b_2)(x) = C_2 x^{a_2-1}(1-x)^{b_2-1}$, with $C_1 = \Gamma(a_1+b_1)/\Gamma(a_1)\Gamma(b_1)$, and similarly for C_2. To calculate the relative entropy in the two dimensional case, we use the formula ([11]

$$\int_0^1 x^{u-1}(1-x)^{v-1} \log x dx = B(u,v)[\Psi(u) - \Psi(u+v)] \tag{39}$$

where $B(u,v)$ is the beta function $B(u,v) = \int_0^1 x^{u-1}(1-x)^{v-1}dx = \Gamma(u)\Gamma(v)/\Gamma(u+v)$ and $\Psi(x)$ is the derivative of the logarithm of the gamma function $\Psi(x) = d(\log\Gamma(x))/dx$. A cross term of the form $F(D_1, D_2)$

$$F(D_1, D_2) = \int_0^1 C_1 x^{a_1-1}(1-x)^{b_1-1}[\log C_2 + (a_2-1)\log x + (b_2-1)\log(1-x)] \tag{40}$$

is equal to

$$F(D_1, D_2) = \log C_2 + (a_2-1)[\Psi(a_1) - \Psi(a_1+b_1)] + (b_2-1)[\Psi(b_1) - \Psi(a_1+b_1)] \tag{41}$$

using the fact that $C_1 B(a_1, b_1) = 1$. In particular, the entropy of a two-dimensional Dirichlet distribution such as D_1 is obtained by taking: $-F(D_1, D_1)$. With some algebra, the Kullback-Liebler divergence between any two Dirichlet distributions is finally given by:

$$\begin{aligned} K(D_1, D_2) &= \log \frac{C_1}{C_2} + (a_1 - a_2)[\Psi(a_1) - \Psi(a_1 + b_1)] + \\ & (b_1 - b_2)[\Psi(b_1) - \Psi(a_1 + b_1)] \end{aligned} \tag{42}$$

With n successes in the sequence D, the posterior is a Dirichlet distribution $D_2(a_2, b_2)$ with [2]

$$a_2 = a_1 + n \quad \text{and} \quad b_2 = b_1 + (N - n) \tag{43}$$

Using this relation between the prior and the posterior, we get the surprise

$$S(D_1, D_2) = \log \frac{C_1}{C_2} + n[\Psi(a_1 + b_1) - \Psi(a_1)] + (N - n)[\Psi(a_1 + b_1) - \Psi(b_1)] \tag{44}$$

Using the general fact that $\Psi(x) - \Psi(y) = \sum_{k=0}^{\infty}(\frac{1}{y+k} - \frac{1}{x+k})$, which implies $\Psi(x + n) - \Psi(x) = \sum_{k=0}^{n-1} \frac{1}{x+k}$ when n is an integer, we get

$$\begin{aligned} S(D_1, D_2) &= \log \frac{C_1}{C_2} + n(\sum_{k=0}^{\infty} \frac{1}{a_1 + k} - \frac{1}{a_1 + b_1 + k}) + \\ & (N - n)(\sum_{k=0}^{\infty} \frac{1}{b_1 + k} - \frac{1}{a_1 + b_1 + k}) \end{aligned} \tag{45}$$

Now we have

$$\sum_{k=0}^{\infty}(\frac{1}{a_1 + k} - \frac{1}{a_1 + b_1 + k}) = \sum_{k=0}^{\lfloor b_1 \rfloor - 1} (\frac{1}{a_1 + k}) + \text{Rest} \tag{46}$$

where

$$
\begin{aligned}
0 \leq \text{Rest} &= \sum_{k=0}^{\infty}\left(\frac{1}{a_1 + \lfloor b_1 \rfloor + k} - \frac{1}{a_1 + b_1 + k}\right) \\
&\leq (b_1 - \lfloor b_1 \rfloor) \sum_{k=0}^{\infty} \frac{1}{(a_1 + \lfloor b_1 \rfloor + k)^2}
\end{aligned} \tag{47}
$$

and similarly for the symmetric term. The rest is exactly 0 when a_1 and b_1 (and hence a_2 abd b_2) are integers, and in general decreases with the size of a_1 and b_1. This yields the approximation

$$
S(D_1, D_2) \approx \log \frac{C_1}{C_2} + n\left(\sum_{k=0}^{\lfloor b_1 \rfloor - 1} \frac{1}{a_1 + k}\right) + (N - n)\left(\sum_{k=0}^{\lfloor a_1 \rfloor - 1} \frac{1}{b_1 + k}\right) \tag{48}
$$

This approximation is *exact* when a_1 and b_1 are integers. Now for $x > 0$ we have $\log((x+n)/x) < \sum_{k=0}^{n-1} 1/(x+k) < \log((x+n-1)/x) + 1/x$ or $0 < \sum_{k=0}^{n-1} 1/(x+k) - \log((x+n)/x) < 1/x$. Thus,

$$
S(D_1, D_2) \approx \log \frac{C_1}{C_2} + n \log \frac{a_1 + b_1}{a_1} + (N - n) \log \frac{a_1 + b_1}{b_1} \tag{49}
$$

Now we have,

$$
\begin{aligned}
\log \tfrac{C_1}{C_2} &= \log \tfrac{\Gamma(a_1+b_1)\Gamma(a_2)\Gamma(b_2)}{\Gamma(a_1)\Gamma(b_1)\Gamma(a_2+b_2)} \\
&= \log \tfrac{(a_1+n-1)(a_1+n-2)\ldots a(b_1+N-n-1)\ldots b_1}{(a_1+b_1+N-1)(a_1+b_1+N-2)\ldots(a_1+b_1)}
\end{aligned} \tag{50}
$$

We can use bounds of the form $\log a + \int_{a_1}^{a_1+n1} \log x dx < \log a_1 + \ldots \log(a_1 + n - 1) \leq \log a_1 \int_{a_1+1}^{a_1+n} \log x dx$ to estimate this term. Alternatively, one can assume that a_1 and b_1 are integers and use binomial coefficient approximations, such as those in [5]. In all cases, neglecting constant terms and terms of order $\log N$, if we let $n = pN$ $(0 < p < 1)$ and N go to infinity we have

$$
\log \frac{C_1}{C_2} \approx -\log \binom{N}{n} \approx -NH(p)) \tag{51}
$$

where $H(p)$ is the entropy of the (p, q) distribution with $q = 1 - p$. Thus when $N \to \infty$, and $n = pN$ with $0 < p < 1$ we have

$$S(D_1, D_2) \approx N[p \log \frac{a_1 + b_1}{a_1} + q \log \frac{a_1 + b_1}{b_1} - H(p)] \approx N\mathrm{K}(p, a_1) \quad (52)$$

where $K(p, a_1)$ is the relative entropy between the empirical distribution (p, q) and the expectation of the prior $(\frac{a_1}{a_1+b_1}, \frac{b_1}{a_1+b_1})$. Thus, asymptotically surprise grows linearly with the number of data points with a proportionality coefficient that depends on the discrepancy between the expectation of the prior and the observed distribution. The same relationship can be expected to be true in the case of a multinomial model.

Symmetric Prior ($a_1 = b_1$)

Consider now the case of a symmetric prior, then

$$S(D_1, D_2) = \log \frac{C_1}{C_2} + N[\Psi(2a_1) - \Psi(a_1)] \quad (53)$$

Using formulas in [11], $\Psi(2a_1) - \Psi(a_1) = \sum_{k=0}^{\infty} \frac{(-1)^k}{2a_1+k} + \log 2$ thus

$$\begin{aligned} S(D_1, D_2)) &= \log \frac{C_1}{C_2} + N \sum_{k=0}^{\infty} \frac{(-1)^k}{2a_1 + k} + \log 2 \\ &\approx N[\sum_{k=0}^{\infty} \frac{(-1)^k}{2a_1 + k} + \log 2 - H(p)] \quad (54) \end{aligned}$$

the approximation being in the regime $n = pN$ and $N \to \infty$. When a_1 is an integer, we also have $\Psi(2a_1) - \Psi(a_1) = \sum_{k=1}^{2a_1-1} (-1)^{k+1}/k = \sum_{k=a_1}^{2a_1-1} 1/k$. Thus when a_1 is an integer

$$S(D_1, D_2) = N[\sum_{k=a_1}^{2a_1-1} \frac{1}{k}] + \log \frac{(2a_1 - 1)\binom{2a_1-2}{a_1-1}}{(2a_1 + N - 1)\binom{N+2a_1-2}{n+a_1-1}} \quad (55)$$

As $N \to \infty$ with $0 < p < 1$

$$S(D_1, D_2) \approx N[\sum_{k=a_1}^{2a_1-1} \frac{1}{k}] - \log \binom{N + 2a_1 - 2}{n + a_1 - 1} \approx N[\sum_{k=a_1}^{2a_1-1} \frac{1}{k}] - \log \binom{N}{n} \quad (56)$$

and therefore

$$S(D_1, D_2) \approx N[\sum_{k=a_1}^{2a_1-1} \frac{1}{k} - H(p)] \tag{57}$$

For instance, when $a_1 = b_1 = 1$, this gives:

$$S(D_1, D_2) = N - \log(N+1) - \log \binom{N}{n} \tag{58}$$

with the asymptotic form

$$S(D_1, D_2) \approx N(1 - H(p)) + \log \frac{\sqrt{2N\pi pq}}{N+1} \approx N(1 - H(p)) \tag{59}$$

With a uniform symmetric prior, the empirical distribution with maximal entropy brings the least information. When $a_1 = b_1 = 5$ this gives $R(D_1, D_2) \approx N[0.746 - H(p)]$. As we increase $a_1 + b_1$, keeping $a_1 = b_1$, the constant $\sum_{a_1}^{2a_1-1}(1/k)$ decreases to its asymptotic value $\log 2$ which corresponds to the asymptotic form $S(D_1, D_2) \approx NK(p, 0.5)$. The stronger the strength of the uniform prior (the larger $a_1 + b_1$), the smaller the surprise created by a die with maximum entropy.

Appendix B: Continuous Case

Unknown Mean/Known Variance

Consider now two Gaussians $G_1(\mu_1, \sigma_1)$ and $G_2(\mu_2, \sigma_2)$. Then, after some algebra, the cross term is given by

$$F(G_1, G_2) = \int_{-\infty}^{+\infty} G_1 \log G_2 dx = -\frac{1}{2}\log(2\pi\sigma_2^2) - \frac{\sigma_1^2 + (\mu_1 - \mu_2)^2}{2\sigma_2^2} \tag{60}$$

here using for simplicity natural logarithms. $F(G, G) = \frac{1}{2}\log[2\pi e\sigma^2] = H(G)$ is the entropy. The Kullback-Liebler divergence can then be obtained

$$K(G_1, G_2) = -\frac{1}{2} + \log\frac{\sigma_2}{\sigma_1} + \frac{\sigma_1^2 + (\mu_1 - \mu_2)^2}{2\sigma_2^2} \tag{61}$$

Consider now a data set with N points $x_1, \ldots, x_N$ with empirical mean $\bar{m}$. If the data has known variance σ^2, then the posterior parameters are given by:

$$\mu_2 = \frac{\frac{\mu_1}{\sigma_1^2} + \frac{N\bar{m}}{\sigma^2}}{\frac{1}{\sigma_1^2} + \frac{N}{\sigma^2}} \quad \text{and} \quad \frac{1}{\sigma_2^2} = \frac{1}{\sigma_1^2} + \frac{N}{\sigma^2} \tag{62}$$

In the general case

$$\begin{aligned} S(G_1, G_2) &= \log \frac{\sigma}{\sqrt{\sigma^2 + N\sigma_1^2}} + N\frac{\sigma_1^2}{2\sigma^2} + \frac{N^2\sigma_1^2(\mu_1 - \bar{m})^2}{2\sigma^2(\sigma^2 + N\sigma_1^2)} \\ &\approx \frac{N}{2\sigma^2}[\sigma_1^2 + (\mu_1 - \bar{m})^2] \end{aligned} \tag{63}$$

when N is large. In the special case where the prior has the same variance has the data $\sigma_1 = \sigma$ then the formula simplifies a little and yields

$$S(G_1, G_2) = \frac{N}{2} - \frac{1}{2}\log(N+1) + \frac{N^2(\mu_1 - \bar{m})^2}{2(N+1)\sigma^2} \approx \frac{N}{2\sigma^2}[\sigma^2 + (\mu_1 - \bar{m})^2] \tag{64}$$

when N is large. In any case, surprise grows linearly with N with a coefficient that is the sum of the prior variance and the square difference between the expected mean and the empirical mean scaled by the variance of the data.

Unknown Variance/Known Mean

In the case of unknown variance and known mean, we have a family $M(\sigma^2)$ of models with a conjugate prior for σ^2 that is a scaled inverse gamma distribution

$$\Gamma_1(\nu_1, s_1) = \frac{(\frac{\nu_1}{2})^{\nu_1/2} s_1^{\nu_1}}{\Gamma(\frac{\nu_1}{2})} (\sigma^2)^{-(\frac{\nu_1}{2}+1)} e^{-\frac{\nu_1 s_1^2}{2\sigma^2}} d\sigma^2 = C_1(\sigma^2)^{-(\frac{\nu_1}{2}+1)} e^{-\frac{\nu_1 s_1^2}{2\sigma^2}} d\sigma^2 \tag{65}$$

with $\nu_1 > 0$ degrees of freedom and scale $s_1 > 0$. F can be computed expanding the integrals and using the fact that $\int_0^{+\infty} x^{\nu/2-1} e^{-x} \log x = \Gamma(\frac{\nu}{2})\Psi(\frac{\nu}{2})$. This yields:

$$F(\nu_1, s_1; \nu_2, s_2) = \log \frac{(\nu_2/2)^{\nu_2/2} s_2^{\nu_2}}{\Gamma(\frac{\nu_2}{2})} + (\frac{\nu_2}{2} + 1)[\Psi(\frac{\nu_1}{2}) + \log \frac{2}{\nu_1 s_1^2}] - \frac{\nu_2 s_2^2}{2s_1^2} \tag{66}$$

The posterior is then a scaled inverse gamma distribution [10] with

$$\nu_2 = \nu_1 + N \quad \text{and} \quad s_2^2 = \frac{\nu_1 s_1^2 + N\bar{\sigma}^2}{\nu_1 + N} \tag{67}$$

where $\bar{\sigma}^2$ is the empirical variance $\bar{\sigma}^2 = \sum_i (x_i - m)^2/N$, based on the known mean m. The surprise is given by

$$S(\Gamma_1, \Gamma_2) = \log \frac{C_1}{C_2} - \frac{N}{2}(\Psi(\frac{\nu_1}{2}) + \log \frac{2}{\nu_1 s_1^2}) + \frac{N\bar{\sigma}^2}{2s_1^2} \tag{68}$$

For large values of N, taking only the leading terms

$$\begin{aligned} S(\Gamma_1, \Gamma_2) \quad &\approx \quad \frac{N}{2}(\frac{\bar{\sigma}^2}{s_1^2} + \log \frac{\nu_1 s_1^2}{2} - \Psi(\frac{\nu_1}{2})) + \log \Gamma(\frac{\nu_1 + N}{2}) \\ &- \quad \frac{\nu_1 + N}{2} \log \frac{\nu_1 + N}{2} - \frac{(\nu_1 + N)}{2} \log \frac{\nu_1 s_1^2 + N\bar{\sigma}^2}{\nu_1 + N} \end{aligned} \tag{69}$$

$$S(\Gamma_1, \Gamma_2) \approx \frac{N}{2}[\frac{\bar{\sigma}^2}{s_1^2} + \log \frac{\nu_1 s_1^2}{2\bar{\sigma}^2} - \Psi(\frac{\nu_1}{2})] \tag{70}$$

Thus surprise information scales linearly with N, with a coefficient of proportionality that typically depends mostly on the ratio of the empirical variance to the scale parameters s_1^2, which is roughly the expectation of the prior [the expectation of the prior is $\nu_1 s_1^2/(\nu_1 - 2)$ provided $\nu_1 > 2$]. The effects of very large of very small values of $\bar{\sigma}$, or ν_1 can also be seen in the formula above. In particular, surprise is largest when the empirical variance $\bar{\sigma}^2$ goes to 0 or infinity, i.e. is very different from the prior expectation.

Unknown Mean/Unknown Variance

In the case of unknown mean and unknown variance, we have a family $M(\mu, \sigma^2)$ of models with a conjugate prior of the form $G_1\Gamma_1 =$

$P(\mu|\sigma^2)P(\sigma^2) = G_1(\mu_1, \sigma^2/\kappa_1)\Gamma_1(\nu_1, s_1)$. Thus the prior has four parameters $(\mu_1, \kappa_1, \nu_1, s_1)$, with $\kappa_1 > 0$, $\nu_1 > 0$, and $s_1 > 0$. The conjugate posterior has the same form, with similar parameters $(\mu_2, \kappa_2, \nu_2, s_2)$ satisfying (see for instance [10])

$$\mu_2 = \frac{\kappa_1}{\kappa_1+N}\mu_1 + \frac{N}{\kappa_1+N}\bar{m} \tag{71}$$

$$\kappa_2 = \kappa_1 + N \tag{72}$$

$$\nu_2 = \nu_1 + N \tag{73}$$

$$\nu_2 s_2^2 = \nu_1 s_1^2 + (N-1)\bar{\sigma}^2 + \frac{\kappa_1 N}{\kappa_1+N}(\bar{m}-\mu_1)^2 \tag{74}$$

with $\bar{m} = \sum x_i/N$ and $\bar{\sigma}^2 = \sum(x_i - \bar{m})^2/(N-1)$. Computation of $F = F(\mu_1, \kappa_1, \nu_1, s_1; \mu_2, \kappa_2, \nu_2, s_2)$ is similar to the two cases treated above and yields:
$F(\mu_1, \kappa_1, \nu_1, s_1; \mu_2, \kappa_2, \nu_2, s_2) =$

$$-\frac{1}{2}[\log\frac{2\pi}{\kappa_2} + \frac{\kappa_2}{\kappa_1} + \log\frac{\nu_1 s_1^2}{2} - \Psi(\frac{\nu_1}{2}) + \kappa_2(\mu_2-\mu_1)^2 s_1^{-2}]$$
$$\frac{+\log(\frac{\nu_2}{2})^{\nu_2/2}s_2^{\nu_2}}{\Gamma(\frac{\nu_2}{2})} + (\frac{\nu_2}{2}+1)[\Psi(\frac{\nu_1}{2}) + \log\frac{2}{\nu_1 s_1^2}] - \frac{\nu_2 s_2^2}{2s_1^2} \tag{75}$$

From Equation 75, we can derive the surprise

$$\begin{aligned} S(G_1\Gamma_1, G_2\Gamma_2) &= \frac{1}{2}[\log\frac{\kappa_1}{\kappa_2} - 1 + \frac{\kappa_2}{\kappa_1} + \kappa_2(\mu_2-\mu_1)^2 s_1^{-2}] + \log\frac{C_1}{C_2} \\ &+ (\frac{\nu_1-\nu_2}{2})[\Psi(\frac{\nu_1}{2}) + \log\frac{2}{\nu_1 s_1^2}] + \frac{\nu_2 s_2^2 - \nu_1 s_1^2}{2s_1^2} \end{aligned} \tag{76}$$

Substituting the value of the posterior parameters

$$\begin{aligned} S(G_1\Gamma_1, G_2\Gamma_2) &= \frac{1}{2}\log\frac{\kappa_1}{\kappa_1+N} + \frac{N}{2\kappa_1} + \frac{\kappa_1+N}{2}\left[\frac{N(\bar{m}-\mu_1)}{(\kappa_1+N)s_1}\right]^2 \\ &+ \log\frac{C_1}{C_2} - \frac{N}{2}[\Psi(\frac{\nu_1}{2}) + \log\frac{2}{\nu_1 s_1^2}] \\ &+ \frac{(N-1)\bar{\sigma}^2 + \frac{\kappa_1 N}{\kappa_1+N}(\bar{m}-\mu_1)^2}{2s_1^2} \end{aligned} \tag{77}$$

For simplicity, we can consider the case where $\mu_1 = \bar{m}$. Then

$$\begin{aligned} S(G_1\Gamma_1, G_2\Gamma_2) &= \frac{1}{2}\log\frac{\kappa_1}{\kappa_1+N} + \frac{N}{2\kappa_1} + \log\frac{C_1}{C_2} \\ &- \frac{N}{2}[\Psi(\frac{\nu_1}{2}) + \log\frac{2}{\nu_1 s_1^2}] + \frac{(N-1)\bar{\sigma}^2}{2s_1^2} \end{aligned} \quad (78)$$

In all cases, for large values of N we always have the approximation

$$S(G_1\Gamma_1, G_2\Gamma_2) \approx \frac{N}{2}[\frac{1}{\kappa_1} + \frac{\bar{\sigma}^2}{s_1^2} + \log\frac{\nu_1 s_1^2}{2\bar{\sigma}^2} - \Psi(\frac{\nu_1}{2}) + \frac{(\bar{m}-\mu_1)^2}{s_1^2}] \quad (79)$$

Surprisee is linear in N with a coefficient that is essentially the sum of the coefficients derived in the unknown mean and unknown variance partial cases.

Acknowledgements

The work of PB is supported by a Laurel Wilkening Faculty Innovation Award and grants from the NIH and Sun Microsystems at UCI.

References

[1] J. Aczel and Z. Daroczy. *On measures of information and their characterizations.* Academic Press, New York, 1975.

[2] P. Baldi and S. Brunak. *Bioinformatics: the machine learning approach.* MIT Press, Cambridge, MA, 2001. Second edition.

[3] J. O. Berger. *Statistical decision theory and Bayesian analysis.* Springer-Verlag, New York, 1985.

[4] R. E. Blahut. *Principles and practice of information theory.* Addison-Wesley, Reading, MA, 1987.

[5] Bela Bollobas. *Random Graphs.* Academic Press, London, 1985.

[6] G. E. P. Box and G. C. Tiao. *Bayesian inference in statistical analysis.* John Wiley and Sons, New York, 1992. (First Edition in 1973).

[7] L. D. Brown. *Fundamentals of Statistical Exponential Families.* Institute of Mathematical Statistics, Hayward, CA, 1986.

[8] T. M. Cover and J. A. Thomas. *Elements of Information Theory.* John Wiley, New York, 1991.

[9] R. T. Cox. Probability, frequency and reasonable expectation. *Am. J. Phys.*, 14:1–13, 1964.

[10] A. Gelman, J. B. Carlin, H. S. Stern, and D. B. Rubin. *Bayesian Data Analysis.* Chapman and Hall, London, 1995.

[11] I. S. Gradshteyn and I. M. Ryzhik. *Table of integrals, series, and products.* Academic Press, New York, 1980.

[12] S.F. Gull. Developments in maximum entropy data analysis. In J. Skilling, editor, *Maximum entropy and Bayesian methods*, pages 53–71. Kluwer, Dordrecht, 1989.

[13] G. E. Hinton, P. Dayan, B. J. Frey, and R. M. Neal. The wake-sleep algorithm for unsupervised neural networks. *Science*, 268(5214):1158–1161, 1995.

[14] E. T. Jaynes. *Probability Theory: The Logic of Science.* 1996. Unpublished.

[15] E.T. Jaynes. Bayesian methods: General background. In J.H. Justice, editor, *Maximum Entropy and Bayesian Methods in Statistics*, pages 1–25. Cambridge University Press, Cambridge, 1986.

[16] G. Jumarie. *Relative information.* Springer Verlag, New York, 1990.

[17] S. Kullback. *Information theory and statistics.* Dover, New York, 1968. (First Edition in 1959).

[18] R. J. McEliece. *The Theory of Information and Coding.* Addison-Wesley Publishing Company, Reading, MA, 1977.

[19] L. J. Savage. *The foundations of statistics.* Dover, New York, 1972. (First Edition in 1954).

[20] C. E. Shannon. A mathematical theory of communication. *Bell System Technical Journal*, 27:379–423, 623–656, 1948.

[21] N. Tishby, F. Pereira, and W. Bialek. The information bottleneck method. In B. Hajek and R. S. Sreenivas, editors, *Proceedings of the 37th Annual Allerton Conference on Communication, Control, and Computing*, pages 368–377. University of Illinois, 1999.

[22] V. Vapnik. *The Nature of Statistical Learning Theory.* Springer Verlag, New York, 1995.

[23] A. Zellner. Optimal information processing and Bayes's theorem. *The American Statistician*, 42(4):278–284, 1998.

Dynamic Group Key Distribution Using MDS Codes

Lihao Xu
Department of Computer Science
Washington University
St. Louis, MO 63130
lihao@cs.wustl.edu

Abstract

Efficient key distribution is an important problem for secure group communications. In this paper, we present a new group key distribution scheme that supports *dynamic* multicast group membership changes. Instead of using encryption algorithms, the scheme employs MDS codes, a class of error control codes. The scheme achieves both forward and backward secrecy of sessions using freshly generated random session keys. The computation complexity of key distribution is significantly reduced using proper MDS codes with efficient encoding and decoding algorithms. The computation load is also more evenly distributed among the group key distribution controller (GC) and all the group members. Easily combined with any key-tree based schemes, this scheme provides much lower computation complexity, while maintaining low and balanced communication complexity and storage complexity for dynamic group key distribution.

Key Words: key distribution, multicast, MDS codes, erasure decoding

1 Introduction

In many applications, multicast is an efficient means of distributing data in terms of resources (such as network bandwidth, server computation and I/O load) usage. The privacy of a multicast communication session is usually ensured using (symmetric) encryption. All the designated receivers or members in a multicast group share a session (encryption) key. In many applications, however, the multicast group membership changes dynamically, i.e., some new members are authorized to join a new multicast session while some old members should be excluded. Thus session keys shall change dynamically to ensure both *forward secrecy* and *backward secrecy* of multicast sessions. The forward secrecy is maintained if an old member who has been excluded from the current session can *not* access the communication of the current session, and the backward secrecy is guaranteed if a new member of the current session can *not* recover the communication of past sessions. This requires each session need a new key that is only known to the current session members, i.e., session keys need to be *dynamically* distributed to authorized session members. In this paper, we adopt the common model where session keys are issued and distributed by a central *group controller* (GC). The resources needed for the GC to distribute session keys to group members include communication, storage and computation resources. The *communication complexity* is usually measured by the number of data bits that need to be transmitted from the GC to group members to convey information of session keys, while the *storage complexity* is measured by the number of data bits that the GC and group members need to store to obtain session keys, and the *computation complexity* is measured by the computation the GC and group member need to perform to derive session keys. Hereafter, the problem of how to effectively use resources to distribute session keys is referred to as the *group key distribution* problem.

The group key distribution problem has been studied extensively. There are two trivial schemes to distribute a session key to a group of n members. The first one is that the GC shares an individual key with each group member, which can be used to encrypt a new group session key. In this scheme, the communication complexity is $O(n)$, the GC needs to store $O(n)$ key information and each member stores $O(1)$ key information, and $O(n)$ encryption and decryption operations are needed. In the second scheme, the GC shares an individual key with each *subset* of the group, which can then be used to multicast a session key to a designated subset of group members. Now the communication complexity and computation complexity reduce to $O(1)$, at the cost of increasing the storage complexity to $O(2^n)$ for both the GC and each group member. It is easy to see neither of the schemes works for practical applications with a reasonable group size n. Thus research efforts have been made to achieve low communication and storage complexity for group key distribution.

Static secret sharing via broadcast channel was studied in [18, 14]. However, this *threshold*-based scheme can distribute a session key to a designated group of members for only one-time use. Once a session key is distributed to the group, any member can calculate the secret information other members in the same group hold. Thus the scheme does not provide forward or backward secrecy. A secure lock method based on the Chinese Remainder Theorem was proposed in [9]. But its prohibitively high communication complexity and computation complexity make it only practical for a very small group with limited number of members. Various theoretical measures and schemes for group key distribution were introduced in [11]. Along the same line, many research efforts have been made on balancing communication complexity and storage complexity of the group key distribution problems, e.g., [7, 12, 5, 6, 20, 21].

For a multicast group with a large number of members, *key-tree* based schemes were introduced to decompose a large group into multiple layers of subgroups with smaller sizes [22, 16, 23, 24]. Using these schemes, a group membership change can be effectively handled in the corresponding subgroups without affecting other ones. Thus communication complexity is reduced at the cost of increase in storage and computation complexity together with extra communication delays. For a group of n members, key-tree based schemes have a communication complexity of $O(logn)$, and storage complexity of $O(n)$ for the GC and $O(logn)$ for each group member. It has been shown that if a group member can store at most $O(logn)$ keys, then the lower bound of communication complexity is $O(logn)$, if a *structure-preserving* protocol is used for group key distribution[8]. Thus the key-tree based schemes are of practical interest for a variety applications because of its balance between communication complexity and storage complexity.

While most research on group key distribution has been on balancing communication complexity and storage complexity, little efforts have been made to reduce computation complexity. It has been long assumed expensive encryption and decryption operations are necessary to distribute group keys. However, in the model where a GC is used to distribute group keys, the computation pattern is *asymmetric*, i.e., the GC needs to perform many more computation operations than each individual group member. Thus besides heavy communication load, the GC can become a bottleneck also because of expensive computation operations it has to perform. Thus it is important to explore approaches of reducing computation complexity by 1) lowering the overall computation cost for both the GC and each individual group member, and 2) distributing some computation load away from the GC to each group member.

In this paper, we propose a new *dynamic* group key distribution scheme that reduces computation complexity and yet maintain at least the same security degree of using symmetric encryption algorithms. In our scheme, information related to session keys is encoded using *error control codes* rather than encryptions. In general, encoding and decoding of a proper error control code have much (at least one order) lower computation complexity than encryption and decryption. Thus computation complexity of key distribution can be significantly reduced. The similar idea of using error control codes to achieve privacy was employed in [18, 14, 3]. The major difference between these schemes and ours is that our scheme allows dynamic group membership changes with very low storage complexity, while the other schemes only work for a predefined static group. The security strength of this scheme will be evaluated, as well as its communication, storage and computation complexity. Besides its low computation complexity, this scheme also has low storage complexity: $O(1)$ for an individual group member, and $O(n)$ for the group controller, where n is the number of group members. It will be also shown this scheme can be easily applied to key-tree based schemes to further reduce its communication complexity from $O(n)$ to $O(logn)$.

This paper is organized as follows: the new scheme based on error control codes is described in Section 2, Section 3 evaluates the security and resources consumption of the scheme. Section 4 extends this basic scheme to key-tree based key distribution schemes. Section 5 concludes the paper.

2 An MDS Code Based Group Key Distribution Scheme

2.1 MDS Codes

MDS (*Maximum Distance Separable*) codes are a class of error control codes that meet the Singleton bound [13]. An (n, k) (block) error control code is a mapping from $GF(q)^k$ to $GF(q)^n$: $E(m) = c$, where $m = m_1m_2\cdots m_k$ is the original message and $c = c_1c_2\cdots c_n$ is its codeword, and $E(\cdot)$ is an encoding function with $k \leq n$. If a decoding function $D(\cdot)$ exists such that $D(c_{i_1}c_{i_2}\cdots c_{i_k}, i_1, i_2, \cdots, i_k) = m$ for $1 \leq i_j \leq n$ and $1 \leq j \leq k$, then this code is called an (n, k) MDS code. For an (n, k) MDS code, the k symbols of the original message can be recovered from *any* k symbols of its codeword. All the symbols are defined over $GF(q)$, and usually $q = 2^m$. The RS (Reed-Solomon) codes [17] are a class of frequently used MDS codes. Notably RS codes and other MDS codes can be used to construct secret-sharing and threshold schemes [18, 14].

2.2 Basic Scheme: Key Distribution Using MDS Codes

For a dynamic multicast group of up to n members, a session key is issued by a *group controller* (GC). Using this session key, the GC can establish a secure multicast channel with the authorized group members. Every time group memberships change because of join or leave of some group members, the GC reissues a new session key which is independent of all the old session keys. This *rekeying* procedure ensures the security of the current session, as well as that of the old sessions, namely, the newly-joined members cannot recover the communications of the old sessions and those old members who left the group cannot access the current session. Thus both backward secrecy and forward secrecy of group communication are maintained. The complexity of the rekeying operation is asymmetric when a new member joins and an old member leaves. When a new member joins, the GC can easily multicast the new session key encrypted by the current session to the current members, followed by a unicast to the new member to send the new session key encrypted by a predetermined encryption key shared between the GC and the new member. Thus join is easy with low communication and computation cost. However, when an old member leaves, the current session key cannot be used to convey the new session key information securely, since it is also known to the old member. So hereafter we will focus on the rekeying operation for member leaves.

Now we describe a new key distribution scheme which consists of three phases: 1) initialization of the GC; 2) join of a new member; and 3) rekeying procedure whenever a group member leaves.

2.2.1 Initialization

Initially, the GC constructs an (L, n) MDS code C over $GF(q)$ ($L \geq 2n - 1$) and a secure *one-way* hash function $H(\cdot)$ whose codomain is $GF(q)$. The domain of $H(\cdot)$ can be an arbitrary large enough space F so that $H(\cdot)$ has a secure one-way property: given any arbitrary $y \in GF(q)$, it is impossible or computationally hard to get $x \in F$ such that $H(x) = y$. Since other strong properties, such as 2nd-preimage resistance[15, Ch.9.2], are not necessary, the hash function H can be implemented more efficiently. The GC then makes public both the MDS code C and the one-way hash function H.

2.2.2 Initial Join

Whenever a new member i is authorized to join the multicast group for the first time, the GC sends it (using a secure unicast) a pair (j_i, s_i), where s_i is a random element in $H(\cdot)$'s domain F, and j_i is a positive integer satisfying 1) $n \leq j_i \leq L$; and 2) $j_i \neq j_k$ for all k's, where k is a current member of the multicast group. The pair (j_i, s_i) will be used as member i's *seed key* (denoted as S_i) and kept in GC's local database as long as member i remains a potential member of the multicast group.

2.2.3 Rekeying

Whenever some new members join or some old members leave the multicast group, the GC needs to distribute a new session key to all the current members. As already discussed, we will focus on the rekeying operation when an old member leaves. The GC executes the rekeying process in the following steps:

1. The GC randomly chooses a *fresh* element r in F, which has not been used to generate previous keys;

2. For each member i of the current group with its seed key (j_i, s_i), the GC constructs an element c_{j_i} in $GF(q)$: $c_{j_i} = H(s_i + r)$, where $+$ is a simple combining operation in F, e.g., string concatenation;

3. Using all the c_{j_i}'s in the above step, the GC constructs a codeword c of the (L, n) MDS code C: set the j_ith symbol of the codeword c to be c_{j_i} and the first $(n - p)$ symbols to be random elements in $GF(q)$, where p is the number of the current members in the multicast group. Since C is an (L, n) MDS code, the codeword c is uniquely determined by its n symbols. Using an efficient erasure decoding algorithm for C, the GC can easily calculate the first $(n - 1)$ symbols $c_1 c_2 \cdots c_{n-1}$ of the codeword c;

4. The GC decodes c using its n symbols to get the corresponding *message word* $M = M_1 M_2 \cdots M_n$. Set the new session key k to be $k = g(M)$, where $g(\cdot)$ is a predefined function whose output is l-bit long, as desired for the session key.

5. The GC multicasts r and $c_1 c_2 \cdots c_{n-1}$.

Upon receiving r and $c_1 c_2 \cdots c_{n-1}$ from the GC, an authorized member i of the current group executes the following steps to obtain the new session key:

1. Calculate $c_{j_i} = H(s_i + r)$ with its seed key (j_i, s_i);
2. Decode the codeword c using its n symbols $c_1 c_2 \cdots c_{n-1} c_{j_i}$ $(n \leq j_i \leq L)$, and recover the corresponding message word $M = M_1 M_2 \cdots M_n$;
3. Recover the new session key k from M: $k = g(M)$.

The MDS property of the code C ensures that each authorized member with its c_{i_j} and $c_1 c_2 \cdots c_{n-1}$ can decode the codeword c to get the *same* message word m as the GC. From the same m, a new session key k can be generated from M using any predetermined algorithm $g(\cdot)$, as long as $g(\cdot)$ generates an output with the desired length. Any unauthorized receiver cannot recover the message word M just from $c_1 c_2 \cdots c_{n-1}$, since it needs one more symbol of the codeword c. Any stale seed key (j_i', s_j') cannot generate a valid symbol of the current codeword c, since the pair (j_i', s_j') is not used in the generation of c.

3 Evaluation of the Basic Scheme

As can be seen from the above basic scheme, the GC generates a new session key by generating a *common* new message word for all the authorized group members. The new message word is decided by a random element r in F, as well as all the seed keys of the current authorized group members. The random element r and the first $(n-1)$ symbols of the codeword corresponding to the message word are multicast in *plaintext* to all the group members, thus computational cost is much lower than using secure point-to-point communications for the rekeying process. The computations needed for this new scheme are encoding and decoding of a chosen MDS code, a one-way hash function and some simple combining functions, all of which can be implemented efficiently and are computationally much cheaper than encryptions.

The following subsections evaluate the details of the new key distribution scheme: security, costs of communication, storage and computation.

3.1 Security

For simplicity, hereafter we assume $q = 2^m$ and the desired new session key is l-bit long. Since r and $c_1 c_2 \cdots c_{n-1}$ are multicast in plaintext and thus known to all interested parties, including unauthorized receivers who attempt to access the current session, the security of the new session key relies on the secrecy of a codeword symbol c_{j_i} that the authorized member i of the current multicast group has. The secrecy of c_{j_i}, in turn, depends on the secrecy of the seed key that member i has. Thus an attacker who attempts to deduce the new session key k has two ways: 1) guess a symbol c_k of the codeword c, which eventually generates the new session key k; 2) guess a current member's seed key. The effort of an attacker needs to make can be measured by the entropy of the information an attacker attempts to recover [10].

If the attacker chooses to guess a symbol c_{j_i} in the codeword directly, then the entropy of c_{j_i} is $\mathcal{H}(c_{j_i}) = log_2\ q = m$, where $\mathcal{H}(X)$ is the entropy of a random variable X. Even though the attacker also needs to get the correct location of a symbol in the codeword to generate the corresponding session key, the attacker can use the fact that for any location j_i, there always exists a correct symbol c_{j_i} which is a symbol of the codeword that generates a new session key. Thus the attacker can first pick up an arbitrary location j_i, where $n \geq j_i \geq L$, then guess the codeword symbol c_{j_i} at that location. So the entropy of getting a correct codeword symbol is $\mathcal{H}(c_{j_i}) = m$.

On the other hand, a seed key consists of two components: an element s_i in F and an integer j_i. Again for simplicity, we assume $F = GF(2^\infty)$, i.e., an input to the one-way function H can be a binary vector of arbitrary length. It is easy to see that $\mathcal{H}(s_i) = t$ if s_i is t-bit long. Also notice that the two components s_i and j_i of a seed key S_i are chosen *independently*, thus $\mathcal{H}(S_i) = \mathcal{H}(s_i) + \mathcal{H}(j_i)$, i.e., the entropy of a seed key is $t + log_2(L - n + 1)$.

Thus the amount of information an attacker needs to guess to obtain the session key k is $min(m, t + log_2(L - n + 1))$. In order to make the one-way hash function H secure, we need $t > m$, thus the entropy of the information an attacker needs to guess is m. Of course, the attacker can simply guess the l-bit session key k itself, whose entropy is $\mathcal{H}(k) = l$. Thus as long as $m \geq l$, the attacker needs to make an effort *no less* than guessing the session key itself.

Finally we examine the security strength of this scheme against conspiracy attack, where some old members cooperate to deduce the current session key. Since its freshness is determined by a fresh random number r, a new session key is totally independent of all the old session keys. This guarantees the *forward secrecy* and *backward secrecy*, i.e., old members cannot access the new session and new members cannot recover communications of all the previous sessions. One possible way for an old member to get a new session key is to calculate the seed key of a current member from old keys. It is easy to compute a symbol c_k of a codeword c from a message word m, once k is known. However, because of the secure one-way property of the hash function used to the generate c, it is impossible (or at least computationally hard) to compute a seed key (s_i, j_i) from c_{j_i} even if c_{j_i} is known. Cooperation among d old members helps reduce neither the number of possible locations for j_i of a current member nor computational hardness to get s_i from c_{j_i}, since j_i can be recycled from old ones as long as the current member i has unique j_i which is different from that of all other current members. Thus this scheme is *resilient* to conspiracy attack.

3.2 Communication, Storage and Computation Complexity

When a new member i is authorized to join the multicast group, the GC assigns a pair of seed keys (j_i, s_i) to it. This pair of seed keys is valid until member i leaves the multicast group. Thus the seed key pair is unicast only *once* to an authorized member. Compared to the rekeying procedure that is needed whenever there is a membership change for the multicast group, this communication cost (in terms of number of the bits transmitted) is negligibly small.

In the rekeying procedure, the GC needs to multicast a fresh random number r and $n-1$ symbols of the new codeword $c_1c_2\cdots c_{n-1}$. Each of the codeword symbols has m bits. The random number r is used to guarantee that the new session key is different from all the old keys used. The length of r determines the total number of sessions the scheme supports. Let r be l_r-bit long, then this scheme can support up to 2^{l_r} sessions. Thus the number of bits that the GC needs to multicast for a rekeying procedure is $l_r + (n-1)m$.

Now we consider storage space needed for both an individual member and the GC. A current member i only needs to store its own pair of seed keys (j_i, s_i). Since $j_i \leq L$ and s_i is t-bit long, member i needs to store $log_2 L + t$ bits locally. On the other hand, the GC needs to store the seed keys of all the current members of the multicast group. So if there are p ($p \leq n$) members in the current multicast group, the GC needs to store $p(log_2 L + t)$ bits. To ensure freshness of the random number l_r-bit r, the GC can simply store r's used in all the previous session so that a new r will be different from old r's. Then the GC needs an additional storage space of Ml_r bits, where M is the number of sessions. But this additional storage space is not necessary, since the GC can choose a well designed random number generator, that generates a sequence of random numbers with a long period, e.g., $2^{l_r}-1$ for l_r-bit random numbers. Such a sequence of random numbers can support up to $2^{l_r}-1$ sessions where each session is guaranteed to have a different random number r from all the other sessions.

Finally we examine computational cost of this scheme. Seed keys for a current member are distributed only once using a secure unicast when the member joins the multicast group. In this join procedure, the GC needs only one encryption operation and the newly-joined member needs only one decryption operation. Most computations are carried in the rekeying procedure. During the rekeying procedure, the GC needs p hashing operations with p current members in the multicast group and one encoding operation. Symmetrically a current member needs one hashing operation and one decoding operation. Thus the GC only performs $(p-1)$ more hashing operations than an individual member. Since hashing operations are usually much less expensive than encryption operations, exact cost of which depends on the hash function chosen, computation load is more evenly distributed from the GC to all the members. In most applications, the rekeying operation is a *blocking* operation, i.e., the operation does not finish until all the members do. Thus computation load balancing is important to reduce the overall execution time of such operations.

Both the encoding and decoding operations for an (L, n) MDS code only need $O(n^2)$ arithmetic operations if standard encoding and erasure decoding algorithms are used. Fast encoding and decoding algorithms only need $O(nlogn)$ operations [13]. Particularly, when a proper MDS code is chosen, both encoding and decoding only need $O(mn^2)$ binary exclusive OR operations [2][4][25]. Notice here that in the encoding operation, the GC only needs to compute n (instead of L) codeword symbols for the (L, n) MDS code. Similarly, in the decoding operation, a current member only needs to recover n original message symbols.

3.3 Proper Parameters

As analyzed above, the security strength of this scheme depends on the lengths of m, r, L and s_i's. Meanwhile, these parameters also decide the cost, namely communication, storage and computation complexity, of this scheme. Naturally one wants to achieve lowest cost while maintaining high security strength. It is easy to see from the above analysis, the smaller the lengths of these parameters are, the lower the cost of this scheme is. Also as been shown, the secrecy of a session key is guaranteed as long as $m \geq l$, i.e., the effort an attacker needs to make to obtain a session key is no less than directly guessing the l-bit session key itself. Thus it should set $m = l$.

It is conjectured that for all (L, n) MDS codes over $GF(q)$, where $q = 2^m$, the largest possible L is $q + 1 = 2^m + 1$ [13]. On the other hand, it is necessary $L \geq 2n - 1$ to ensure that each group member receives enough information to obtain the session key. Thus for a session key of l-bit, this scheme can support up to $2^{l-1} + 1$ members in one session. Since L doesn't affect communication and computation (encoding and decoding) complexity, while the storage space needed is proportional to $log_2 L$, we suggest L be maximized, i.e., $L = q + 1 = 2^m + 1 = 2^l + 1$, so that n, the number of members in one session the scheme can support, is maximized to $2^{l-1} + 1$.

Both r and s_i's, together with the hash function $H(\cdot)$, decide the freshness and unpredictability of a new session key. Assume that all the s_i's are of the same length t-bit, and r is of l_r-bit, then to guarantee the one-way hashing property of $H(\cdot)$, we need $t + l_r \gg m$. While t decides storage complexity, l_r affects the communication complexity of the rekeying procedures. Also l_r decides the maximum number of sessions where each session can have a different r. Thus we can make trade-offs between t and l_r, depending on the needs of applications. In most applications, setting $l_r = l$ for l-bit session keys should ensure the freshness of r and support up to $2^l - 1$ sessions, without significantly increasing the communication complexity in the rekeying procedures. Then t should be set to as least l.

Since the communication complexity is proportional to n, and the encoding and decoding complexity is proportional to n^2, then n, the maximal number of multicast group members in one session, should be kept as small as possible. One way of reducing n is to set $n = p$, where p is the actual number of members in the current session. Then of course, instead of using one fixed (L, n) MDS code, a class of (L, n_k) MDS codes should be used for different sessions. L can still be fixed to $L = 2^l + 1$ for all sessions, while n_k is set to the actual number of members in the kth session.

As described in the initial join operation, the GC only sends a user its seed key once when it is authorized to join the group for the first time. This seed key will be valid as long as the user remains to be a group member of a series of sessions which are not necessary consecutive. This reduces the communication complexity in many applications, where a user is authorized to join session C again, after it was a member of session A and stayed out of session B, where $A < B < C$. The cost of this communication and computation reduction is the need to increase storage space at the GC, since the GC needs to store the seed keys of all the possible participants, not just of the members in the current session. Since each participant i needs an unique j_i as one of its seed keys, and $n \leq j_i \leq L$, the maximum number of distinguished participants in the all session is $L-n+1 \leq \frac{L+1}{2}$, considering the factor that $L \geq 2n-1$. Thus for l-bit session keys, this scheme can support up to $2^{l-1}+1$ different participants for all the sessions.

4 Extending the Basic Scheme to Key Trees

The basic key distribution scheme reduces computation complexity by replacing computation expensive encryption and decryption operations with more efficient encoding and decoding operations of MDS codes. However, this basic scheme has the same communication complexity as the first trivial key distribution scheme.

To reduce the communication complexity of rekeying operation, a key-tree based scheme and many of its variations have been proposed [22, 16, 23, 24, 19]. This scheme reduces the communication complexity of rekeying operation to $O(logn)$, while each member needs to store $O(logn)$ keys and the GC needs to store $O(nlogn)$ keys, where n is the multicast group size. This is the most practical key distribution scheme, which balances the communication and storage complexity for dynamic multicast key distribution.

Here we briefly describe a basic key-tree based scheme for the rekeying operation.

4.1 Key-Tree Based Rekeying Scheme

The main idea to reduce the rekeying communication complexity of this scheme is to have the GC distribute *subgroup* keys in addition to individual member keys and the group session key. These keys are arranged in a *logical* tree hierarchy, where the group session key serves as the root, the individual member keys are the leaves, and the subgroup keys correspond to intermediate nodes. Each member stores all the keys along the path from the corresponding leaf to the root in the tree. Then each subgroup key can be used to securely multicast to the members that are leaves of the corresponding subtree. During the rekeying process, the

GC can thus securely multicast to a subgroup of members using their shared subgroup key instead of individual member keys.

Figure 1 shows a key-tree for a 9-member group. K_i $(1 \leq i \leq 9)$ is the individual key of member i. K_{1_9} is the group session key that is shared by all the members. Finally, K_{1_3}, K_{4_6} and K_{7_9} are 3 subgroup keys for the 3 corresponding subgroups respectively, e.g., K_{1_3} is shared by members 1 through 3, who form the first subgroup.

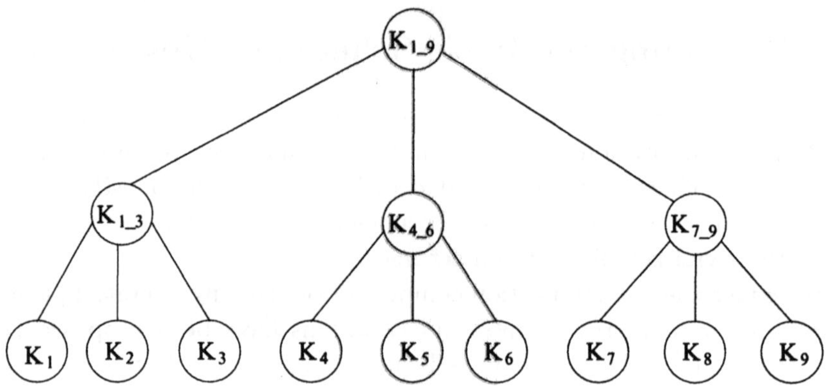

Figure 1: A Key Tree for a 9-member Group

Now, suppose member 9 leaves the group. Then a rekeying operation is needed to change the current group session key K_{1_9} and the corresponding subgroup key K_{7_9}. This can be achieved by the GC multicasting the following 5 messages (of equal size) to the remaining 8 members: $E_{K_7}(K_{7\_8})$, $E_{K_8}(K_{7\_8})$, $E_{K_{7\_8}}(K_{1\_8})$, $E_{K_{1\_3}}(K_{1\_8})$, $E_{K_{4\_6}}(K_{1\_8})$, where K_{1_8} is the new group session key, K_{7_8} is the new subgroup key for the new subgroup formed by members 7 and 8, and $E_k(m)$ is the encrypted version of message m using key k.

While the key-tree can be arbitrary, it has been shown that a balanced tree is required to achieve low communication complexity of the rekeying operation. In general, it is easy to show that the communication complexity of a rekeying operation on a balanced d-ary key tree is $O(dlog_d n)$, and each member needs to store $O(log_d n)$ keys. It is not hard to show $dlog_d n$ is minimized when d is 3[19]. The only cost of this key-tree-based scheme is the storage increase at the GC: the GC now needs to store all the additional subgroup keys as well. The total number of the keys stored on the GC increases to $\frac{dn-1}{d-1}$ from $n+1$ for a d-ary balanced key tree.

4.2 MDS Code Based Rekeying on A Key-Tree

The basic key distribution scheme based on MDS codes discussed in the previous section replaces expensive encryption and decryption operations with much cheaper encoding and decoding operations. Thus this basic scheme can be readily applied on a key-tree to further reduce the computation complexity of the rekeying operation, simply by replacing encryption operations with proper encoding operations.

As in other key-tree based rekeying schemes, MDS codes are used to rekey from bottom (leaves) up, so that child subtrees have their keys for use when their parent node is rekeyed. In Figure 1, when member 9 leaves, the new subgroup key K_{7_8} is rekeyed before the server changes the new group session key to K_{1_8}. When MDS codes are used for the rekeying process, each node (leaf or intermediate) key becomes a pair of (j_i, s_i) as discussed in the previous section. The GC stores all the key pairs on the key-tree. Whenever encryptions are needed for rekeying a subgroup key, a new MDS codeword is constructed from all the key pairs ((j_i, s_i)'s) of the corresponding immediate child nodes and then multicast by the GC.

In Figure 1, when member 9 leaves the group, the GC first uses the key pairs $K_7 = (j_7, s_7)$ and $K_8 = (j_8, s_8)$, together with a fresh random r, to construct a codeword of an $(L, 2)$ MDS code, and then follow the rekeying procedure of the basic scheme as described in the previous section. After proper decoding, member 7 and 8 share a new subgroup key K_{7_8} that is only known to them, besides the GC. This finishes the rekeying of K_{7_8}. Next, the GC constructs another codeword of an $(L, 3)$ MDS code from the subgroup keys K_{1_3}, K_{4_6} and K_{7_8}, and the decoding output from this codeword produces a new group session key K_{1_8}, which is shared by all the remaining group members.

Note that when the key tree is a d-ary balanced tree, only an $(L, d-1)$ MDS code is needed to rekey the immediate subgroup key shared by old member. Then an (L, d) MDS code is needed for rekeying all the other subgroup keys and the new group session key. Since the rekeying scheme based on MDS codes does *not* change the communication and storage complexity of the underlining key-tree-based rekeying scheme, the communication complexity still remains $O(dlog_d n)$, As pointed out in the previous subsection, d needs to be 3 to minimize the rekeying communication complexity. Thus the MDS codes needed are $(L, 2)$ and $(L, 3)$ codes, and their encoding and decoding are significantly simpler than general (L, n) MDS codes, where $n \gg 3$. Thus the MDS-Code-based scheme is particular suitable and practical for rekeying operations on key trees.

5 Conclusion

We have presented a dynamic group key distribution scheme using MDS codes. The computation complexity of key distribution is greatly reduced by employing only encoding and erasure decoding operations instead of more expensive encryption and decryption computations. The overall execution time of group key distribution is further lessened by balancing the computation load among the GC and all the group members. Easily combined with key trees, this scheme provides much lower computation complexity, while maintaining low and balanced communication complexity and storage complexity for dynamic group key distribution. This scheme is thus practical for many applications, including Internet multicast.

References

[1] Michel Abdalla, Yuval Shavitt, and Avishai Wool, "Towards Making Broadcast Encryption Practical", IEEE/ACM Transactions on Networking, 8(4):443–454, August 2000.

[2] M. Blaum, J. Bruck, A. Vardy, "MDS Array Codes with Independent Parity Symbols", *IEEE Trans. on Information Theory*, 42(2), 529-542, March 1996.

[3] R. Blom, "An Optimal Class of Symmetric Key Generation Systems", *Advances in Cryptology*, EUROCRYPT'84, LNCS 209, 335-338, 1984.

[4] J. Bloemer, M. Kalfane, M. Karpinski, R. Karp, M. Luby and D. Zuckerman, "An XOR-Based Erasure-Resilient Coding Scheme", ICSI Technical Report, TR-95-048, August 1995.

[5] C. Blundo, A. Cresti, "Space Requirement for Broadcast Encryption", *Advances in Cryptology*, EUROCRYPT'94, LNCS 950, 287-298, 1995.

[6] C. Blundo, A. De Santis, A. Herzberg, S. Kutten, U. Vaccaro and M. Yung, "Perfectly Secure Key Distribution in Dynamic Conferences", *Advances in Cryptology*, EUROCRYPT'92, LNCS 740, 471-486, 1993.

[7] C. Blundo, L. A. Frota Mattos, D. R. Stinson, "Trade-offs between Communication and Storage in Unconditionally Secure Schemes for Broadcast Encryption and Interactive Key Distribution", *Advances in Cryptology*, CRYPTO'96, LNCS 1109, 387-400, 1996.

[8] R. Canetti, T. Malkin, K. Nissim, "Efficient Communication-Storage Tradeoffs for Multicast Encryption", *Advances in Cryptology*, EUROCRYPT 1999, May 1999.

[9] G. H. Chou and W. T. Chen, "Secure Broadcasting Using the Secure Lock", *IEEE Trans. on Software Engineering*, 15(8), 929-934, Aug. 1989.

[10] T. M. Cover and J. A. Thomas, *Elements of Information Theory*, John Wiley & Son, 1991.

[11] A. Fiat and M. Naor, "Broadcast Encryption", *Advances in Cryptology*, CRYPTO'93, LNCS 773, 480-491, 1994.

[12] M. Luby and J. Staddon, "Combinatorial Bounds for Broadcast Encryption", *Advances in Cryptology*, EUROCRYPT'98, LNCS 1403, 512-526, 1998.

[13] F. J. MacWilliams and N. J. A. Sloane, *The Theory of Error Correcting Codes*, Amsterdam: North-Holland, 1977.

[14] R. J. McEliece and D. V. Sarwate, "On Sharing Secrets and Reed-Solomon Codes", *Communications of ACM*, 26(9), 583-584, Sep. 1981.

[15] A. J. Menezes, P. C. van Oorschot and S. A. Vanstone, *Handbook of Applied Cryptography*, CRC Press, 4th Printing, 1999.

[16] S. Mittra, "Iolus: A Framework for Scalable Secure Multicasting", *Proc. of ACM SIGCOMMi'97*, 277-288, Sep. 1997.

[17] I. S. Reed and G. Solomon, "Polynomial Codes over Certain Finite Fields", *J. SIAM*, 8(10), 300-304, 1960.

[18] A. Shamir, "How to Share a Secret", *Communications of ACM*, 24(11), 612-613, Nov. 1979.

[19] J. Snoeyink, S. Suri and G. Varghese, "A Lower Bound for Multicast Key Distribution", *Proc. IEEE INFOCOM 2001*, Apr. 2001.

[20] D. R. Stinson, "On Some Methods for Unconditionally Secure Key Distribution and Broadcast Encryption", *Designs, Codes and Cryptography*, 12, 215-243, 1997.

[21] D. R. Stinson and T. van Trung, "Some New Results on Key Distribution Patterns and Broadcast Encryption", *Designs, Codes and Cryptography*, 14, 261-279, 1998.

[22] M. Waldvogel, G. Caronni, D. Sun, N. Weiler and B. Plattner, "The VersaKey Framework: Versatile Group Key Management", *IEEE Journal on Selected Areas in Communications*, 7(8), 1614-1631, August 1999.

[23] D. M. Wallner, E. J. Harder and R. C. Agee, "Key Management for Multicast: Issues and Architectures", *Internet Draft* (work in progress), draft-wallner-key-arch-01.txt, Sep. 15, 1998.

[24] C. K. Wong, M. Gouda and S. S. Lam, "Secure Group Communications Using Key Graphs", *Proc. ACM SIGCOMM'98*, Sep. 1998.

[25] L. Xu and J. Bruck, "X-Code: MDS Array Codes with Optimal Encoding," *IEEE Trans. on Information Theory*, 45(1), 272-276, Jan., 1999.

Worst-Case Mutual Information Trajectories in Concatenated Codes with Asymptotic Interleavers

Dariush Divsalar
Sequoia Communications,Los Angeles,
and Jet Propulsion Laboratory
4800 Oak Grove Drive
Pasadena, CA 91109
darius@sequoia-communications.com
dariush@shannon.jpl.nasa.gov

Shlomo Shamai
Department of Electrical Engineering,
Technion -Israel Institute of Technology,
Haifa 32000, Israel
sshlomo@ee.technion.ac.il

Abstract

In this paper we present preliminary results on extremal (worst and best) average mutual information values carried by the extrinsic loglikelihood under the constraint of a given mean and variance while accounting for the consistency feature of loglikelihoods. This is done in an effort to gain insight into the iterative decoding procedure without resorting to the classical "Gaussian approximation".

Keywords: Mutual Information, Extrinsic information, Concatenated codes.

1. Introduction

In recent contributions [1]-[4] an approximate method based on a Gaussian approximation for the a posteriori extrinsic probability has been invoked in an effort to gain insight into the major parameters affecting the performance of iterative decoding of concatenated interleaved coding systems. While [1] focuses on signal-to-noise ratio transfer, the studies in [2]-[6] adopt the mutual information measure. All results in [1]-[4] rely on the Gaussian approximation in the sense of [7], that is the extrinsic [8] probability function modeled as Gaussian and hence characterized by its average and variance. See [7] for further details and justifications.

In this work, we wish to relax the Gaussian approximation. Specifically, we shall use the mutual information framework as in [3]-[4], but rather than invoking the Gaussian assumption, we shall introduce the worst case associated mutual information which still satisfies the discrimination (the expectation of the loglikelihood ratio) and a variance constraint as well as the consistency property [7], which is satisfied by the a posteriori extrinsic loglikelihood ratio (interpreted here as a random variable). To be addressed also the maximum of the associated mutual information under the same average and variance constraints of the extrinsic loglikelihood ratio.

It is emphasized that the parameters of the a posteriori extrinsic loglikelihood ratio variable, such as its average (discrimination) and variance should be evaluated via probability propagation [9], not adhering to any Gaussian approximations.

2. Preliminaries

Consider a random variable y which carries information on a binary symmetric random variable x taking on the values -1 and 1 with equal probability $1/2$. Assume that y takes on values in some observation space $\mathcal{Y}$, and is characterized by the conditional probability measure $p(y|x)$.

We now define a symmetry property as follows:

Symmetry Definition [11]. For all $y \in \mathcal{Y}$, there corresponds a $y' \in \mathcal{Y}$, such that the Radon-Nykodim derivatives satisfy:

$$dp(y|1)/dp(y'|-1) = dp(y|-1))/dp(y'|1) = 1\,. \tag{2.1}$$

A special case of this definition is the standard symmetric transition probability property where $y' = -y$.

In the following it is useful to introduce to loglikelihood variable

$$\lambda = \log \frac{dp(y|1)}{dp(y|-1)}\,, \tag{2.2}$$

which is in fact a transformation of the random variable y.

Following [9],[10] it is elementary to show the consistency of the random variable λ. Assume that the effective probability measure is $p(y|1)$ (interpreted as assuming that $x = 1$, was transmitted). This induces a probability measure on λ (2.2) designated by $q_1(\lambda)$, and respectively $q_{-1}(\lambda)$ is induced by $p(y|-1)$.

Consistency: For a symmetric conditional measure we have [11]

$$q_1(-\lambda) = e^{-\lambda} q_1(\lambda) \tag{2.3}$$

and correspondingly,

$$q_{-1}(\lambda) = e^{-\lambda} q_{-1}(-\lambda)\,. \tag{2.4}$$

2.1. Mutual Information

Consider now the average mutual information:

$$I = I(y;x) = \mathrm{E} \log \frac{dp(y|x)}{dp(y)} , \tag{2.5}$$

where E designates here the expectation operator. It follows straightforwardly that

$$I = \log 2 - \frac{1}{2} \int_{-\infty}^{\infty} dp(y|1) \log \left(1+e^{-\lambda}\right) - \frac{1}{2} \int_{-\infty}^{\infty} dp(y|-1) \log \left(1+e^{\lambda}\right) , \tag{2.6}$$

where λ and y are interrelated by (2.2).

Invoking the symmetry condition (2.1) yields immediately that

$$\begin{aligned} I &= \log 2 - \int_{-\infty}^{\infty} dp(y|1) \log \left(1+e^{-\lambda}\right) \\ &= \log 2 - \int_{-\infty}^{\infty} dq_1(\lambda) \log \left(1+e^{-\lambda}\right) . \end{aligned} \tag{2.7}$$

Since

$$I = H(x) - H(x|y) ,$$

where H stands for entropies or conditional entropies, by (2.7) it is clear that the equivocation

$$H(x|y) \triangleq H_{x|\lambda} \triangleq \int dq_1(\lambda) \log \left(1+e^{-\lambda}\right) . \tag{2.8}$$

3. Optimization Problem

We are now in the position to set up the optimization problem. We first wish to minimize I or to maximize the equivocation $H_{x|\lambda}$.

$$\max_{q_1(\lambda)} \int_{-\infty}^{\infty} dq_1(\lambda) \log\left(1 + e^{-\lambda}\right) \tag{3.1}$$

over all distributions of λ that satisfy

$$\int_{-\infty}^{\infty} dq_1(\lambda) = 1 , \tag{3.2}$$

$$\int_{-\infty}^{\infty} \lambda \, dq_1(\lambda) = \bar{\lambda} , \tag{3.3}$$

$$\int_{-\infty}^{\infty} \lambda^2 \, dq_1(\lambda) = \sigma^2 + (\bar{\lambda})^2 , \tag{3.4}$$

$$\frac{dq_1(-\lambda)}{dq_1(\lambda)} = e^{-\lambda} . \tag{3.5}$$

Constraints (3.3),(3.4) are interpreted as a given predetermined mean ($\bar{\lambda}$) and variance (σ^2) while (3.5) designates the consistency property.

We now use the consistency property (3.4) and the evident equality

$$\int_{-\infty}^{0} dq_1(\lambda) \, G(\lambda) = \int_{0}^{\infty} dq_1(-\lambda) \, G(-\lambda) = \int_{0}^{\infty} dq_1(\lambda) \, e^{-\lambda} \, G(-\lambda) \tag{3.6}$$

for any $G(\lambda)$, to transform the optimization problem originally stated (3.1)-(3.5) in the double infinite interval $(-\infty, \infty)$ to the positive infinite interval $(0, \infty)$. This yields

$$\max_{\substack{q_1(\lambda) \\ \lambda \geq 0}} \int_0^\infty dq_1(\lambda) \left[\log\left(1+e^{-\lambda}\right) + e^{-\lambda} \log\left(1+e^{\lambda}\right)\right] , \tag{3.7}$$

with the constraints,

$$\int_0^\infty dq_1(\lambda) \left(1+e^{-\lambda}\right) = 1 , \tag{3.8}$$

$$\int_0^\infty dq_1(\lambda)\, \lambda\left(1-e^{-\lambda}\right) = \bar{\lambda} , \tag{3.9}$$

$$\int_0^\infty dq_1(\lambda)\, \lambda^2\left(1+e^{-\lambda}\right) = \sigma^2 + (\bar{\lambda})^2 . \tag{3.10}$$

We can further transform the optimization into an optimization with respect to probability measure $\mu(\lambda)$ where

$$d\mu(\lambda) = dq_1(\lambda)\left(1+e^{-\lambda}\right) \quad , \quad \lambda \geq 0 , \tag{3.11}$$

which reads,

$$\begin{aligned} \max_{\mu(\lambda), \lambda \geq 0} \quad & \int_0^\infty d\mu(\lambda) \left[\frac{\log\left(1+e^{-\lambda}\right) + e^{-\lambda} \log\left(1+e^{\lambda}\right)}{1+e^{-\lambda}}\right] \\ = & \int_0^\infty d\mu(\lambda) \left[\log\left(1+e^{-\lambda}\right) + \frac{\lambda e^{-\lambda}}{1+e^{-\lambda}}\right] , \end{aligned} \tag{3.12}$$

$$\int_0^\infty d\mu(\lambda) = 1 , \tag{3.13}$$

$$\int_0^\infty d\mu(\lambda)\, \lambda \left(\frac{1-e^{-\lambda}}{1+e^{-\lambda}}\right) = \bar{\lambda} , \tag{3.14}$$

$$\int_0^\infty \lambda^2 \, d\mu(\lambda) = \sigma^2 + (\bar{\lambda})^2 \, . \tag{3.15}$$

Assertion (tentative solution):

The maximizing distribution for (3.12) under constraints (3.13)-(3.15) is given by:

$$d\mu(\lambda) = (1-a)\,\delta(0)\,d\lambda + a\delta(\lambda - \lambda_0)\,d\lambda \, , \tag{3.16}$$

or equivalently in terms of:

$$dq_1(\lambda) = (1-a)\,\delta(0)\,d\lambda + a\left(\frac{\delta(\lambda-\lambda_0)}{1+e^{-\lambda_0}} + \frac{e^{-\lambda_0}\,\delta(\lambda+\lambda_0)}{1+e^{-\lambda_0}}\right) d\lambda \, , \tag{3.17}$$

where

$$\lambda_0 = \left(\sigma^2 + (\overline{\lambda})^2\right)/x \tag{3.18}$$

and where

$$x \triangleq a\lambda_0 \tag{3.19}$$

is the solution of the equation

$$x \left(\frac{1 - e^{-\frac{(\sigma^2+\overline{\lambda}^2)}{x}}}{1 + e^{-\frac{(\sigma^2+\overline{\lambda}^2)}{x}}} \right) = \overline{\lambda} \, . \tag{3.20}$$

By (3.18), (3.20) we have,

$$a = \frac{x^2}{\left(\sigma^2 + (\overline{\lambda})^2\right)} \, . \tag{3.21}$$

The corresponding average mutual information is

$$I = \frac{-x^2}{\sigma^2 + \overline{\lambda}^2} \log\left(\frac{1 + e^{-\left(\frac{\sigma^2+\overline{\lambda}^2}{x}\right)}}{2} \right) - \frac{x\,e^{-\left(\frac{\sigma^2+\overline{\lambda}^2}{x}\right)}}{1 + e^{-\left(\frac{\sigma^2+\overline{\lambda}^2}{x}\right)}} \, . \tag{3.22}$$

Results (3.20),(3.22) are presented in a parametric form. That is for a fixed value of $\frac{\bar{\lambda}^2}{\sigma^2}$, $\bar{\lambda}$ and $I_{\min}$ are given in a parametric form as a function of $x \in \left(0, \sqrt{\sigma^2 + (\bar{\lambda})^2}\right)$. It might be useful to introduce another parameter a (3.21) which takes on values in $a \in (0, 1]$.

Assertion (tentative solution):

The minimization in (3.12) over $\mu(\lambda)$ under constraints (3.13)-(3.15) is achieved by

$$d\mu(\lambda) = \lim_{\varepsilon \to 0} \left\{ (1-\varepsilon)\delta(\lambda - \lambda_0)\, d\lambda + \varepsilon\delta\left(\lambda - \frac{\sqrt{\sigma^2 + \bar{\lambda}^2 - (1-\varepsilon)\lambda_0^2}}{\sqrt{\varepsilon}}\right) \right\} d\lambda\,, \tag{3.23}$$

where

$$\lambda_0 \left(\frac{1 - e^{-\lambda_0}}{1 + e^{-\lambda_0}} \right) = \bar{\lambda}\,.$$

In terms of $q_1(\lambda)$ (3.23) at the limit $\varepsilon = 0$ transforms into

$$dq_1(\lambda) = \frac{1}{1 + e^{-\lambda_0}}\, \delta(\lambda - \lambda_0)\, d\lambda + \frac{e^{-\lambda_0}}{1 + e^{-\lambda_0}}\, \delta(\lambda + \lambda_0)\, d\lambda\,. \tag{3.24}$$

The corresponding mutual information is given by

$$I_{\max} = -\log\left(\frac{1 + e^{-\lambda_0}}{2} \right) - \frac{\lambda_0\, e^{-\lambda_0}}{1 + e^{-\lambda_0}}\,. \tag{3.25}$$

Note that $I_{\max}$ is independent of the value of σ^2 and is fully determined by the sole constraint $\bar{\lambda}$.

4. Conclusions

In this paper we have identified the minimum (and maximum) average mutual information associated with the log likelihood consistent random variable, standing for the extrinsic information. This is done in an effort to relax the Gaussian approximation while propagating mutual information expression through the iteration process in a similar manner as done in [2]-[4]. To that end the a posteriori average $\bar{\lambda}$ and variance σ^2 should be exactly evaluated [10] without resorting to a Gaussian approximation [7]. The minimum mutual information is then, a function of $\bar{\lambda}$ and σ^2. The maximal value of the mutual information depends on $\bar{\lambda}$ only. Our mathematical results can also be used to propagate the densities resulting from maximum and minimum mutual information. This research is in progress and the results are being extended by using higher moments of the extrinsic information.

References

[1] D. Divsalar, S. Dolinar and F. Pollara, "Iterative Turbo Decoder Analysis Based on Density Evolution," IEEE JSAC, May 2001.

[2] S. ten Brink, "Convergence Behavior of iterative Decoded Parallel Concatenated Codes," IEEE Trans. Commun. Vol. 49, No. 10, pp. 1727-1737, October 2001.

[3] S. ten Brink, "Design of Serially Concatenated Codes Based on Iterative Decoding Convergence," *Proc. 2nd and International Symposium on Turbo Codes*, pp. 319–322, Brest, France, Sept. 4–7, 2000.

[4] S. ten Brink, "Designing Iterative Decoding Schemes with Extrinsic Information Transfer Chart," AEU-International Journal of Electronics and Communications, Vol. 54, No. 6, pp. 389-398, 2000.

[5] S. ten Brink, "Exploiting the Chain Rule of Mutual Information for the Design of Iterative Decoding Schemes," 39th Ann. Allerton Conf. Com. Contr. and Computing, Monticello, Illinois, October 2001.

[6] A. Ashimkin, G. Kramer and S. ten Brink, "Code Rate the Area under Extrinsic Information Transfer Curves," submission to: IEEE Trans. Inform. Theory, Lausanne, Switzerland, June 30-July 5, 2002.

[7] S.-Y. Chung, T. Richardson and R. Urbanke, "Analysis of Sum Product Decoding of Low Density Parity Check Codes using Gaussian Approximations," *IEEE Trans. Inform. Theory*,Feb. 2001.

[8] J. Hagenauer, E. Offer and L. Papke, "Iterative Decoding of Binary Block and Convolutional Codes," *IEEE Trans. Inform. Theory*, Vol. 42, No. 2, pp. 429–445.

[9] T. Richardson and R. Urbanke, "Analysis and Design of Iterative Decoding Systems," *1999 IMA Summer Program: Codes Systems and Graphical Models*, Minnesota, USA, August 2–6, 1999.

[10] T. Richardson and R. Urbanke, "An Introduction to the Analysis of Iterative Coding Systems," preprint.

[11] M. Peleg, "Consistency with General Symmetry," *Personal Communication*.

Results to get Maximal Quasihermitian Curves. New possibilities for AG Codes

Robert J. McEliece*
Department of Electrical Engineering
California Institute of Technology
Pasadena, CA 91125
rjm@systems.caltech.edu

M.C. Rodríguez-Palánquex†
Escuela Universitaria de Estadística
Universidad Complutense de Madrid
28040 Madrid, Spain
mcrodri@mat.ucm.es

Abstract

For *Quasihermitian curves* defined over $\mathbf{F}_2$, with genus $g \geq 1$, we present sufficient conditions for getting maximal curves on $\mathbf{F}_{2^{2g}}$. This shows a way to obtain good AG Codes from this class of curves.

Keywords: Maximal curves, zeta function.

*R.J. McEliece's contribution to this paper was supported by NSF grant no. CCR9804793, and grants from the Sony Corp., Qualcomm, Caltech's Lee Center for Advanced Networking.

†M.C. Rodríguez-Palánquex was supported by DGICYT (*"Dirección General de Investigación del Ministerio de Ciencia y Tecnología"*) under grant TIC2000-0735. During this work M.C. Rodríguez-Palánquex was with the *Department of Electrical Engineering at California Institute of Technology.* This author would like to express her appreciation to the *"Programa Complutense del Amo"* (Universidad Complutense de Madrid), for providing her a grant for this stay.

1 Introduction

By the *Hasse-Weil Theorem*, the number of rational points of a *smooth projective curve* C, of genus g, defined over a finite field $K = \mathbf{F}_q$ is bounded by

$$\#C(\mathbf{F}_q) \leq q + 1 + 2g\sqrt{q} \tag{1}$$

C is *maximal* when $\#C(\mathbf{F}_q)$ attains this upper bound.

Curves with many rational points are interesting in AG Coding Theory to get good codes. This is the case of the *Quasihermitian Curves* [4,5], recently studied. From them good AG Goppa Codes have been defined [1]. The results that we present in this paper, are consequence of the observations from many examples of maximal Quasihermitan curves over the finite field $\mathbf{F}_{2^{2g}}$. We wanted to answer the question: Which are the conditions on g to get maximal Quasihermitian curve over $\mathbf{F}_{2^{2g}}$?

As is well known, compute $\#C(\mathbf{F}_{q^r})$, for all $r \leq g$ is necessary to determine $\#C(\mathbf{F}_{q^r})$, for all $r > g$. From this study, we have a faster way to distinguish maximal curves C over $\mathbf{F}_{2^{2g}}$, and, therefore, interesting consequences in AG Coding Theory.

2 Properties of $\mathbf{Z}(\mathbf{C}, z)$. Examples for Quasihermitian Curves

Let C be a smooth projective curve over the finite field $\mathbf{F}_2$, with genus g.

We can determine the number of rational points of C over the finite fields $\mathbf{F}_{2^r}$, $\#C(\mathbf{F}_{2^r})$, for all $r > g$, if we compute previously $\#C(\mathbf{F}_{2^r})$, $\forall r = 1, ..., g$. From the definition and properties of *zeta function* ([2], Part II, Sect.IV), we know that

$$\#C(\mathbf{F}_{2^r}) = 2^r + 1 - \sum_{i=1}^{g}(\alpha_i^r + \overline{\alpha_i}^r) = 2^r + 1 - 2\sum_{i=1}^{g}\mathrm{Re}(\alpha_i^r) \tag{2}$$

with $\alpha_1, ..., \alpha_g$ complex numbers, $|\alpha_i| = \sqrt{2}$, $\quad \alpha_i \cdot \overline{\alpha_i} = 2$, and being then, the *zeta function*

$$Z(C, z) = \frac{\prod_{i=1}^{g}(1 - \alpha_i z)(1 - \overline{\alpha_i} z)}{(1 - z)(1 - 2z)} \tag{3}$$

We go to apply these properties to some examples with *Quasihermitian curves.*

Quasihermitian curves are defined over $\mathbf{F}_{2^r}$, $\forall a, b \in \mathbf{Z}$, being $a \geq 2$, $b > -a$, $\beta_1, \beta_2 \in \mathbf{F}_{2^r} - \{0\}$, by the affine equation

$$y^a + \beta_1 y + \beta_2 x^{a+b} = 0 \tag{4}$$

So, this class of curves are given by the homogeneus equations

$$\begin{array}{l} C_{a,a+b} : Y^a Z^b + \beta_1 Y Z^{a+b-1} + \beta_2 X^{a+b} = 0 \quad \text{if } b \geq 0 \\ \text{or} \\ D_{a,a+b} : Y^a + \beta_1 Y Z^{a-1} + \beta_2 X^{a+b} Z^{-b} = 0 \quad \text{if } 0 > b \end{array} \tag{5}$$

Example 2.1 We consider the Quasihermitian curve $C_{3,5} : Y^3Z^2 + YZ^4 + X^5 = 0$, whose genus is 2 (its singular points are: $(0, 1, 0)$ and $(0, 0, 1)$, each one of them with only a branch over $\mathbf{F}_{2^r}$). Over $\mathbf{F}_2$ there are three points as one can easily check, and over $\mathbf{F}_{2^2}$ one finds 2 more points, so $\#C_{3,5}(\mathbf{F}_2) = 3$ and $\#C_{3,5}(\mathbf{F}_{2^2}) = 5$. According to (2), we have that

$$\left\{ \begin{array}{l} \alpha_1 + \overline{\alpha_1} + \alpha_2 + \overline{\alpha_2} = 0 \\ \alpha_1^2 + \overline{\alpha_1}^2 + \alpha_2^2 + \overline{\alpha_2}^2 = 0 \\ \alpha_1 \cdot \overline{\alpha_1} = 2 \\ \alpha_2 \cdot \overline{\alpha_2} = 2 \end{array} \right. \Bigg| \Rightarrow \left\{ \begin{array}{l} \alpha_1 = 1 + j \\ \alpha_2 = -1 + j \end{array} \right.$$

being, then, the zeta function

$$Z(C_{3,5}, z) = \frac{(1 - 2z + 2z^2)(1 + 2z + 2z^2)}{(1 - z)(1 - 2z)} = \frac{1 + 4z^2}{(1 - z)(1 - 2z)}$$

Therefore,

$$\begin{aligned} \#C_{3,5}(\mathbf{F}_{2^r}) &= 2^r + 1 - 2\sum_{i=1}^{2} \mathrm{Re}(\alpha_i^r) = \\ &= 2^r + 1 - 2\sqrt{2^r}\left(\cos\frac{r\pi}{4} + \cos\frac{3r\pi}{4}\right) \end{aligned}$$

We have, for example, that

$$\#C_{3,5}(\mathbf{F}_{2^4}) = 33 \text{ and } \#C_{3,5}(\mathbf{F}_{2^{12}}) = 4,353$$

$C_{3,5}$ is maximal over $\mathbf{F}_{2^4}$ and $\mathbf{F}_{2^{12}}$.

Example 2.2 For the Quasihermitian curve $C_{4,6} : Y^4Z^2+YZ^5+X^6 = 0$, whose genus is 3, we have that $\#C_{4,6}(\mathbf{F}_2) = 3$, $\#C_{4,6}(\mathbf{F}_{2^2}) = 5$ and $\#C_{4,6}(\mathbf{F}_{2^3}) = 9$, being, then, the zeta function

$$Z(C_{4,6}, z) = \frac{1+8z^6}{(1-z)(1-2z)}$$

being $C_{4,6}$ maximal over $\mathbf{F}_{2^6}$ with 113 points.

3 Maximal Quasihermitian Curves over $\mathbf{F}_{2^{2g}}$

Let C be a smooth projective curve over a finite field $\mathbf{F}_2$, with genus g whose zeta function is

$$Z(C, z) = \frac{\prod_{i=1}^{g}(1-\alpha_i z)(1-\overline{\alpha_i} z)}{(1-z)(1-2z)} \tag{6}$$

First of all, we observe the following (from the Hasse-Weil bound)

$$C \text{ is maximal over } \mathbf{F}_{2^{2g}} \Leftrightarrow \#C(\mathbf{F}_{2^{2g}}) = 2^{2g} + 1 + 2^{g+1}g$$

By (2)

$$\#C(\mathbf{F}_{2^{2g}}) = 2^{2g} + 1 - \sum_{i=1}^{g}(\alpha_i^{2g} + \overline{\alpha}_i^{2g}) \tag{7}$$

Then

$$C \text{ is maximal over } \mathbf{F}_{2^{2g}} \Leftrightarrow \sum_{i=1}^{g}(\alpha_i^{2g} + \overline{\alpha}_i^{2g}) = -2^{g+1}g$$

From here we present the new results. For them, we consider

$$A = \{\{w_1, \overline{w}_1\}, \{w_2, \overline{w}_2\}, ..., \{w_g, \overline{w}_g\}\} \tag{8}$$

set of g pairs of conjugated complex numbers of modulus 1:

$$|w_i|^2 = |\overline{w}_i|^2 = w_i\overline{w}_i = 1, \forall i = 1, ..., g \tag{9}$$

and for each integer $k \geq 1$, we define the kth *power sum* for A as follows

$$S_k = \sum_{i=1}^{g}(w_i^k + \overline{w}_i^k) \tag{10}$$

Theorem 3.1 Let $\sigma_A(z) = \prod_{i=1}^{g}(1 - w_i z)(1 - \overline{w}_i z) = 1+\sigma_1 z+...+\sigma_{2g-1}z^{2g-1}+z^{2g}$. If $S_1 = S_2 = ... = S_g = 0$ then $S_{2g} = -2g$.

Proof. Our basic tool is called the *Newton Identity* ([3], Ch.8, Sect.6). It relates an arbitrary polinomial of the form

$$p(z) = \prod_{i=1}^{n}(1 - \beta_i z) = 1 + p_1 z + ... + p_n z^n \tag{11}$$

to the corresponding power sums

$$S_k = \sum_{i=1}^{n} \beta_i^k \tag{12}$$

via the power sum generating function

$$S(z) = \sum_{k\geq 1} S_k z^{k-1}$$

Then, the *Newton Identity* says that[1]

$$S(z) = -\frac{p'(z)}{p(z)} \tag{13}$$

So, for $\sigma_A(z)$ we have

$$\sigma_A(z)S_A(z) = -\sigma'_A(z) \tag{14}$$

where $S_A(z) = \sum_{k\geq 1} S_k z^{k-1}$ with $S_k = \sum_{i=1}^{g}(w_i^k + \overline{w}_i^k)$

Then

$$\begin{aligned}&(1 + \sigma_1 z + ... + \sigma_{2g-1}z^{2g-1} + z^{2g})(S_1 + S_2 z + S_3 z^2 + ...) =\\ &= -[\sigma_1 + 2\sigma_2 z + ... + (2g-1)\sigma_{2g-1}z^{2g-2} + 2g z^{2g-1}]\end{aligned}$$

therefore

$$\begin{aligned}&S_k + \sigma_1 S_{k-1} + ... + \sigma_{k-1}S_1 + k\sigma_k = 0, \quad \forall k = 1, 2, ..., 2g-1\\ &S_{2g} + \sigma_1 S_{2g-1} + ... + \sigma_{2g-1}S_1 + 2g = 0\end{aligned} \tag{15}$$

[1]For example, if $p(z) = 1+z^4$ then by (13) $S(z) = -4\frac{z^3}{1+z^4} = -4z^3+4z^7-4z^{11}+...$ So $S_k = -4$ if $k \equiv 4$ (mod8).

and

$$S_k + \sigma_1 S_{k-1} + ... + \sigma_{2g-1} S_{k-2g+1} + S_{k-2g} = 0, \quad \forall k \geq 2g+1$$

Thus $\sigma_A(z)$ can be determined from the first $(2g-1)$ terms of the sequence (S_k), and given $S_1, ..., S_{2g-1}$, and $\sigma_A(z)$, the remaining terms $S_{2g}, S_{2g+1}, ...$ can be determined by linear recursion.

Our goal is to find the coefficients $\sigma_1, ..., \sigma_g, \sigma_{g+1}, ..., \sigma_{2g-1}$ of $\sigma_A(z)$ knowing $S_1, S_2, ..., S_g$. By (15) we have that

$$\begin{cases} \sigma_1 = -S_1 \\ \sigma_2 = -\frac{1}{2}(S_2 + \sigma_1 S_1) \\ \vdots \\ \sigma_g = -\frac{1}{g}(S_g + \sigma_1 S_{g-1} + ... + \sigma_{g-1} S_1) \end{cases} \tag{16}$$

But, unfortunately, that leaves the remaining coefficients $\sigma_{g+1}, ..., \sigma_{2g-1}$ of $\sigma_A(z)$ undetermined. But, it is resolved by the following observation.

Let $\widetilde{\sigma}_A(z)$ denote the monic polynomial whose roots are reciprocal to those of $\sigma_A(z)$, then

$$\widetilde{\sigma}_A(z) = z^{2g}\sigma_A\left(\frac{1}{z}\right)$$

so

$$\widetilde{\sigma}_A(z) = z^{2g} + \sigma_1 z^{2g-1} + ... + \sigma_{2g-1} z + 1 \tag{17}$$

But, since the reciprocals of the w_i's are the $\overline{w_i}'$s, $\sigma_A(z) = \widetilde{\sigma}_A(z)$. Comparing $\sigma_A(z)$ to (17), $\sigma_i = \sigma_{2g-i}$ for $i = 1, ..., g-1$. The polynomial $\sigma_A(z)$ is *palindromic*, i.e.,$\sigma_A(z)$ reads the same forwards or backwards.

Combining this with (16), we see that $\sigma_1, ..., \sigma_{2g-1}$ can all be determined recursively from $S_1, ..., S_g$, as follows

$$\begin{array}{l} \sigma_1 = \sigma_{2g-1} = -S_1 \\ \sigma_2 = \sigma_{2g-2} = -\frac{1}{2}(S_2 + \sigma_1 S_1) \\ \vdots \\ \sigma_{g-1} = \sigma_{g+1} = -\frac{1}{g-1}(S_{g-1} + \sigma_1 S_{g-2} + ... + \sigma_{g-2} S_1) \\ \sigma_g = -\frac{1}{g}(S_g + \sigma_1 S_{g-1} + ... + \sigma_{g-1} S_1) \end{array} \tag{18}$$

In particular, if $S_1 = S_2 = ... = S_g = 0$, then by (18) $\sigma_1 = \sigma_2 = ... = \sigma_{2g-1} = 0$, and so $\sigma_A(z) = 1 + z^{2g}$.

Thus, by (14)

$$S_A(z) = -2g\frac{z^{2g-1}}{1+z^{2g}} = -2g(z^{2g-1} - z^{4g-1} + z^{6g-1} - \ldots$$

which proves the theorem.

Corollary 3.1 If $\sum_{i=1}^{g}(\alpha_i^r+\overline{\alpha}_i^r) = 0, \forall r \le g$, then $\sum_{i=1}^{g}(\alpha_i^{2g}+\overline{\alpha}_i^{2g}) = -2^{g+1}g$

Proof. Let $\alpha_i = \sqrt{2}w_i$, $\forall i = 1, \ldots, g$, with $|w_i| = 1$. Then, by Theorem 3.1,

$$\sum_{i=1}^{g}(\alpha_i^{2g} + \overline{\alpha}_i^{2g}) = -2^{g+1}g$$

In the following Theorem we have the answer to the question stated in the Introduction of this paper.

Theorem 3.2 Let g be the genus of C, curve defined over $\mathbf{F}_2$.

$$\#C(\mathbf{F}_{2^r}) = 2^r + 1, \forall r \le g \Rightarrow C \text{ is maximal over } \mathbf{F}_{2^{2g}}$$

Proof. We know that

$$\#C(\mathbf{F}_{2^r}) = 2^r + 1 - \sum_{i=1}^{g}(\alpha_i^r + \overline{\alpha_i}^r) \tag{19}$$

with $\alpha_1, \ldots, \alpha_g$ complex numbers, $|\alpha_i| = \sqrt{2}$, $\alpha_i \cdot \overline{\alpha}_i = 2$, and being then, the *zeta function*

$$Z(C, z) = \frac{\prod_{i=1}^{g}(1-\alpha_i z)(1-\overline{\alpha_i} z)}{(1-z)(1-2z)} \tag{20}$$

Let $S_r = \sum_{i=1}^{g}(\alpha_i^r + \overline{\alpha}_i^r)$, then, by hypothesis,

$$S_1 = S_2 = \ldots = S_g = 0$$

and by the Corollary 3.1

$$S_{2g} = -2^{g+1}g$$

Therefore, C is maximal over $\mathbf{F}_{2^{2g}}$. Applying these results to Quasihermitian curves, we have examples as

With $g = 2, C_{9,10} : Y^9Z + YZ^8 + X^{10} = 0$
maximal over $\mathbf{F}_{2^4}$ (33 points)
With $g = 3, C_{4,12} : Y^4Z^8 + YZ^{11} + X^{12} = 0$
maximal over $\mathbf{F}_{2^6}$ (113 points)
With $g = 5, C_{2,11} : Y^2Z^9 + YZ^{10} + X^{11} = 0$
maximal over $\mathbf{F}_{2^{10}}$ (1,345 points)

4 Conclusion

We have obtained an easy way to determine maximal Quasihermitian curves over $\mathbf{F}_{2^{2g}}$. This result together with the known algorithm for computing the minimum distance [1] makes the construction of good algebraic-geometric codes easier.

References

[1] García-Villalba, L.J. & Rodríguez-Palánquex M.C. & Montoya-Vitini, F. *An Algorithm for Computing the Minimum Distance.* Electronics Letters, Vol. 35, No. 18, pp. 1534-1535. 1999.

[2] van Lint, J.H. & van der Geer, G. *Introduction to Coding Theory and Algebraic Geometry.* Birkhäuser Verlag, Basel. 1988.

[3] MacWilliams, F.J. & Sloane, N.J.A. *The Theory of Error Correcting Codes.* Amsterdam, North-Holland. 1977.

[4] Rodríguez-Palánquex, M.C. *Aritmética de Curvas Cuasihermíticas. Aplicaciones a Códigos Geométricos de Goppa.* Ph. D. Dissertation Universidad Complutense de Madrid, 1996.

[5] Rodríguez-Palánquex, M.C. & García-Villalba, L.J. & Luengo-Velasco, I. *Computing the Genus of a Class of Curves.* Berlin: Springer-Verlag, 2001. pp. 182-191.

On Asymmetric Error Detection with Feedback [1]

Paul Oprisan Bella Bose
Department of Computer Science
Oregon State University
{paulo,bose}@cs.orst.edu

Abstract

The asymmetric nature of bit errors in optical communication provides the grounds for efficient error control techiques; the Z-channel and systematic asymmetric error detection (AED) codes can be succesfully combined with ARQ protocols for feedback error control enhancement. The throughput and the accepted packet error rate for such systems are derived from the codes construction. The results show that only a few check bits can provide a great deal of error protection, almost error free transmission in optical networks.

Keywords: ARQ protocols, asymmetric error detection, throughput efficiency, Z-channel.

[1]This research is supported by NSF grant CCR-0105204

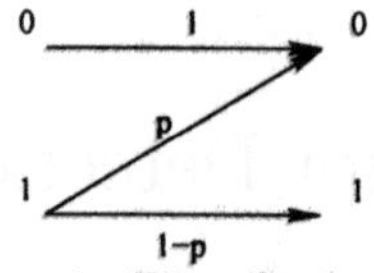

Figure 1: The Z-channel.

1 Introduction

Error detecting codes are effective in providing protection against transient, intermittent and permanent faults in computer and communication systems [1],[2], [3]. Many of the binary codes developed over the last five decades use the binary symmetric channel model, with equal bit error probabilities. In optical communication systems, 1's are represented by the presence of photons and 0's by their absence. Upon transmission, photons may fade or decay, but new photons cannot be generated [4], [5]. Furthermore, in optical storage there is a significant asymmetry between written and non-written data. Particularly, in write-once and erasable optical recording, because the incoming data stream of 1's and 0's is translated into pits and lands (or vice versa), most of the errors due to scratches are of only one type. In these types of systems, the observed errors are asymmetric and the error characteristics can be modeled by the Z-channel [4], [6], [5], [7] as shown in Fig. 1. The probability of an $1 \to 0$ transition is p (the probability of $1 \to 1$ is $1 - p$, consequently) and $0 \to 1$ transitions are not possible – they have 0 probability.

There are situations in which the communication channel may be affected by short bursts of noise. In such situations, a great deal of extra protection can be provided by the ARQ protocols. This means that the system detects the error packet, discards it and requests a retransmission. It is desired that the noise duration is shorter than the retransmission time, such that the retransmitted packet will pass undisturbed across the channel, thus avoiding multiple retransmission attempts. However, the rate of retransmission requests increases as the channel quality deteriorates which affects the efficiency of the system.

Some analysis procedures of the ARQ protocols for binary symmetric channels using symmetric error detecting codes are given in [6], [2] and [1]. In the present paper we estimate and analyze the specific parameters of ARQ protocols implemented with Bose-Lin unidirectional/asymmetric error detecting codes [8].

2 Asymmetric Error Detecting Codes

A code C is capable of detecting t asymmetric errors if and only if for all the codewords X, Y in C, either X and Y are unordered[2] or $d(X, Y) \geq t+1$ when one covers the other, where $d(X, Y)$ is the Hamming distance between X and Y [8], [4].

A code capable of detecting t asymmetric errors is also capable of detecting t unidirectional errors (asymmetric errors are of only one type (say $1 \rightarrow 0$) in all the data words, while unidirectional errors can be of both types in a message, but not in the same data word).

The check symbols for detecting 2 and 3 unidirectional errors are $w_0 \bmod 2^2$ and $w_0 \bmod 2^3$, respectively, where w_0 is the number of 0's in the information word. So, the 2 and 3 unidirectional error detecting codes use 2 and 3 check bits, respectively.

When $r \geq 4$ check bits are used, the code designed by Method 1 in [8] can detect up to $2^{r-2} + r - 2$ errors. In order to get the check, the first $r-1$ bits are derived as $(b_{r-2} \ldots b_1 b_0) = w_0 \bmod 2^{r-1}$. Then, the check symbol is $(a_{r-1}a_{r-2}a_{r-3} \ldots a_1 a_0) = (b_{r-2}\bar{b}_{r-2}b_{r-3} \ldots b_1 b_0)$.

When $r \geq 5$ check bits are used, the following alternate method of encoding, with better detecting capabilities (in terms of the number of errors detected) is used. First, calculate $(b_{r-2}b_{r-3} \ldots b_1 b_0) = w_0 \bmod 3 \cdot 2^{r-3}$. Then, the 3 most significant bits, $(b_{r-2}b_{r-3}b_{r-4})$, which can be one of the patterns in the set {000, 001, 010, 011, 100, 101}, are mapped into a 2-out-of-4 code. A simple example of mapping function is $f(i) \leq f(j)$ for $i < j$, $0 \leq i, j \leq 5$ i.e. $f(000) = 0011$, $f(001) = 0101$, $f(010) = 0110$, $f(011) = 1001$, $f(100) = 1010$ and $f(101) = 1100$. The concatenation of such a symbol with the remaining $r-4$ bits: $(b_{r-5} \ldots b_1 b_0)$, gives the check symbol. This code can detect up to $5 \cdot 2^{r-4} + r - 4$ errors.

In Bose-Lin code design it is assumed that $k \geq 2^r$. Otherwise, one can use the Berger codes [7], which can detect all asymmetric errors. The check symbol for this code is w_0 in binary and so the code uses $\lceil log_2(k+1) \rceil$ check bits.

[2] $X = (x_1, x_2, \ldots, x_n)$ covers $Y = (y_1, y_2, \ldots, y_n)$ if $y_i \leq x_i$ for all $i = 1, 2, \ldots, n$. If neither covers the other, they are called unordered.

3 ARQ Parameters

We measure the performance of an ARQ protocol by:

- the *accepted packet error rate* $P(E)$, which is the percentage of erroneous packets accepted by the receiver and
- the *throughput* η, which is the average number of data packets accepted by the receiver in the time it takes the transmitter to send one k-bit block.

The probability of detected errors P_d, (which is the probability that a retransmission request is generated) is defined as the value to which the ratio of blocks with detected errors (N_d) to the number of blocks transmitted over the channel (N) converges in probability [6]:

$$\lim_{N\to\infty} P\left\{\left|\frac{N_d}{N} - P_d\right| > \epsilon\right\} = 0, \quad \forall \quad \epsilon > 0. \tag{1}$$

The probability of undetected errors P_u is similarly defined:

$$\lim_{N\to\infty} P\left\{\left|\frac{N_u}{N} - P_u\right| > \epsilon\right\} = 0, \quad \forall \quad \epsilon > 0. \tag{2}$$

where N_u is the number of blocks with undetected errors.

In the following we will assume that the probabilities of the source generating 1's and 0's are equal. In order to determine the probability of detecting errors in the received word, we consider the double error detecting codes first and then generalize the result.

3.1 Double Error Detection

Errors detected, or retransmission request, means that the received word has one or two errors. Thus,

$$P_d \cong P_1 + P_2, \tag{3}$$

where P_1 and P_2 are the probabilities of one error and two error events, respectively. We use '$\cong$' in the above relation because multiple errors may be detected sometimes; for analysis purposes we restrict the number of errors to 2. The bit errors can occur either in the information part of the data word or in the check part. However, they occur independently, and so we can write:

$$P_1 = P_{1_{inf}} P_{0_{cs}} + P_{0_{inf}} P_{1_{cs}} \quad \text{and} \tag{4}$$

$$P_2 = P_{2_{inf}} P_{0_{cs}} + P_{1_{inf}} P_{1_{cs}} + P_{0_{inf}} P_{2_{cs}}, \tag{5}$$

where the sub-indices indicate the error location. For the information part, the probability that the received word contains one bit in error can be calculated by counting all the possibilities of error location. Any '1' bit can be in error, independently and equally likely:

$$P_{1_{inf}} = \frac{2^k - 1}{2^k} \sum_{j=0}^{k-1} (k-j) p (1-p)^{k-j-1}. \tag{6}$$

Similarly, the probability that the received word contains two bits in error is given by:

$$P_{2_{inf}} = \frac{2^k - k - 1}{2^k} \sum_{j=0}^{k-2} \binom{k-j}{2} p^2 (1-p)^{k-j-2}. \tag{7}$$

By using the equality $(n-m)\binom{n}{m} = n\binom{n-1}{m}$, we can rewrite (6) as:

$$P_{1_{inf}} = \frac{1}{2} p \sum_{j=0}^{k-1} \binom{k-1}{j} \frac{1}{2^j} \frac{1}{2^{(k-1)-j}} (1-p)^{(k-1)-j}, \tag{8}$$

which yields:

$$P_{1_{inf}} = \frac{1}{2} k p \left(1 - \frac{p}{2}\right)^{k-1}. \tag{9}$$

Similarly, equation (7) becomes:

$$P_{2_{inf}} = \frac{1}{8} k(k-1) p^2 \sum_{j=0}^{k-2} \binom{k-2}{j} \frac{1}{2^j} \frac{1}{2^{(k-2)-j}} (1-p)^{(k-2)-j}, \tag{10}$$

which is:

$$P_{2_{inf}} = \frac{1}{8} k(k-1) p^2 \left(1 - \frac{p}{2}\right)^{k-2}. \tag{11}$$

The probability of no error in the information part is:

$$P_{0_{inf}} = \left(1 - \frac{p}{2}\right)^k. \tag{12}$$

The error distribution for the check symbol depends on the information part. This is because the number of '1' bits in the check symbol is given by the information part of the codeword. The events no error, one error and two errors have their probabilities given by the following equations:

$$P_{0_{cs}} = p_{00} + p_{01}(1-p) + p_{10}(1-p) + p_{11}(1-p)^2, \tag{13}$$

$$P_{1_{cs}} = p_{01}p + p_{10}p + 2p_{11}p(1-p) \quad \text{and} \tag{14}$$

$$P_{2_{cs}} = p_{11}p^2, \tag{15}$$

where p_{00}, p_{01}, p_{10} and p_{11} are the probabilities that the codewords have the check symbols as shown by the indices. The check symbol for double error detecting codes is generated by taking the number of 0's modulo 4. It can be also determined by using the relation $3-(w \bmod 4)$, where w is the weight of the information part of the codeword. Let j, $j = 0, \ldots, 3$ be the decimal representation of the indices of p_{00}, p_{01}, p_{10} and p_{11}. These probabilities can be estimated by counting all the codewords with the corresponding weights:

$$p_j = P(w_0 \bmod 4 = j) = \frac{1}{2^k}\sum_{i=0}^{m/4}\binom{k}{4i+j} \tag{16}$$

where m is the largest multiple of 4 such that $m \leq k-3$.

3.2 Multiple Error Detection

Equations (6) and (7) can be generalized for an e error detecting code:

$$P_{e_{inf}} = \frac{p^e}{2^e}\binom{k}{e}(1-\frac{p}{2})^{k-e}. \tag{17}$$

This relation can be replaced in the general form of the probability of e errors in the received word (for Method 2 in Section 3):

$$P_e = \sum_{i=0}^{min\{r-2,e\}} P_{i_{cs}} P_{(e-i)_{inf}} \quad (18)$$

for all $e = 0, 1, \ldots, 5 \cdot 2^{r-4} + r - 4$.

Although it seems very laborious to calculate the check symbol error probabilities, the task is actually simplified by the code construction. For example, looking at the 4 bits check symbol construction in Section 3, we see that these probabilities are the same as for the 3 bits check symbol: $p_{0100} = p_{000}$, $p_{0101} = p_{001}$, $p_{0110} = p_{010}$, $p_{0111} = p_{011}$, $p_{1000} = p_{100}$, $p_{1001} = p_{101}$, $p_{1010} = p_{110}$ and $p_{1011} = p_{111}$. The second factor in (18) is given by the following relations:

$$P_{0_{cs}} = (p_{100} + p_{000})(1-p) + (p_{101} + p_{110} + p_{001} + p_{010})(1-p)^2 + (p_{111} + p_{011})(1-p)^3, \quad (19)$$

$$P_{1_{cs}} = (p_{100} + p_{000})p + 2(p_{101} + p_{110} + p_{001} + p_{010})p(1-p) + 3(p_{111} + p_{011})p(1-p)^2, \quad (20)$$

$$P_{2_{cs}} = (p_{101} + p_{110} + p_{001} + p_{010})p^2 + +3(p_{111} + p_{011})p^2(1-p), \quad (21)$$

$$P_{3_{cs}} = (p_{111} + p_{011})p^3. \quad (22)$$

Thus, only 8 check symbol error probabilities are needed. Further, for the 5 bits check symbol, derived using the second method presented in Section 3, there are only 12 check symbol error probabilities needed due to the 2-out-of-4 mapping. Moreover, they are based on the 3 bits check symbol again. Thus, considering the mapping given as an example in Section 3, note that the check symbol can contain at most three 1's and the first 4 bits of the check contain at most two 1's. Hence:

$$P_{0_{cs}} = P\{LSB = 0\}(1-p)^2 + P\{LSB = 1\}(1-p)^3, \quad (23)$$
$$P_{1_{cs}} = 2P\{LSB = 0\}p(1-p) + 3P\{LSB = 1\}p(1-p)^2, \quad (24)$$
$$P_{2_{cs}} = P\{LSB = 0\}p^2 + 3P\{LSB = 1\}p^2(1-p), \quad (25)$$
$$P_{3_{cs}} = P\{LSB = 1\}p^3. \quad (26)$$

$log_{10}p$	P_u			
	# of check bits			
	2	3	5	0 (uncoded)
-9	$2.08 \cdot 10^{-20}$	$2.61 \cdot 10^{-27}$	$5.06 \cdot 10^{-85}$	$5.00 \cdot 10^{-7}$
-8	$2.08 \cdot 10^{-17}$	$2.61 \cdot 10^{-23}$	$5.06 \cdot 10^{-73}$	$4.99 \cdot 10^{-6}$
-7	$2.08 \cdot 10^{-14}$	$2.61 \cdot 10^{-19}$	$5.06 \cdot 10^{-61}$	$4.99 \cdot 10^{-5}$
-6	$2.08 \cdot 10^{-11}$	$2.61 \cdot 10^{-15}$	$5.06 \cdot 10^{-49}$	$4.99 \cdot 10^{-4}$
-5	$2.08 \cdot 10^{-8}$	$2.60 \cdot 10^{-11}$	$5.03 \cdot 10^{-37}$	$4.98 \cdot 10^{-3}$
-4	$2.01 \cdot 10^{-5}$	$2.51 \cdot 10^{-7}$	$4.81 \cdot 10^{-25}$	$4.87 \cdot 10^{-2}$
-3	$1.44 \cdot 10^{-2}$	$1.76 \cdot 10^{-3}$	$3.08 \cdot 10^{-13}$	$3.93 \cdot 10^{-1}$
-2	0.8768	0.7377	0.0055	0.9933
-1	1	1	0.9999	1

Table 1: Probability of undetected error for $k = 1000$ information bits

Note that:

$$P\{LSB = 0\} = p^0_{000} + p^0_{001} + p^0_{010} + p^0_{011} + p^0_{100} + p^0_{101}, \tag{27}$$

and

$$P\{LSB = 1\} = p^1_{000} + p^1_{001} + p^1_{010} + p^1_{011} + p^1_{100} + p^1_{101}, \tag{28}$$

where the lower indices represent the first 3 check bits before mapping and the upper indices indicate the LSB's of the check symbol. These probabilities are calculated as follows:

$$p^0_j = P(w_0 \bmod 12 = j) = \frac{1}{2^k} \sum_{i=0}^{m/12} \binom{k}{12i + j} \tag{29}$$

$$p^1_l = P(w_0 \bmod 12 = l) = \frac{1}{2^k} \sum_{i=1}^{m/12} \binom{k}{12i + l}, \tag{30}$$

where j, $j = 0, 2, \ldots, 10$ is 2 times the decimal representation of the binary indices of $p_{000}, \ldots, p_{101}$, $l = j + 1$, and m is the largest multiple of 12 such that $m \leq k - 11$. The 6 or more check bits cases can be treated similarly.

$log_{10}p$	P_E		
	# of check bits		
	2	3	5
-9	$2.14 \cdot 10^{-23}$	$2.76 \cdot 10^{-31}$	$4.65 \cdot 10^{-97}$
-8	$2.14 \cdot 10^{-20}$	$2.76 \cdot 10^{-27}$	$4.65 \cdot 10^{-85}$
-7	$2.14 \cdot 10^{-17}$	$2.76 \cdot 10^{-23}$	$4.65 \cdot 10^{-73}$
-6	$2.14 \cdot 10^{-14}$	$2.76 \cdot 10^{-19}$	$4.65 \cdot 10^{-61}$
-5	$2.14 \cdot 10^{-11}$	$2.76 \cdot 10^{-15}$	$4.65 \cdot 10^{-49}$
-4	$2.15 \cdot 10^{-8}$	$2.76 \cdot 10^{-11}$	$4.65 \cdot 10^{-37}$
-3	$2.17 \cdot 10^{-5}$	$2.79 \cdot 10^{-7}$	$4.67 \cdot 10^{-25}$
-2	0.0241	0.0031	$4.94 \cdot 10^{-13}$
-1	0.9940	0.9933	0.3705

Table 2: Accepted packet error rate for $k = 100$ information bits

4 Performance Evaluation

For analysis purposes, we consider a perfect feedback channel, which is modeled by a noiseless channel, independent of the forward channel.

There are three basic retransmission protocols, stop-and-wait (SW-ARQ), go-back-N (GBN-ARQ) and selective repeat (SR-ARQ) [1], [2].

$P(E)$ can be computed by adding together the probabilities of all the events which result in the acceptance of an erroneous packet: $P(E) = P_u + P_d P_u + P_d^2 P_u + P_d^3 P_u + \dots$, as the erroneous packet can be accepted on the first transmission, or on the second one, third one and so on. The series converges to:

$$P(E) = \frac{P_u}{1 - P_d} = \frac{1 - P_0 - P_d}{1 - P_d}. \tag{31}$$

Table 1 shows values of P_u for 1000 information bits. These values confirm an error free transmission, for all practical purposes, at least in the fiber optics range of p. The accepted packet error rates for 100 information bits are shown in Table 2.

The throughput is defined as the average number of encoded packets accepted by the receiver in the time it takes the transmitter to send one information word (k-bit packet). In a feedforward system, it is equal to the code rate, k/n. In an ARQ system, it is a function of the average number of times a data packet has to be transmitted before it is accepted. The explicit form of this function depends on the protocol [1], [2], [6].

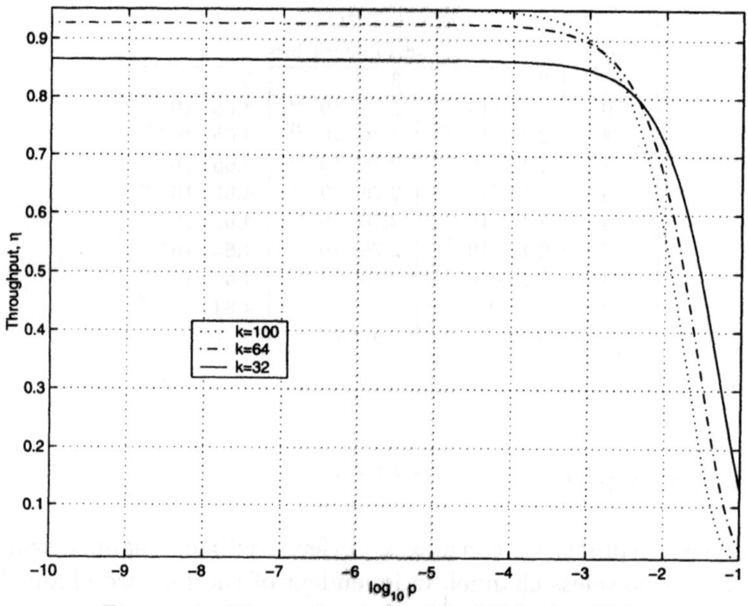

Figure 2: Throughput for SR-ARQ/11AED.

The throughput is a good measure for describing the delivery time characteristics (with the exception of the protocols which require the transmitted data to be delivered within a deadline). For a SR-ARQ protocol, the throughput is given by:

$$\eta = \frac{k}{n}(1 - P_d). \tag{32}$$

Throughput plots for 5 check bits Bose-Lin AED codes are shown in Figure 2. The throughputs for SW-ARQ and GBN-ARQ can be also calculated, given some specific parameters of the system, like the forward propagation delay, the receiver processing delay and the feedback propagation delay [1], [6].

A better throughput for longer information sequences with the same number of check bits and small values of p is due to the better code rate, k/n in equation (32). However, the throughput values drop for larger k's because of the increasing rate of retransmission requests; the steeper decreasing slope of the throughput curves in the 10^{-2} region of p, which corresponds to a noisier channel (typically p is less than 10^{-5} for optical channels), confirms the above statement.

5 Conclusions

The asymmetric transmission channel (or the Z-channel) is a common model for optical communications. In this paper, we study the properties of the Bose-Lin asymmetric error detecting codes within the feedback systems frame.

The accepted packet error rate and the throughput for ARQ protocols which use these codes for error detection are derived from the code statistical properties. The results provide efficient and simple error control in optical communication.

References

[1] S.B. Wicker. *Error Control Systems for Digital Communication and Storage.* Prentice Hall, New Jersey, 1995.

[2] S. Lin and D. J. Costello. *Error Control Coding, Fundamentals and Applications.* Prentice Hall, New Jersey, 1983.

[3] R. J. McEliece. *The Theory of Information and Coding.* Addison-Wesley, Reading, MA, 1977.

[4] J.M. Borden. Optimal asymmetric error detecting codes. *Inform. Control*, pages 66–73, 1982.

[5] R. J. McEliece. The effectiveness of Turbo-like codes on non-standard channel models. *IEEE Int. Symposium. on Inform. Theory*, Plenary session talk, June 2001.

[6] T. Kløve and V. Korzhik. *Error Detecting Codes. General Theory and their Application in Feedback Communication Systems.* Kluwer Academics, Boston/London/Dordrecht, 1995.

[7] M. Blaum. *Codes for Detecting and Correcting Unidirectional Errors.* IEEE Computer Society Press, Los Alamitos, California, 1993.

[8] B. Bose and D.J. Lin. Systematic unidirectional error-detecting codes. *IEEE Trans. Computers.*, C-34:1026–1032, November 1985.

Cryptanalysis of block ciphers and weight divisibility of some binary codes

Anne Canteaut, Pascale Charpin and Marion Videau
INRIA - projet CODES
B.P. 105 - 78153 Le Chesnay Cedex, France
{Anne.Canteaut, Pascale.Charpin, Marion.Videau}@inria.fr

Abstract

The resistance of an iterated block cipher to most classical attacks can be quantified by some properties of its round function. The involved parameters (nonlinearity, degrees of the derivatives...) for a function F from $\mathbf{F}_2^m$ into $\mathbf{F}_2^m$ are related to the weight distribution of a binary linear code $\mathcal{C}_F$ of length $2^m - 1$ and dimension $2m$. In particular, the weight divisibility of $\mathcal{C}_F$ appears as an important criterion in the context of linear cryptanalysis and of higher-order differential attacks. When the round function F is a power permutation over $\mathbf{F}_{2^m}$, the associated code $\mathcal{C}_F$ is the dual of a primitive cyclic code with two zeroes. Therefore, McEliece's theorem provides a powerful tool for evaluating the resistance of some block ciphers to linear and higher-order differential attacks.

Keywords: block ciphers, cryptanalysis, Boolean functions, almost bent functions, cyclic codes.

1 Introduction

This paper focuses on a large class of symmetric block ciphers called *iterated block ciphers.* In such systems the ciphertext is obtained by iteratively applying a keyed function, called the *round function*, to the plaintext. The underlying idea of this construction is that many iterations of a cryptographically weak round function are expected to lead to a cryptographically strong encryption function.

A round function can in general be split up into three steps: a key dependant function (usually a bitwise xor), a non linear part called the *confusion function* and a linear permutation.

The design of such ciphers relies on the development of their cryptanalysis. It is particularly true since the publication of two generic attacks: the *differential* and the *linear* cryptanalysis. They are at the origin of a new approach to define the security of a cipher by some mathematical properties of the round function. This leads to the concept of *provable security* [34]. However those properties imply strong requirements on the algebraic structures of the round function and it appears that it can weaken the cipher.

In this paper we show how optimal functions in regard to differential and linear cryptanalysis suffer from weaknesses against *higher-order differential* attacks. The explanation has to be found the high *weight divisibility* of the code associated to the round function. From this property we have derived a new upper bound for the degree of some composed function. Then, it has been used to mount a generic attack on any 5-round *Feistel cipher* using an *almost bent* function as a round function and to explain the origin of a weakness in a reduced version of MISTY1 presented in [38, 1]. Finally we sum up the requirements, known up to now, a confusion function must verify to ensure the security of an iterated block cipher.

2 Cryptanalysis of iterated block ciphers

To define an iterated block cipher more formally, we consider a family $(F_k)_{k\in\mathcal{K}}$ of permutations of the set of m-bit words, $\mathbf{F}_2^m$, indexed by a value $k \in \mathcal{K}$. In a cryptographic context, the round function F is public; F depends of a parameter k and each value of k provides a permutation F_k. The set $\mathcal{K}$ is called the *round key space.*

The encryption function of the iterated block cipher with block size m, with r rounds and with round function F is the keyed permutation of $\mathbf{F}_2^m$

defined by

$$x_i = F_{k_i}(x_{i-1}) \quad \text{for } 1 \le i \le r ,$$

where x_0 is the plaintext and x_r is the ciphertext. The vector $(k_1, \ldots, k_r)$ is called *the key* and its components are the *round keys*. The round keys may be derived from a unique master key which is shorter than the concatenation of all round keys.

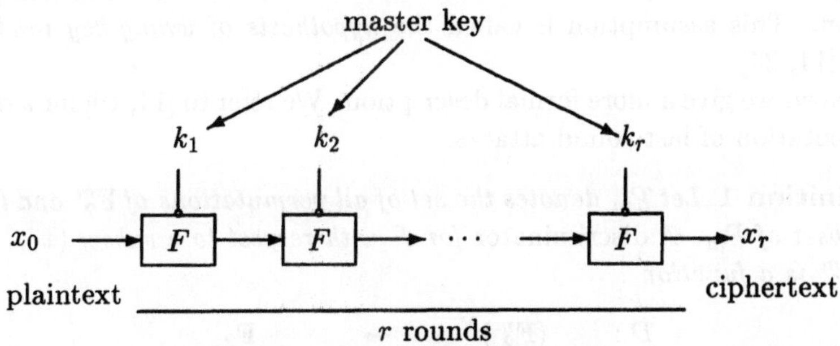

Figure 1: Iterated block cipher

2.1 Last-round attacks

Most attacks on iterated block ciphers are divide-and-conquer techniques which recover the *last round key* k_r from the knowledge of some pairs of plaintexts and ciphertexts. The other round keys $(k_1, \ldots, k_{r-1})$ (or the entire master key) may then be recovered either directly from k_r (e.g. by exhaustive search) or one after another by successively applying the last round attack on the cipher obtained by removing the last round.

In a *last round attack*, we consider the *reduced cipher*, i.e., the cipher obtained by removing the final round of the original cipher. The reduced cipher corresponds to the function

$$G_{(k_1,\ldots,k_{r-1})} = F_{k_{r-1}} \circ \ldots \circ F_{k_1} . \tag{1}$$

The key point in a last-round attack is to be able to distinguish the reduced cipher from a random permutation for all possible values of the first $(r-1)$ round keys $k_1, \ldots, k_{r-1}$. Some information on k_r can be recovered by applying a discriminator to all functions

$$H_k : x_0 \mapsto F_k^{-1}(x_r) = F_k^{-1}\left(F_{k_r} \circ G_{(k_1,\ldots,k_{r-1})}(x_0)\right) , \quad k \in \mathcal{K}$$

(k describes here the set of possible values of k_r). If the guess k matches the actual last round key k_r, then F_k^{-1} inverts the last encryption round and H_k corresponds to the reduced cipher. On the contrary, when k is a wrong guess, we get

$$H_k = F_k^{-1} \circ F_{k_r} \circ F_{k_{r-1}} \circ \ldots \circ F_{k_1} .$$

Since it essentially corresponds to the reduced cipher followed by two more encryption rounds, this function is supposed to act like a random permutation. This assumption is called the *hypothesis of wrong-key randomization* [14, 24].

Now, we give a more formal description. We refer to [14, 19] for a detailed presentation of last-round attacks.

Definition 1 *Let $\mathcal{P}_m$ denotes the set of all permutations of $\mathbf{F}_2^m$ and let $\mathcal{F}$ be a subset of $\mathcal{P}_m$. A* discriminator *for $\mathcal{F}$ with respect to a subset $(x_1, \ldots, x_N)$ of $\mathbf{F}_2^m$ is a function*

$$\begin{array}{cccc} \mathcal{D}: & (\mathbf{F}_2^m)^N & \to & \mathbf{F}_2 \\ & (y_1, \ldots, y_N) & \mapsto & \mathcal{D}(y_1, \ldots, y_N) \end{array}$$

for which there exists $\varepsilon > 0$ such that

$$\begin{array}{rcl} \varepsilon & < & \mid Pr_{f \in \mathcal{F}} [\mathcal{D}(f(x_1), \ldots, f(x_N)) = 1] \\ & & - \ Pr_{\pi \in_R \mathcal{P}_m} [\mathcal{D}(\pi(x_1), \ldots, \pi(x_N)) = 1] \mid . \end{array}$$

Now, the existence of a discriminator $\mathcal{D}$ for the family of reduced ciphers,

$$\mathcal{G} = \{G_{\mathbf{k}}, \ \mathbf{k} = (k_1, \ldots, k_{r-1}) \in \mathcal{K}^{r-1}\}$$

with respect to a set $(x_1, \ldots, x_N)$ leads to a last-round attack. The discriminator $\mathcal{D}$ should satisfy the *hypothesis of fixed-key equivalence*, i.e., it should return the same value for almost all reduced ciphers in $\mathcal{G}$ [14, 15, 24]. This hypothesis obviously holds when the round key is introduced by addition, i.e., $F_k(x) = F(x + k)$. This situation occurs in many ciphers, like DES, AES... The last-round attack derived from $\mathcal{D}$ is as follows:

INPUT: $(c_1, \ldots, c_N)$: the N ciphertexts corresponding to the plaintexts $(x_1, \ldots, x_N)$.

OUTPUT: A set of candidates for the last-round key k_r.

For all $k \in \mathcal{K}$
  For i from 1 to N do $y_i \leftarrow F_k^{-1}(c_i)$
  If $\mathcal{D}(y_1, \ldots, y_N) = 1$ then return k.

The attack requires the knowledge of N pairs of plaintexts-ciphertexts. Its average cost is $\#\mathcal{K}\times(NT_{F^{-1}}+T_{\mathcal{D}})$, where $\#\mathcal{K}$ is the size of the round key space, $T_{\mathcal{D}}$ is the average cost of the discriminator and $T_{F^{-1}}$ is the average number of operations required for evaluating F_k^{-1}. Notice that the costs of the most commonly used discriminators are proportional to N. If the attack returns several round keys, it can be repeated with another discriminator.

2.2 Basic properties of Boolean functions

Several specific properties of the reduced cipher may yield a discriminator. Now, we define some basic notions related to Boolean functions, which appear in the most commonly used last-round attacks.

A *Boolean function f of m variables* is a function from $\mathbf{F}_2^m$ into $\mathbf{F}_2$. It can be expressed as a polynomial in $x_1, \ldots, x_m$, called its *algebraic normal form.* The *degree* of f, denoted by $deg(f)$, is the degree of its algebraic normal form.

Differential and higher order differential attacks involve the derivatives of the reduced cipher.

Definition 2 [26] *Let F be a function from $\mathbf{F}_2^m$ into $\mathbf{F}_2^m$. For any $a \in \mathbf{F}_2^m$, the* derivative of F with respect to a *is the function*

$$D_aF(x) = F(x+a) + F(x) \ .$$

For any t-dimensional subspace V of $\mathbf{F}_2^m$, the t-th derivative of F with respect to V *is the function*

$$D_VF = D_{a_1}D_{a_2}\ldots D_{a_t}F \ ,$$

where $(a_1, \ldots, a_t)$ is any basis of V.

Linear cryptanalysis has concern with the Walsh spectrum of the reduced cipher. In the following, the usual dot product between two vectors x and y is denoted by $x \cdot y$. For any $\alpha \in \mathbf{F}_2^m$, φ_α denotes the linear function of m variables: $x \mapsto \alpha\cdot x$. For any Boolean function f of m variables, we denote by $\mathcal{F}(f)$ the following value related to the Walsh (or Fourier) transform of f:

$$\mathcal{F}(f) = \sum_{x\in\mathbf{F}_2^m}(-1)^{f(x)} = 2^m - 2wt(f) \ ,$$

where $wt(f)$ is the Hamming weight of f, i.e., the number of $x \in \mathbf{F}_2^m$ such that $f(x) = 1$.

Definition 3 *The* Walsh spectrum *of a Boolean function f of m variables f is the multiset*

$$\{\mathcal{F}(f + \varphi_\alpha), \alpha \in \mathbf{F}_2^m\} .$$

The Walsh spectrum *of a vectorial function F from $\mathbf{F}_2^m$ into $\mathbf{F}_2^m$ consists of the Walsh spectra of all Boolean functions $\varphi_\alpha \circ F : x \mapsto \alpha \cdot F(x)$, $\alpha \neq 0$. Therefore, it corresponds to the multiset*

$$\{\mathcal{F}(\varphi_\alpha \circ F + \varphi_\beta), \ \ \alpha \in \mathbf{F}_2^m \setminus \{0\}, \beta \in \mathbf{F}_2^m\} .$$

Definition 4 *The* nonlinearity *of a function F from $\mathbf{F}_2^m$ into $\mathbf{F}_2^m$ is the Hamming distance between all $\varphi_\alpha \circ F, \alpha \in \mathbf{F}_2^m$, $\alpha \neq 0$, and the set of affine functions. It is given by*

$$2^{m-1} - \frac{1}{2}\mathcal{L}(F) \quad \textit{where} \quad \mathcal{L}(F) = \max_{\alpha \in \mathbf{F}_2^m \setminus \{0\}} \max_{\beta \in \mathbf{F}_2^m} |\mathcal{F}(\varphi_\alpha \circ F + \varphi_\beta)| .$$

In the following, we focus on three classes of last-round attacks: differential cryptanalysis, linear cryptanalysis and higher-order differential cryptanalysis. There exist some other attacks on iterated block cipher. For example, a last-round attack can be performed when the reduced cipher, seen as a univariate polynomial in $\mathbf{F}_{2^m}[X]$, is close to a low-degree polynomial [18]. But, the mathematical nature of the property exploited by the latter attack is quite different.

2.3 Differential cryptanalysis

Differential cryptanalysis was introduced by Biham and Shamir [2]. It can be applied when the reduced cipher has a derivative which is not uniformly distributed. More precisely, assume that there exist two nonzero elements a and b in $\mathbf{F}_2^m$ such that for any $\mathbf{k} = (k_1, \ldots, k_{r-1})$ the reduced cipher $G_\mathbf{k}$ (as defined by 1) satisfies:

$$\#\{x \in \mathbf{F}_2^m, \ \ D_a G_\mathbf{k}(x) = b\} \simeq A ,$$

for a large integer A. This property leads to a discriminator $\mathcal{D}$ for the family $\mathcal{G}$ of reduced ciphers with respect to any subset $(x_{2i}, x_{2i} + a \ ; \ 0 \leq i < N/2)$, where the x_{2i} form a set of $N/2$ randomly chosen elements in $\mathbf{F}_2^m$ and $N \geq \frac{2^{m+1}}{A-1}$.

Set $J = \mathcal{D}(y_0, y_1, \ldots, y_{2i}, y_{2i+1}, \ldots)$ with i in the range $[0, N/2[$, where y_{2i} and y_{2i+1} correspond respectively to the unknown values $G_\mathbf{k}(x_{2i})$ and

$G_{\mathbf{k}}(x_{2i}+a)$ for a correct guess of the last-round key. Then

$$\begin{aligned} J &= 1 \text{ if } \#\{i,\ 0 \le i < N/2,\ y_{2i}+y_{2i+1}=b\} \simeq \frac{AN}{2^{m+1}}, \\ &= 0 \text{ if } \#\{i,\ 0 \le i < N/2,\ y_{2i}+y_{2i+1}=b\} \simeq \frac{N}{2^{m+1}}\ . \end{aligned}$$

It follows that a cipher is resistant to differential cryptanalysis if each $G_{\mathbf{k}}$, $\mathbf{k} \in \mathcal{K}^{r-1}$, is such that for any nonzero $a \in \mathbf{F}_2^m$, the output distribution of $x \mapsto D_a G_{\mathbf{k}}(x)$ is close to the uniform distribution. A necessary security condition is that the round function satisfies this property; it may be a sufficient condition for some ciphers, e.g. for Feistel ciphers [34]. Therefore, the round function F of an iterated cipher should satisfy the following requirement: for any $k \in \mathcal{K}$,

$$\delta_{F_k} = \max_{a\neq 0,\ b} \#\{x \in \mathbf{F}_2^m,\ F_k(x+a)+F_k(x)=b\}$$

should be small. As the number of solutions $x \in \mathbf{F}_2^m$ of $D_a F_k(x) = b$ is even (because x_0 is a solution if and only if $x_0 + a$ is a solution), we can deduce

Proposition 1 [34] *For any function F from $\mathbf{F}_2^m$ into $\mathbf{F}_2^m$, we have*

$$\delta_F \ge 2\ .$$

In case of equality, F is said to be almost perfect nonlinear (APN).

Note that the terminology APN comes from the general bound

$$\delta_F \ge 2^{n-m}$$

for a function from $\mathbf{F}_2^n$ into $\mathbf{F}_2^m$, where the functions achieving this bound are called *perfect nonlinear functions* [31]. Such functions only exist when n is even and $n \ge 2m$ [32].

All known APN functions are functions of an odd number of variables. Actually, it is conjectured that, for any function F from $\mathbf{F}_2^m$ into $\mathbf{F}_2^m$ with m even, we have

$$\delta_F \ge 4\ .$$

This statement is proved for some particular cases, most notably for power functions [3, 9].

2.4 Linear cryptanalysis

Linear cryptanalysis exploits the existence of a linear combination of the m output bits of the reduced cipher which is close to an affine function [27, 28]. Let us assume that there exists two nonzero elements a and b in $\mathbf{F}_2^m$ such that for any $\mathbf{k} = (k_1, \ldots, k_{r-1})$

$$|\mathcal{F}(\varphi_a \circ G_{\mathbf{k}} + \varphi_b)| \simeq A ,$$

for a large integer A. This property leads to a discriminator $\mathcal{D}$ for $\mathcal{G}$ with respect to any subset$(x_1, \ldots, x_N)$ of randomly chosen elements:

$$\begin{aligned} \mathcal{D}(y_1, \ldots, y_N) &= 1 \text{ if } \Big|\sum_{i=1}^{N}(-1)^{a \cdot y_i + b \cdot x_i}\Big| \simeq \frac{AN}{2^m}, \\ &= 0 \text{ if } \Big|\sum_{i=1}^{N}(-1)^{a \cdot y_i + b \cdot x_i}\Big| \simeq 0 . \end{aligned}$$

The security criterion corresponding to linear cryptanalysis is that all functions $\varphi_a \circ G_{\mathbf{k}}$, $a \neq 0$ should be far away from all affine functions. Therefore, a necessary condition is that all F_k, $k \in \mathcal{K}$, have a high nonlinearity, i.e. a high value for $2^{m-1} - \frac{1}{2}\mathcal{L}(F)$.

Proposition 2 [37, 8] *For any function* $F : \mathbf{F}_2^m \rightarrow \mathbf{F}_2^m$,

$$\mathcal{L}(F) \geq 2^{\frac{m+1}{2}} .$$

In case of equality F *is called* almost bent (AB).

For a function F from $\mathbf{F}_2^n$ into $\mathbf{F}_2^m$, we have

$$\mathcal{L}(F) \geq 2^{\frac{n}{2}}$$

where the functions achieving this bound were called *bent* functions in [32], as a generalization of the famous Boolean bent functions.

The minimum value of $\mathcal{L}(F)$ where F is a function from $\mathbf{F}_2^m$ into $\mathbf{F}_2^m$ can only be achieved when m is odd. For even m, some functions with $\mathcal{L}(F) = 2^{\frac{m}{2}+1}$ are known and it is conjectured that this value is the minimum [12, 36].

2.5 Higher-order differential cryptanalysis

Higher-order differential cryptanalysis was introduced by Knudsen in [23]. It exploits the fact that the reduced cipher $G_{\mathbf{k}}$, defined by (1), has a constant

t-th derivative. Assume that there exists a t-dimensional subspace $V \subset \mathbf{F}_2^m$ such that for any $\mathbf{k} = (k_1, \ldots, k_{r-1})$ we have

$$D_V G_{\mathbf{k}} = c$$

where $c \in \mathbf{F}_2^m$ is a constant which does not depend on $\mathbf{k}$. In accordance with Definition 2, we have:

$$D_V G_{\mathbf{k}}(x) = \sum_{v \in V} G_{\mathbf{k}}(x + v) \text{ for all } x \in \mathbf{F}_2^m ,$$

where the sum is an addition over $\mathbf{F}_2^m$. So we derive the following discriminator for $\mathcal{G}$ with respect to the set $(x_1, \ldots, x_{2^t})$ of elements of any coset of V, $a + V$ with $a \in \mathbf{F}_2^m$ (here y_i corresponds to $G_{\mathbf{k}}(x_i)$ for a correct guess of the last-round key):

$$\mathcal{D}(y_1, \ldots, y_{2^t}) = 1 \text{ if and only if } \sum_{i=1}^{2^t} y_i = c .$$

A natural candidate for V arises when the degree of the reduced cipher is known; this comes from the next proposition whose proof can be found in [26].

Definition 5 *The* degree *of a function F from $\mathbf{F}_2^m$ into $\mathbf{F}_2^m$ is the maximum degree of its Boolean components:*

$$deg(F) = \max_{1 \le i \le m} deg(\varphi_{e_i} \circ F)$$

where $(e_1, \ldots, e_m)$ denotes the canonical basis of $\mathbf{F}_2^m$.

Actually, we have

Proposition 3 *Let F be a function from $\mathbf{F}_2^m$ into $\mathbf{F}_2^m$ of degree d. Then, for any $(d+1)$-dimensional subspace $V \subset \mathbf{F}_2^m$, we have*

$$D_V F(x) = 0 \quad \text{for all } x \in \mathbf{F}_2^m .$$

Note that the dimension of the smallest subspace V satisfying $D_V F = 0$ may be smaller than $\deg(F) + 1$. Since

$$\max_{\mathbf{k} \in \mathcal{K}^{r-1}} \deg(G_{\mathbf{k}}) \le \left(\max_{k \in \mathcal{K}} \deg(F_k) \right)^{r-1} ,$$

it obviously follows that a cipher is vulnerable to higher-order differential cryptanalysis when its round function has a low degree. This property was used by Jakobsen and Knudsen [20] for breaking a cipher example proposed in [34], whose round function is a quadratic permutation. However, this condition is not sufficient and a stronger requirement on the round function will be exhibited in Section 4.2.

All three properties involved in differential, linear and higher-order differential attacks are invariant under both right and left composition by a linear permutation of $\mathbf{F}_2^m$ [33]. Then, they only concern the *confusion part* of the round function.

3 Optimal cryptographic functions over $\mathbf{F}_2^m$ and weight distributions of some $[2^m - 1, 2m]$ binary codes

Carlet, Charpin and Zinoviev have pointed out that both APN and AB properties can be expressed in terms of error-correcting codes [7]. In the following, F denotes a function from $\mathbf{F}_2^m$ into $\mathbf{F}_2^m$. Since both APN and AB properties are invariant under translation, we here only consider the functions F such that $F(0,\ldots,0) = 0$. We consider $\mathbf{F}_2^m$ as the ordered set

$$\{\ 0,\ \alpha_1,\ \ldots,\ \alpha_{2^m-1}\ \}\ .$$

Now, the linear binary code $\mathcal{C}_F$ of length $(2^m - 1)$ and dimension $2m$ is defined by its generator matrix:

$$G_F = \begin{pmatrix} \alpha_1 & \alpha_2 & \alpha_3 & \ldots & \alpha_{2^m-1} \\ F(\alpha_1) & F(\alpha_2) & F(\alpha_3) & \ldots & F(\alpha_{2^m-1}) \end{pmatrix}, \tag{2}$$

where each entry in $\mathbf{F}_2^m$ is viewed as a binary column vector of length m. It clearly appears that any codeword in $\mathcal{C}_F$ corresponds to a vector $(a \cdot \alpha_i + b \cdot F(\alpha_i), 1 \leq i < 2^m)$. Therefore, its Hamming weight is given by

$$\#\{i, 1 \leq i < 2^m,\ a \cdot \alpha_i + b \cdot F(\alpha_i) = 1\} = 2^{m-1} - \frac{1}{2}\mathcal{F}(\varphi_b \circ F + \varphi_a)\ .$$

Moreover, a vector $(c_1, \ldots, c_{2^m-1})$ belongs to the dual code $\mathcal{C}_F^\perp$ if and only if it satisfies:

$$\sum_{i=1}^{2^m-1} c_i\alpha_i = 0 \text{ and } \sum_{i=1}^{2^m-1} c_iF(\alpha_i) = 0\ .$$

Then, we have that the minimum distance of $C_F^\perp$ is at least 3. Moreover, there exist three different indexes i_1, i_2, i_3 such that

$$F(\alpha_{i_1}) + F(\alpha_{i_2}) + F(\alpha_{i_3}) + F(\alpha_{i_1} + \alpha_{i_2} + \alpha_{i_3}) = 0$$

if and only if $C_F^\perp$ contains a codeword of Hamming weight 4 (or 3 if $\alpha_{i_1} + \alpha_{i_2} + \alpha_{i_3} = 0$). Therefore, we obtain the following correspondence:

Theorem 1 [7] *Let F be a function from $\mathbf{F}_2^m$ into $\mathbf{F}_2^m$ with $F(0) = 0$. Let C_F be the linear binary code of length $2^m - 1$ with generator matrix G_F described by (2). Then,*

(i) $\mathcal{L}(F) = \max_{c \in C_F, c \neq 0} |2^m - 2wt(c)|$.

In particular, for odd m, F is AB if and only if for any non-zero codeword $c \in C_F$,

$$2^{m-1} - 2^{\frac{m-1}{2}} \leq wt(c) \leq 2^{m-1} + 2^{\frac{m-1}{2}} .$$

(ii) *F is APN if and only if the code $C_F^\perp$ has minimum distance 5.*

When the vector space $\mathbf{F}_2^m$ is identified with the finite field $\mathbf{F}_{2^m}$, the function F can be expressed as a unique polynomial of $\mathbf{F}_{2^m}[X]$. Now, we focus on power functions F, i.e., $F(x) = x^s$ over $\mathbf{F}_{2^m}$. Let α be a primitive element of $\mathbf{F}_{2^m}$. Any cyclic code C of length $(2^m - 1)$ can be defined by its generator polynomial whose roots are called the zeros of the code. The *defining set* of C is then the set

$$I(C) = \{i \in \{0, \cdots, 2^m - 2\} |\ \alpha^i \text{ is a zero of } C\} .$$

Since C is a binary code, its defining set is a union of 2-cyclotomic cosets modulo $(2^m - 1)$, $Cl(a)$, where $Cl(a) = \{2^j a \bmod (2^m - 1)\}$. From now on the defining set of a binary cyclic code of length $(2^m - 1)$ is identified with the representatives of the corresponding 2-cyclotomic cosets modulo $(2^m - 1)$. Then, when F is a power function $x \mapsto x^s$ over $\mathbf{F}_{2^m}$, the code $C_F^\perp$ is the cyclic code of length $(2^m - 1)$ with defining set $\{1, s\}$ [7].

In the following, we investigate the weight divisibility of the code C_F associated to a function F from $\mathbf{F}_2^m$ to $\mathbf{F}_2^m$.

Definition 6 *A binary code C is said* 2^ℓ-divisible *if the weight of any of its codewords is divisible by 2^ℓ. Moreover C is said* exactly 2^ℓ-divisible *if, additionally, it contains at least one codeword whose weight is not divisible by $2^{\ell+1}$.*

The following theorem due to McEliece reduces the determination of the exact weight divisibility of binary cyclic codes to a combinatorial problem:

Theorem 2 [30] *A binary cyclic code is exactly 2^{ℓ}-divisible if and only if ℓ is the smallest number such that $(\ell + 1)$ nonzeros of C (with repetitions allowed) have product 1.*

For a cyclic code with two nonzeros, McEliece's theorem can be reformulated as follows [5].

Corollary 1 *The dual of the cyclic code of length $(2^m - 1)$ with defining set $\{1, s\}$ is exactly 2^{ℓ}-divisible if and only if for all u such that $0 \leq u \leq 2^m - 1$,*

$$w_2(A(u)) \leq w_2(u) + m - 1 - \ell$$

where $A(u) = us \bmod (2^m - 1)$ and $w_2(u)$ corresponds to the number of 1s in the 2-adic expansion of u.

4 Optimal cryptographic functions and weight divisibility of some $[2^m - 1, 2m]$ binary codes

We have pointed out that the cryptographic properties of a round function over $\mathbf{F}_2^m$ which guarantee a high resistance to linear and differential attacks have concern with the weight distributions of some $[2^m - 1, 2m]$ linear binary codes. Now, we show that the weight divisibility of the associated code plays a major role in the context of linear cryptanalysis and of higher order differential attacks.

4.1 Characterization of almost bent functions

First, we give some general results on the weight distributions of linear binary codes with parameters $[2^m - 1, 2m]$. These properties are derived from Pless power moment identities [35] and from some ideas due to Kasami [21, th. 13]. The proofs of the results given in this section can be found in [5, 7].

Theorem 3 *Let C be a $[2^m - 1, 2m]$ linear code which does not contain the all-one vector $\mathbf{1} = (1, \cdots, 1)$. Assume that the minimum distance of the dual code $C^{\perp}$ is at least 3. Let $A = (A_0, \cdots, A_{2^m-1})$ (resp. $B = (B_0, \cdots, B_{2^m-1})$) be the weight enumerator of C (resp. $C^{\perp}$). Then, we have*

(i) *If w_0 is such that for all $0 < w < w_0$*

$$A_w = A_{2^m - w} = 0,$$

then

$$6(B_3 + B_4) \leq (2^m - 1)\left[(2^{m-1} - w_0)^2 - 2^{m-1}\right]$$

where equality holds if and only if

$$A_w = 0 \quad \text{for all} \quad w \notin \{0, w_0, 2^{m-1}, 2^m - w_0\} .$$

(ii) *If w_1 is such that for all $w_1 < w < 2^{m-1}$*

$$A_w = A_{2^m - w} = 0,$$

then

$$6(B_3 + B_4) \geq (2^m - 1)\left[(2^{m-1} - w_1)^2 - 2^{m-1}\right]$$

where equality holds if and only if

$$A_w = 0 \quad \text{for all} \quad w \notin \{0, w_1, 2^{m-1}, 2^m - w_1\} .$$

This theorem gives an upper bound on the value of w_0 for which all nonzero weights of C lie between w_0 and $2^m - w_0$:

Corollary 2 *Let C be a $[2^m - 1, 2m]$ linear code which does not contain the all-one vector $\mathbf{1} = (1, \cdots, 1)$. Assume that the minimum distance of the dual code $C^\perp$ is at least 3. Let w_0 be the smallest w such that $0 < w < 2^{m-1}$ and*

$$A_w + A_{2^m - w} \neq 0$$

Then

$$w_0 \leq 2^{m-1} - 2^{\frac{m-1}{2}}$$

and equality holds if and only if the weight of every codeword in C belongs to $\{0, 2^{m-1}, 2^{m-1} \pm 2^{\frac{m-1}{2}}\}$. In this case the weight distribution of C is the same as the weight distribution of the dual of the 2-error-correcting BCH code, i.e.

Weight: w	*Number of words: A_w*
0	1
$2^{m-1} - 2^{(m-1)/2}$	$(2^m - 1)(2^{m-2} + 2^{(m-3)/2})$
2^{m-1}	$(2^m - 1)(2^{m-1} + 1)$
$2^{m-1} + 2^{(m-1)/2}$	$(2^m - 1)(2^{m-2} - 2^{(m-3)/2})$

Then we can deduce a fundamental result for characterizing AB functions by using divisibility properties. Assume that m is odd and consider a code C which satisfies the hypothesis of Theorem 3. Suppose that the minimum distance of the dual code $C^\perp$ is 5 and C is $2^{\frac{m-1}{2}}$-divisible. Then, with notation of Theorem 3, we have $B_3 = B_4 = 0$ and $A_w = A_{2^m-w} = 0$ for all w in the range $[2^{m-1} - 2^{\frac{m-1}{2}}, 2^{m-1}]$. It follows from **(ii)** of Theorem 3 that the nonzero weights of C are 2^{m-1} and $2^{m-1} \pm 2^{\frac{m-1}{2}}$ only. According to the previous corollary, we obtain:

Corollary 3 *The equality $w_0 = 2^{m-1} - 2^{\frac{m-1}{2}}$ holds in Corollary 2 if and only if C is $2^{\frac{m-1}{2}}$-divisible and its dual has minimum distance 5.*

Corollary 4 *Let m be odd. Let F be a function from $\mathbf{F}_2^m$ into $\mathbf{F}_2^m$ and let C_F be the code defined in Theorem 1. Assume that C_F does not contain the all-one vector.*

Then F is AB if and only if F is APN and the code C_F is $2^{\frac{m-1}{2}}$-divisible.

Note that the divisibility condition on C_F equivalently means that all Walsh coefficients of F are divisible by $2^{\frac{m+1}{2}}$. Moreover, C_F does not contain the all-one vector means $\mathcal{L}(F) \neq 2^m$.

When F is a power function, $F : x \mapsto x^s$, the corresponding code C_F is the dual of a binary cyclic code of length $(2^m - 1)$ with defining set $\{1, s\}$. Its weight divisibility can therefore be obtained by applying McEliece's theorem, as expressed in Corollary 1. This leads to the following characterization of AB power functions:

Corollary 5 *Let $m = 2t + 1$. Assume that the power function $F : x \mapsto x^s$ on $\mathbf{F}_{2^m}$ satisfies $\mathcal{L}(F) \neq 2^m$. Then F is AB on $\mathbf{F}_{2^m}$ if and only if F is APN on $\mathbf{F}_{2^m}$ and*

$$\forall u,\ \ 1 \leq u \leq 2^m - 1,\ \ w_2(A(u)) \leq t + w_2(u) \tag{3}$$

where $A(u) = us \bmod (2^m - 1)$.

This result provides a fast algorithm for checking whether an APN power function is AB, and then for finding all AB power functions on $\mathbf{F}_{2^m}$. There are roughly $2^{m-1}/m$ cyclotomic representatives u such that $w_2(u) \leq (m-1)/2$ and each test requires one modular multiplication on m-bit integers and two weight computations. Condition (3) can then be checked with around 2^m elementary operations and at no memory cost.

Moreover, the determination of the values of s such that $x \mapsto x^s$ is almost bent on $\mathbf{F}_{2^m}$ is now reduced to a combinatorial problem, and this technique was directly used to prove that some power functions are AB [4, 17]. Moreover, it leads to a very efficient method for proving that a given power function is not AB. For example, the APN power function $x \mapsto x^s$ over $\mathbf{F}_{2^{5g}}$ with $s = 2^{4g} + 2^{3g} + 2^{2g} + 2^g - 1$ does not satisfy the condition of Corollary 5 [5].

These recent results lead to the following list (up to equivalence) of known AB permutations (Table 1).

Table 1: Known AB power permutations x^s on $\mathbf{F}_{2^m}$, m odd

exponents s	m	
$2^j + 1$, $\gcd(j, m) = 1$, $1 \leq j \leq \frac{m-1}{2}$		[13, 33]
$2^{2j} - 2^j + 1$, $\gcd(j, m) = 1$, $2 \leq j \leq \frac{m-1}{2}$		[22]
$2^{\frac{m-1}{2}} + 3$		[4]
$2^{\frac{m-1}{2}} + 2^{\frac{m-1}{4}} - 1$	$m \equiv 1 \bmod 4$	[17]
$2^{\frac{m-1}{2}} + 2^{\frac{3m-1}{4}} - 1$	$m \equiv 3 \bmod 4$	[17]

4.2 Weight divisibility and resistance to higher-order differential attacks

Now, we show that the weight divisibility of the codes $\mathcal{C}_{F_k}$ associated to the round function F of an iterated block cipher has concern with its resistance to higher-order differential attacks. Recall that a higher-order differential attack can be performed when the reduced cipher has a low degree. The degree of the round function (*i.e.*, of the functions F_k) provides a trivial bound on the degree of the reduced cipher

$$\max_{\mathbf{k} \in \mathcal{K}^{r-1}} \deg(G_{\mathbf{k}}) \leq \left(\max_{k \in \mathcal{K}} \deg(F_k)\right)^{r-1},$$

but this bound can be exploited only when the degree of the round function is very low. Indeed another approach has to be used when the degree of the round function exceeds $\sqrt{m}$, since $(deg(F))^{r-1} > m$ for any $r \geq 3$.

Here, we focus on the degree of the function obtained by composing two permutations F and F'. We show that the trivial bound

$$deg(F' \circ F) \leq deg(F')deg(F)$$

can be improved when the weight divisibility of the code C_F associated with F is high.

First, we assume that F is a power function $F : x \mapsto x^s$ over $\mathbf{F}_{2^m}$. Now, any function F' from $\mathbf{F}_2^m$ into $\mathbf{F}_2^m$ can be written as a univariate polynomial in $\mathbf{F}_{2^m}[X]$:

$$F'(X) = \sum_{u=0}^{2^m-1} a_u X^u ,$$

where the degree of F' is: $\max\{w_2(u),\ a_u \neq 0\}$. Thus, we deduce that $F' \circ F(x) = \sum_{u=0}^{2m-1} a_u X^{us \bmod (2^m-1)}$. Therefore,

$$deg(F' \circ F) \leq \max_{u, a_u \neq 0} w_2(us \bmod (2^m - 1)) ,$$

and McEliece's theorem directly provides a new bound on the degree of $F' \circ F$(see Corollary 1).

Theorem 4 *Let $F : x \mapsto x^s$ be a power function over $\mathbf{F}_{2^m}$ such that the associated code C_F defined in Theorem 1 is 2^ℓ-divisible. Then, for any function F' from $\mathbf{F}_2^m$ into $\mathbf{F}_2^m$, we have*

$$deg(F' \circ F) \leq m - 1 - \ell + deg(F') .$$

In particular, when F is an almost bent power function over $\mathbf{F}_{2^m}$, we obtain

$$deg(F' \circ F) \leq \frac{m-1}{2} + deg(F') .$$

The previous theorem leads to a general explanation for the weakness in the cipher MISTY1 [29] reported in [38] and [1]. The proposed attack relies on the fact that for any quadratic function Q from $\mathbf{F}_2^7$ into $\mathbf{F}_2^7$ we have $\deg(Q \circ S_7) \leq 5$, where S_7 is the almost bent power permutation $x \mapsto x^{81}$ over $\mathbf{F}_{2^7}$. Theorem 4 provides a generalization of this property to any AB power function over $\mathbf{F}_{2^7}$.

We can prove that Theorem 4 is not specific to power functions: it is also valid for any function F from $\mathbf{F}_2^m$ into $\mathbf{F}_2^m$. The proof of the next theorem, very technical, has to be found in [6].

Theorem 5 *Let F be a function from $\mathbf{F}_2^m$ into $\mathbf{F}_2^m$ such that the associated code C_F is 2^ℓ-divisible. Then, for any function F' from $\mathbf{F}_2^m$ into $\mathbf{F}_2^m$, we have*

$$deg(F' \circ F) \leq m - 1 - \ell + deg(F') .$$

Now, we show how the weight divisibility of the code associated with the round function can be exploited in a practical situation. As an example, we consider a 5-round Feistel cipher. In a Feistel cipher with block size $2m$, the round function is defined by

$$\begin{array}{rccc} F_k: & \mathbf{F}_2^m \times \mathbf{F}_2^m & \rightarrow & \mathbf{F}_2^m \times \mathbf{F}_2^m \\ & (L,R) & \mapsto & (R, L + S_k(R)) \end{array}$$

where S_k is a function from $\mathbf{F}_2^m$ into $\mathbf{F}_2^m$ called the confusion function. In the following, L_i (resp. R_i) denotes the left part (resp. right part) of the output of the i-th round. In a 5-round Feistel cipher, the right part of the output of the third round, R_3, can be derived from the ciphertext (L_5, R_5) and the last-round key:

$$R_3 = R_5 + S_{k_5}(L_5) .$$

Moreover, when we consider any plaintext (x, c_0) whose right part is a given constant c_0, R_3 can be computed from x by only two iterations of the confusion function :

$$R_3(x) = x + c_1 + S_{k_3}(c_0 + S_{k_2}(x + c_1))$$

where x stands for the left half of the plaintext, c_0 and c_1 being some constants.

When the confusion function S_k is such that all codes $\mathcal{C}_{S_k}$ are 2^ℓ-divisible, we can apply Theorem 5. Then, we obtain the following upper bound for the degree of R_3:

$$deg(R_3) \leq m - 1 - \ell + \max_{k \in \mathcal{K}} \deg(S_k).$$

Let $\delta = \min(\max_{k \in \mathcal{K}} \deg(S_k)^2 + 1, m - \ell + \max_{k \in \mathcal{K}} \deg(S_k))$. We have exhibited a higher-order differential attack on any 5-round Feistel cipher using a confusion function S_k which satisfies $\delta \leq m$:

INPUT: $(L_i, R_i)_{1 \leq i \leq 2^\delta}$: the ciphertexts corresponding to the plaintexts $(x_i, c_0)_{1 \leq i \leq 2^\delta}$, where c_0 is any fixed element of $\mathbf{F}_2^m$ and $(x_i)_{1 \leq i \leq 2^\delta}$ is a δ-dimensional subspace of $\mathbf{F}_2^m$:

OUTPUT: A set of candidates for the last-round key k_5.

For all $k \in \mathcal{K}$
  For i from 1 to 2^δ do $y_i \leftarrow R_i + S_k(L_i)$
  If $\sum_{i=1}^{2^\delta} y_i = 0$ then return k.

For example, if all S_k are almost bent, the previous higher order differential attack can be performed except when $\max_k \deg(S_k) = (m+1)/2$, i.e., when S_k is an almost bent function of maximum degree.

4.3 Highly nonlinear round functions having a low divisibility

The previous results point out that the use of an almost bent round function (or confusion function) is not suitable, even if the corresponding iterated block cipher is provably-secure against differential and linear cryptanalysis. An almost bent function may make the cipher vulnerable to higher-order differential cryptanalysis because of the high weight-divisibility of the associated code.

When m is even, the smallest known value of $\mathcal{L}(F)$ for a function F from $\mathbf{F}_2^m$ into $\mathbf{F}_2^m$ is $\mathcal{L}(F) = 2^{m/2+1}$. The only known functions (up to equivalence) achieving this bound are power functions. Since power permutations cannot be APN when m is even, that the security criteria corresponding to differential cryptanalysis and to linear cryptanalysis are not so strongly related. Moreover, the divisibility of the codes associated with these highly nonlinear functions varies. In particular, the degree of such a function is not upper-bounded since there is no requirement on the weight divisibility. All known functions satisfying this property are equivalent to one of the power functions given in Table 2 (or to one of their inverses) [12].

Table 2: Known power permutations x^s on $\mathbf{F}_{2^m}$, m even, with the highest nonlinearity and exact weight divisibility of the associated code

exponents s	m	divisibility	
$2^{m-1}-1$	$m \equiv 0 \bmod 2$	2	[25]
2^j+1, $\gcd(j,m)=2$, $j<\frac{m}{2}$	$m \equiv 2 \bmod 4$	$2^{\frac{m}{2}}$	[13]
$2^{2j}-2^j+1$, $\gcd(j,m)=2, j<\frac{m}{2}$	$m \equiv 2 \bmod 4$	$2^{\frac{m}{2}}$	[22]
$2^{\frac{m}{2}}+2^{\frac{m+2}{4}}+1$	$m \equiv 2 \bmod 4$	$2^{\frac{m}{2}}$	[10]
$2^{\frac{m}{2}}+2^{\frac{m}{2}-1}+1$	$m \equiv 2 \bmod 4$	$2^{\frac{m}{2}}$	[10]
$\sum_{i=0}^{m/2} 2^{ij}$, $\gcd(j,m)=1, j<\frac{m}{2}$	$m \equiv 0 \bmod 4$	$2^{\frac{m}{2}-1}$	[12]
$2^{\frac{m}{2}}+2^{\frac{m}{4}}+1$	$m \equiv 4 \bmod 8$	$2^{\frac{m}{2}-1}$	[12]

It appears that all known optimal functions for m even are such that the associated codes are either $2^{\frac{m}{2}-1}$-divisible or $2^{\frac{m}{2}}$-divisible, except the inverse function. The inverse function is very specific in this context. The corresponding code C_F is the dual of the Melas code. Its weights are all even integers w such that $|w - \frac{2^m-1}{2}| \leq 2^{\frac{m}{2}}$ [25]. Therefore, it has the smallest possible weight divisibility when F is a permutation.

Moreover, $x \mapsto x^{2^{m-1}-1}$ is the only power permutation F of $\mathbf{F}_{2^m}$ (up to equivalence) which corresponds to an exactly 2-divisible code.

Proposition 4 [16] *Let m and s be two positive integers such that* $\gcd(s, 2^m - 1) = 1$. *Let C_s be the dual of binary cyclic code of length $(2^m - 1)$ with defining set $\{1, s\}$. Then, C_s is exactly 2-divisible if and only if $w_2(s) = m - 1$.*

Thus, the inverse function is the only confusion function which is optimal with respect to all resistance criteria; it opposes the best resistance to differential, linear and higher-order differential attacks. This function is used in the new block cipher standard AES [11].

The search for other confusion functions which are "almost optimal" with respect to these criteria leads to the following open problem.

Open problem 1 *Find all permutations F of $\mathbf{F}_2^m$ such that the associated code C_F defined in Theorem 1 satisfies:*

(i) $\max_{c \in C_F, c \neq 0} |2^m - 2wt(c)|$ *is close to* $2^{\frac{m-1}{2}}$;

(ii) *The number of codewords of weight 3 and 4 in the dual code $C_F^{\perp}$ is small;*

(iii) *The exact weight divisibility of C_F is low.*

For power functions, McEliece theorem provides a useful tool for finding such functions. For instance, the exponents s such that the code associated with $x \mapsto x^s$ is exactly 4-divisible are known.

Proposition 5 [5, Prop. 5.3] *Let m and s be two positive integers such that* $\gcd(s, 2^m - 1) = 1$. *Let C_s be the dual of the binary cyclic code of length $(2^m - 1)$ with defining set $\{1, s\}$. Then, C_s is exactly 4-divisible if and only if either $w_2(s) = m - 2$ or $w_2(s^{-1}) = m - 2$ where s^{-1} is the only integer in $\{0, \ldots, 2^m - 1\}$ such that* $s^{-1}s \equiv 1 \bmod (2^m - 1)$.

A necessary condition on s for obtaining an exactly 8-divisible code can also be found in [5, Prop. 5.4].

References

[1] S. Babbage and L. Frisch. On MISTY1 Higher Order Differential Cryptanalysis. In *Proceedings of ICISC 2000*, number 2015 in Lecture Notes in Computer Science, pages 22–36. Springer-Verlag, 2000.

[2] E. Biham and A. Shamir. Differential cryptanalysis of DES-like cryptosystems. *Journal of Cryptology*, 4(1):3–72, 1991.

[3] A. Canteaut. Differential cryptanalysis of Feistel ciphers and differentially uniform mappings. In *Selected Areas on Cryptography, SAC'97*, pages 172–184, Ottawa, Canada, 1997.

[4] A. Canteaut, P. Charpin, and H. Dobbertin. Binary m-sequences with three-valued crosscorrelation: A proof of Welch conjecture. *IEEE Trans. Inform. Theory*, 46(1):4–8, 2000.

[5] A. Canteaut, P. Charpin, and H. Dobbertin. Weight divisibility of cyclic codes, highly nonlinear functions on GF(2^m) and crosscorrelation of maximum-length sequences. *SIAM Journal on Discrete Mathematics*, 13(1):105–138, 2000.

[6] A. Canteaut and M. Videau. Weakness of block ciphers using highly nonlinear confusion functions. research report, INRIA, 2001. to appear.

[7] C. Carlet, P. Charpin, and V. Zinoviev. Codes, bent functions and permutations suitable for DES-like cryptosystems. *Designs, Codes and Cryptography*, 15:125–156, 1998.

[8] F. Chabaud and S. Vaudenay. Links between differential and linear cryptanalysis. In *Advances in Cryptology - EUROCRYPT'94*, number 950 in Lecture Notes in Computer Science, pages 356–365. Springer-Verlag, 1995.

[9] P. Charpin, A. Tietäväinen, and V. Zinoviev. On binary cyclic codes with minimum distance $d = 3$. *Problems of Information Transmission*, 33(4):287–296, 1997.

[10] T. Cusick and H. Dobbertin. Some new 3-valued crosscorrelation functions of binary m-sequences. *IEEE Transactions on Information Theory*, 42:1238–1240, 1996.

[11] J. Daemen and V. Rijmen. AES proposal: the Rijndael block cipher. Available at `http://csrc.nist.gov/encryption/aes/rijndael/`, 1999.

[12] H. Dobbertin. One-to-one highly nonlinear power functions on $GF(2^n)$. *Appl. Algebra Engrg. Comm. Comput.*, 9(2):139–152, 1998.

[13] R. Gold. Maximal recursive sequences with 3-valued recursive crosscorrelation functions. *IEEE Transactions on Information Theory*, 14:154–156, 1968.

[14] C. Harpes. *Cryptanalysis of iterated block ciphers*, volume 7 of *ETH Series in Information Processing*. Hartung-Gorre Verlag, Konstanz, 1996.

[15] C. Harpes, G. Kramer, and J. L. Massey. A generalization of linear cryptanalysis and the applicability of Matsui's piling-up lemma. In *Advances in Cryptology - EUROCRYPT'95*, number 921 in Lecture Notes in Computer Science, pages 24–38. Springer-Verlag, 1995.

[16] T. Helleseth. Some results about the cross-correlation function between two maximal linear sequences. *Discrete Mathematics*, 16:209–232, 1976.

[17] H. Hollman and Q. Xiang. A proof of the Welch and Niho conjectures on crosscorrelations of binary m-sequences. *Finite Fields and Their Applications*, 7(2):253–286, 2001.

[18] T. Jakobsen. Cryptanalysis of block ciphers with probabilistic non-linear relations of low degree. In *Advances in Cryptology - CRYPTO'98*, number 1462 in Lecture Notes in Computer Science, pages 212–222. Springer-Verlag, 1998.

[19] T. Jakobsen. *Higher-order cryptanalysis of block ciphers.* PhD thesis, Technical University of Denmark, 1999.

[20] T. Jakobsen and L.R. Knudsen. The interpolation attack on block ciphers. In *Fast Software Encryption 97*, number 1267 in Lecture Notes in Computer Science. Springer-Verlag, 1997.

[21] T. Kasami. Weight distributions of Bose-Chaudhuri-Hocquenghem codes. In *Proceedings of the conference on combinatorial mathematics and its applications*, pages 335–357. The Univ. of North Carolina Press, 1968.

[22] T. Kasami. The weight enumerators for several classes of subcodes of the second order binary Reed-Muller codes. *Information and Control*, 18:369–394, 1971.

[23] L. R. Knudsen. Truncated and higher order differentials. In *Fast Software Encryption - Second International Workshop*, number 1008 in Lecture Notes in Computer Science, pages 196–211. Springer-Verlag, 1995.

[24] Z. Kukorelly. *On the validity of certian hypotheses used in linear cryptanalysis*, volume 13 of *ETH Series in Information Processing*. Hartung-Gorre Verlag, Konstanz, 1999.

[25] G. Lachaud and J. Wolfmann. The weights of the orthogonal of the extended quadratic binary Goppa codes. *IEEE Transactions on Information Theory*, 36(3):686–692, 1990.

[26] X. Lai. Higher order derivatives and differential cryptanalysis. In *Proc. "Symposium on Communication, Coding and Cryptography", in honor of J. L. Massey on the occasion of his 60'th birthday*, 1994.

[27] M. Matsui. Linear cryptanalysis method for DES cipher. In *Advances in Cryptology - EUROCRYPT'93*, number 765 in Lecture Notes in Computer Science. Springer-Verlag, 1994.

[28] M. Matsui. The first experimental cryptanalysis of the Data Encryption Standard. In *Advances in Cryptology - CRYPTO'94*, number 839 in Lecture Notes in Computer Science. Springer-Verlag, 1995.

[29] M. Matsui. New Block Encryption Algorithm MISTY. In *Proceedings of the Fourth International Workshop of Fast Software Encryption*, number 1267 in Lecture Notes in Computer Science, pages 54–68. Springer-Verlag, 1997.

[30] R.J. McEliece. Weight congruence for p-ary cyclic codes. *Discrete Mathematics*, 3:177–192, 1972.

[31] W. Meier and O. Staffelbach. Nonlinearity criteria for cryptographic functions. In *Advances in Cryptology - EUROCRYPT'89*, number 434 in Lecture Notes in Computer Science, pages 549–562. Springer-Verlag, 1990.

[32] K. Nyberg. Perfect nonlinear S-boxes. In *Advances in Cryptology - EUROCRYPT'91*, number 547 in Lecture Notes in Computer Science, pages 378–385. Springer-Verlag, 1991.

[33] K. Nyberg. Differentially uniform mappings for cryptography. In *Advances in Cryptology - EUROCRYPT'93*, number 765 in Lecture Notes in Computer Science, pages 55–64. Springer-Verlag, 1993.

[34] K. Nyberg and L.R. Knudsen. Provable security against differential cryptanalysis. In *Advances in Cryptology - CRYPTO'92*, number 740 in Lecture Notes in Computer Science, pages 566–574. Springer-Verlag, 1993.

[35] V. Pless. Power moment identities on weight distributions in error-correcting codes. *Info. and Control*, 3:147–152, 1963.

[36] D.V. Sarwate and M.B. Pursley. Crosscorrelation properties of pseudorandom and related sequences. *Proceedings of the IEEE*, 68(5):593–619, 1980.

[37] V.M. Sidelnikov. On mutual correlation of sequences. *Soviet Math. Dokl.*, 12:197–201, 1971.

[38] H. Tanaka, K. Hisamatsu, and T. Kaneko. Strength of MISTY1 without FL function for Higher Order Differential Attack. In *Applied Algebra, Algebraic Algorithms and Error-Correcting Codes*, number 1719 in Lecture Notes in Computer Science, pages 221–230. Springer-Verlag, 1999.

Sloppy Alice attacks! Adaptive chosen ciphertext attacks on the McEliece Public-Key Cryptosystem

Eric R. Verheul, Jeroen M. Doumen and Henk C.A. van Tilborg
Department of Math. and Comp. Sc.,
P.O. Box 513, Eindhoven University of Technology,
5600 MB Eindhoven, The Netherlands.
Eric.Verheul@pobox.com, doumen@win.tue.nl, henkvt@win.tue.nl

Abstract

We introduce new adaptive chosen ciphertext attacks, called *Sloppy Alice Attacks*, in which a malicious sender or an adaptive eavesdropper Eve has an oracle which allows her to find out whether a sent encrypted message does, or does not, decrypt properly. From this information she can extract the plaintext that was encrypted. In this paper we show that the McEliece public–key cryptosystem is susceptible to Sloppy Alice attacks.

Keywords: Ciphertext Only Adaptive Cryptanalysis, Decryption Problems, Differential Cryptanalysis, Differential Fault Analysis, McEliece Public–Key Cryptosystem

1 Introduction

In the last decade, several forms of attacks have been published where some of the inputs of an encryption system with a secret fixed key are adaptively chosen. By letting each new input (either plaintext or ciphertext) depend on the previous outputs and by looking at certain aspects of the resulting output at each step, additional secret information of the cryptosystem (e.g. the fixed key) may be determined. Among the studied aspects of the output are:

- the differences in the output when the differences in the plain inputs are known, see [4], [15];
- some statistical or number–theoretic aspects of the output of the cryptosystem when errors are inflicted to the cryptosystem itself (e.g. by radiation), see for instance [5], [6];
- the execution time when the precise complexity of the underlying cryptographic process is known, see [13].

In this paper, we will look at a different setting. Here an attacker, called Eve, has access (for instance by interception) to one or more encrypted messages (called ciphertexts) for a receiver Bob.

We also suppose that Eve has access to an oracle that can tell her whether a ciphertext deciphers correctly or not. This oracle scenario was already studied in a more general form by Goldwasser, Micali and Tong [9]. Here we will present an efficient algorithm against the McEliece cryptosystem.

Note that in practice such an oracle might be easy to obtain: Eve may have access to Bob's decryption device or Eve might be an active eavesdropper. Another possibility is that Bob's decryption device is in fact automated, and will send a reply if the decryption somehow went wrong, thus asking for a retransmission. This reply can then be intercepted by Eve.

Now the general idea of this attack is based on the following components:

1. Eve alters the ciphertext slightly in such a way that there is a reasonable probability that the message still deciphers correctly and sends the altered message to her oracle.

2. Knowledge on whether the message still deciphers correctly or not reveals new information and opens interesting new possibilities for adapting the ciphertext.

Eve will continue to alter messages in this way (called a *round*), until she has retrieved enough secret information. It is very likely that Eve will have to send a considerable number of altered messages.

In the case that Eve uses Bob as her oracle (she sends the messages to him and checks for a retransmission request), this means that Bob (or the protocol he is using) has to be willing to ask the presumed sender Alice over and over again for retransmission. This is only likely if Bob accepts sloppy (or faulty) behavior of Alice (or her computer). That is why we refer to this type of attacks as *Sloppy Alice attacks* (SA–attacks).

In [19] we discuss Sloppy Alice attacks on key–recovery schemes. Here, we will concentrate on SA attacks on the McEliece [16] public–key cryptosystem, which is based on algebraic coding theory and more in particular on *Goppa codes.* So we assume that Eve has a validly encrypted McEliece message for Bob which she can alter and for which she is able to find out (e.g. by using her oracle) if it still is a validly encrypted McEliece message.

The outline of this paper is as follows: in Section 2 we will describe the McEliece public–key cryptosystem – in a general form – and we will determine the value of the *maximum error correcting capability* that is attained by two most important decoding (deciphering) methods for Goppa codes: the Berlekamp–Massey algorithm and Euclid's algorithm. The proof of this value is given in Appendix A. Section 3 is the main part of this paper; there we will describe an effective SA–attack on the McEliece public–key cryptosystem based on the maximum error correcting property of the above two decoding algorithms. In Section 4, we consider some countermeasures against the described SA–attacks. Finally, we summarize our findings in Section 5.

Related Work

The attack described here differs from the one in [2] where it is assumed that the (original) sender, instead of an eavesdropper, sends the same message more than once using different random error vectors.

Also, please note that since the initial submission of this paper in 1998, an independent description of a similar algorithm has appeared in [10].

2 The McEliece Public–Key Cryptosystem

We shall give some introductory remarks on error–correcting codes. Let $V_n(q)$ denote an n–dimensional vector space over the finite field $GF(q)$ (where q is a prime power). Let $\mathcal{C}$ be a k dimensional linear subspace of $V_n(q)$. It will be called a linear $[n, k]$ *code.* The elements in $\mathcal{C}$ are similarly called *codewords.* A *generator matrix* G for $\mathcal{C}$ is a $k \times n$ (q–ary) matrix whose rows span $\mathcal{C}$. This means that each codeword $\mathbf{c}$ can be written as $\mathbf{c} = \mathbf{m}G$ (in $\mathcal{C}$) with $\mathbf{m} \in V_k(q)$. One can describe this by saying that the information vector $\mathbf{m}$ is encoded in the codeword $\mathbf{c}$. The quantity $n - k$ is the *redundancy* of $\mathcal{C}$. It gives the number of excess symbols in $\mathbf{c}$ with regards to $\mathbf{m}$.

Now $\mathbf{c}$ is sent over an (unreliable) channel and certain errors may be inflicted on $\mathbf{c}$: the received vector is $\mathbf{r} = \mathbf{c} + \mathbf{e}$ where $\mathbf{e}$ is the so called *error vector.* Let the Hamming weight $w_H(\underline{x})$ of a vector $\mathbf{x}$ simply count the number of non–zero coordinates of $\mathbf{x}$. If the weight of $\mathbf{e}$ is not too large, the received vector $\mathbf{r}$ coincides on many coordinates with $\mathbf{c}$ and so $\mathbf{c}$ can be recovered.

To this end, the *(Hamming) distance* $d_H(\mathbf{x}, \mathbf{y})$ between vectors $\mathbf{x}$ and $\mathbf{y}$ is defined as the number of coordinates where $\mathbf{x}$ and $\mathbf{y}$ differ. Note that $d_H(\mathbf{x}, \mathbf{y}) = w_H(\mathbf{x} - \mathbf{y})$. The *minimum distance* d of $\mathcal{C}$ is defined as the minimum Hamming distance between different codewords in $\mathcal{C}$. Since $\mathcal{C}$ is linear, an equivalent definition would be that d is the minimum non–zero weight in $\mathcal{C}$. A code with minimum distace d will be denoted as a $[n, k, d]$ code. The number $t = \lfloor (d-1)/2 \rfloor$ is called the *error–correcting capability* of $\mathcal{C}$. It follows from the triangle inequality that for each element $\mathbf{r}$ in $V_n(q)$ there can be at most one element $\mathbf{c}$ in $\mathcal{C}$ at distance $\leq t$ to it. So, in principle, one can correct up to t errors inflicted to an element in $\mathcal{C}$ by finding the nearest point in $\mathcal{C}$. However, in practice the process of determining the nearest point (called *decoding*) is often very complex. To illustrate, the problem for general linear codes on deciding on whether there exists a point in $\mathcal{C}$ at a given distance of a given point $x \in V_n(q)$ is known to be in the class NP–complete, see [3]. Fortunately there are certain classes of linear codes where decoding can be done quite effectively, see below.

In 1978, McEliece [16] proposed a public–key cryptosystem based on the general difficulty of decoding. Consider a generator matrix G, generating a q–ary code $\mathcal{C}$ with parameters $[n, k, d]$, which is constructed by a user Bob. Let $0 \leq t \leq e = \lfloor (d-1)/2 \rfloor$ and let $\mathcal{A}_t$ be an effective decoding algorithm for $\mathcal{C}$ that can correct up to t errors.

Now, to use this in a cryptographic setting, Bob generates a random, invertible, q–ary, $k \times k$ matrix S and a random permutation matrix P of size $n \times n$. The public key of Bob is $G' = SGP$ together with the value of t. The matrices, S, G, P are kept secret. The idea is that G', although it generates a codespace $\mathcal{C}'$ which is equivalent to $\mathcal{C}$, behaves like a "random" generator matrix for which the decoding problem is hard.

Now suppose another user, say Alice, wants to encrypt a message $\mathbf{m} \in V_k(q)$ for Bob. Then, she generates a random error vector $\mathbf{e}$ of weight $\leq t$ and she forms:[1]

$$\mathbf{r} = \mathbf{m}G' + \mathbf{e} \ \ (= \mathbf{m}SGP + \mathbf{e}) \tag{1}$$

On delivery, Bob calculates

$$\mathbf{r}P^{-1} = (\mathbf{m}S)G + \mathbf{e}P^{-1}.$$

As $\mathbf{e}P^{-1}$ has the same weight as $\mathbf{e}$, Bob can determine $\mathbf{m}S$ (and $\mathbf{e}P^{-1}$) from $\mathbf{r}P^{-1}$ by means of algorithm $\mathcal{A}_t$. Since S is a invertible matrix, Bob can easily determine $\mathbf{m}$, for instance by the method of Gaussian elimination.

More in particular, in his scheme McEliece proposed to use binary (i.e. $q = 2$), irreducible Goppa codes, with $n = 1024$, $k \approx 524$ and $t = 50$. There exist many (different) codes of these parameters, they are easy to generate (randomly) and efficient decoding algorithms for them are easy to find. McEliece's construction can be extended to larger classes of codes (for instance non–binary Goppa codes). We will not go into detail here as that is not necessary for our attack; it suffices to mention the following bounds on the security–related parameters of the system.

[1] In some variants of McEliece the weight of the error vector e is always exactly equal to t.

Assumption 2.1 (The security of McEliece cryptosystem) *The following observations can be made on the parameters of a McEliece public-key cryptosystem:*

Sec–1 $k \approx n/2 \geq 512$: *this makes "syndrome decoding" as well as an exhaustive search for finding the nearest codeword to the received word infeasible;*

Sec–2 $50 \leq t \leq 100$: *this makes all kinds of techniques that are based on guessing/finding k (almost) error free coordinates less time consuming than the methods in Sec–1, but still infeasible (see [1, 7, 16]).*

We shall now formulate a property of the decoding algorithm $\mathcal{A}_t$. This property will play a crucial role in the proposed SA Attack.

Property 2.1 (Maximum Error Property (MEP)) *On input of a vector* $\mathbf{r} \in V_n(q)$, *the decoding algorithm* $\mathcal{A}_t$ *will either return a codeword* $\mathbf{c}$ *in* C *at distance* $\leq t$ *to* $\mathbf{r}$ *(if such a codeword exists) or it will return an error–message.*

The MEP property states that the decoding algorithm $\mathcal{A}_t$, on input $\mathbf{r}$, never returns an element $\mathbf{c} \in C$ at distance more than t from $\mathbf{r}$. It is important to realize that if too many transmission errors have occurred, the received vector $\mathbf{r}$ may be at distance $\leq t$ form another codeword than the transmitted one. In this case $\mathcal{A}_t$ will not return an error message. The probability that this occurs, will be of importance in the analysis of our SA attack and will be discussed in Section 3.

We end this section with stating that two relevant decoding algorithms for Goppa codes have the MEP; the proof of this will be given in Appendix A.

Proposition 2.1 *Let* C *be a Goppa code with designed distance* t, *and let* $\mathcal{A}_t$ *be either Euclid's algorithm or the Berlekamp–Massey algorithm for decoding* C. *Then* $\mathcal{A}_t$ *has the MEP.*

3 A SA–Attack on the McEliece Cryptosystem

We shall now describe the SA–attack on the McEliece cryptosystem. Let $\mathbf{g}'_i$ denote the i–th column of the (public) generator matrix G' ($1 \leq i \leq n$).

Algorithm 3.1 (SA–attack on McEliece) *Assume that the decoding algorithm $\mathcal{A}_t$ used in a McEliece cryptosystem has the Maximal Error Property. Let* $\mathbf{r}$ *be the ciphertext sent by Alice and intercepted by Eve (it is of the form $\mathbf{r}_A = \mathbf{m}G' + \mathbf{e}_A$). Then Eve does the following:*

Step 1. Increase the number of errors made by Alice to exactly t.

In order to increase the number of errors to the maximum, Eve continues changing a random coordinate arbitrarily (though each coordinate is selected at most once) and sending the resulting codeword to Bob until an error message is returned, i.e. the message is not accepted as a valid McEliece ciphertext. Once this occurs, Eve knows that this message contains one error too many, and thus the previous message she sent to Bob has the maximum number of errors. She now goes on to step 2 with this message $\mathbf{r}'$.

Step 2. Determine enough error–free coordinates.

Once Eve knows she has a message with exactly t errors, she can start probing a random coordinate (different from all preceeding choices, including those made in step 1) by changing this arbitrarily in $\mathbf{r}$*', and sending the mutated message to Bob. If an error message is returned, this coordinate was error–free. Once enough error–free coordinates are determined, Eve can determine the plaintext in step 3.*

Step 3. Determine the plaintext.

Once Eve knows enough error–free coordinates, she can solve the matrix equation $r' = mG$ for the plaintext m by using Gaussian elimination on the columns corresponding with the (known) error–free coordinates.

Before formulating Theorem 3.1, we recall the following definition. Let A_w, $0 \leq w \leq n$ be the number of codewords in $\mathcal{C}$ of weight w. The weight distribution of an $[n, k, d]$ code C will be called *approximately binomial* if (in the context of the problem here) the weight distribution may be approximated as follows:

$$A_w \approx \frac{\binom{n}{w}(q-1)^w}{q^{n-k}}, \quad d \leq w \leq n, \tag{2}$$

or, in other words, the cardinality of codewords with weight d is approximately $\binom{n}{w}(q-1)^w/q^{n-k}$.

Note that in this case,

$$\sum_{w=0}^{n} A_w \approx \sum_{w=0}^{n} \frac{\binom{n}{w}(q-1)^w}{q^{n-k}} = \frac{(1+(q-1))^n}{q^{n-k}} = q^k = |C|,$$

as it should be. Certainly the weight distribution of $V_n(q)$ itself is binomial (approximation (2) is actually an equality). We are not familiar with any result on how well the weight distribution of Goppa codes can be approximated by the binomial distribution, but based on [11, 12] it seems very reasonable to make that assumption here.

We will also assume that the minimum distance of the used code is odd, i.e. $d = 2t + 1$.

Theorem 3.1 *With the notation of Section 2, let C be a McEliece–like cryptosystem based on a q–ary code with approximately binomial weight distribution and with a decoding algorithm that has the maximum error property, for instance a Goppa code. Then Algorithm 3.1 is a SA attack on C returning the plaintext* $\mathbf{m}$.

Let the number of times the loop in step 1 is executed be S_1. We have that $S_1 \leq 2t + 1$ times. With large probability the loop in step 2 will be executed at most $k + \log_q(k) + X + 1$ times, where X is equal to $\min(t, 2t + 1 - S_1)$.

So, with large probability Algorithm 3.1 is a SA attack of at most $k + \log_q(k) + 2t + 2$ rounds.

We remark that if a binary code is used, then any error–coordinate is a one, so Algorithm 3.1 can be slightly improved to a SA attack of at most $k + 2t + 1$ rounds. We also remark, that if in step 2 of Algorithm 3.1 t rejections are encountered, then all errors introduced by Alice have been found and all remaining coordinates are error–free. In this case deciphering can be done much faster. Of course, the probability that this occurs is negligible.

Before proving Theorem 3.1, we need two lemmas. We first recall that the *binary entropy function* $h(x)$ is defined by: $h(0) = h(1) = 0$ and $h(x) = -x\log_2(x) - (1-x)\log_2(1-x)$ $(0 < x < 1)$.

Lemma 3.1 *In the notation of Section 2, let* $\mathbf{e} \in V_n(q)$ *be an 'error-vector' of weight* $t+1$*. Then the following holds.*

i) *There is at most one vector* $\mathbf{f} \in V_n(q)$ *of weight* $\leq t$ *such that* $\mathbf{e}+\mathbf{f} \in C$*. Also, if such an* $\mathbf{f}$ *exists, then the weight of* $\mathbf{f}$ *is exactly equal to* t*, the supports (the sets of non-zero coordinates) of* $\mathbf{e}$ *and* $\mathbf{f}$ *are disjoint and* $d = 2t+1$*.*

ii) *If* $\mathbf{e}$ *is chosen uniformly random, then the probability* P *that a vector* $\mathbf{f}$ *of weight at most* t *exists such that* $\mathbf{e}+\mathbf{f} \in C$ *is given by*

$$P = \frac{A_{2t+1}\binom{2t+1}{t+1}}{\binom{n}{t+1}(q-1)^{t+1}}. \tag{3}$$

iii) *If the weight distribution of* C *is approximately binomial, then*

$$P \approx \frac{\binom{n-(t+1)}{t}(q-1)^t}{q^{n-k}} \leq \frac{2^{(n-(t+1))h(\frac{t}{n-(t+1)})}(q-1)^t}{q^{n-k}}.$$

iv) *Assume that the weight distribution of the Goppa codes is indeed approximately binomial [11, 12]. If a binary Goppa code is used in a McEliece cryptosystem, the above probability* P *is negligible. To be more precies, when the parameters originally proposed by McEliece in [16] are used, we have* $P \leq 2^{-215}$*. When the improved parameters as mentioned in Assumptions 2.1 are used, we have that* $P \leq 2^{-54}$*.*

Proof: i) Suppose that two distinct candidates for $\mathbf{f}$ – as mentioned in the first part of the lemma – exist, say $\mathbf{f}$ and $\ddot{\mathbf{e}}$. Then

$$d_H(\mathbf{e}+\mathbf{f}, \mathbf{e}+\ddot{\mathbf{e}}) = d_H(\mathbf{f}, \ddot{\mathbf{e}}) = w_H(\mathbf{f}-\ddot{\mathbf{e}}) \leq w_H(\mathbf{f}) + w_H(\ddot{\mathbf{e}}) = \leq 2t < d.$$

As $\mathbf{e}+\mathbf{f}$ and $\mathbf{e}+\ddot{\mathbf{e}}$ are two distinct members of C at distance less than d, we arrive at a contradiction. A similar argument shows that the weight of $\mathbf{f}$ must be equal to t and that $d = 2t+1$.

ii) Let

$$B = \{(\mathbf{e},\mathbf{f}) \in V_n(q) \times V_n(q) \mid \mathbf{e}+\mathbf{f} \in C,\ w_H(\mathbf{e}) = t+1,\ w_H(\mathbf{f}) = t\}.$$

Let $\mathbf{c}$ be a codeword in C of weight $2t+1$. Then each $(t+1)$–subset of the support of $\mathbf{c}$ gives rise to a unique pair $(\mathbf{e},\mathbf{f}) \in B$ (change the remaining t non–zero coordinates of $\mathbf{c}$ into a zero, resp. the $t+1$ coordinates themselves). Conversely, each element $(\mathbf{e},\mathbf{f}) \in B$ can be obtained this way. Hence it follows that $|B| = A_{2t+1}\binom{2t+1}{t+1}$.

The total number of 'error–vectors' of weight $t+1$ in $V_n(q)$ is $\binom{n}{t+1}(q-1)^{t+1}$. The probability P is the quotient of $|B|$ and this number.

iii) By assumption

$$A_{2t+1} \approx \frac{\binom{n}{2t+1}(q-1)^d}{q^{n-k}}.$$

It follows from ii) that

$$P \approx \frac{\binom{n}{2t+1}(q-1)^{2t+1}\binom{2t+1}{t+1}}{q^{n-k}\binom{n}{t+1}(q-1)^{t+1}} = \frac{\binom{n-(t+1)}{t}(q-1)^t}{q^{n-k}}.$$

To show the inequality in iii), we note that the binomial theorem implies for each $0 \le m \le n$ the inequality

$$n^n = (m+(n-m))^n = \sum_{i=0}^{n}\binom{n}{i} m^i (n-m)^{n-i} \ge \binom{n}{m} m^m (n-m)^{n-m}.$$

This can be rewritten as

$$\binom{n}{m} \le \frac{n^n}{m^m(n-m)^{n-m}} = \frac{2^{n\log_2 n}}{2^{m\log_2 m}2^{(n-m)\log_2(n-m)}} =$$
$$2^{-m\log_2(m/n)-(n-m)\log_2((n-m)/n)} = 2^{nH(m/n)}.$$

iv) With the assumption that the weight distributions of the Goppa codes are indeed approximately binomial, the probability P mentioned in ii) can be approximated using iii). If we take the parameters proposed by McEliece, i.e. $n = 1024$, $k = 524$, $t = 50$, we obtain

$$P \leq \frac{2^{975*H(50/975)}}{2^{500}} \approx \frac{2^{285}}{2^{500}} = 2^{-215},$$

which is a negligible. Similarly, if we take the parameters as in Assumption 2.1 we obtain that $P \leq 2^{-54}$. So the same holds for general McEliece cryptosystems, provided of course the weight distribution of code is approximately binomial.

□

The following observation may be of interest to the reader. It is well known that for a perfect t–error correcting code $\binom{2t+1}{t} A_{2t+1} = \binom{n}{t+1}(q-1)^{t+1}$ (e.g. see [18, Problem 3.4.9]). Substitution of this relation into (3) gives $P = 1$, as it should be for a perfect code: each word at distance $t+1$ from one codeword lies at distance t from exactly one other codeword. Thus a Sloppy Alice attack on McEliece–like cryptosystem which uses a perfect code will not work, since each vector can be decoded.

Lemma 3.2 *The probability that a random $k \times (k + \log_q k)$ q–ary matrix has rank k is at least $1 - \frac{1}{k}$.*

Proof: Let $P(k, m)$, $m \geq k$, denote the probability that a random $k \times m$ binary matrix A has rank k. Looking at the rows of A we observe that the first row of A should be non–zero, the second row should be independent of the first, etc. This argument leads to

$$\begin{aligned} P(k,m) &= \frac{\prod_{i=0}^{k-1}(q^m - q^i)}{\prod_{i=0}^{k-1} q^m} = \prod_{i=m-k+1}^{m} (1 - \frac{1}{q^i}) \\ &\overset{(*)}{\geq} 1 - \sum_{i=m-k+1}^{m} \frac{1}{q^i} = 1 - \frac{1 - \frac{1}{q^k}}{(q-1)q^{m-k}} \geq 1 - \frac{1}{q^{m-k}}, \end{aligned}$$

where $(*)$ follows quite easily with an induction argument.

Now substituting $m = k + \log_q k$ in the above relation gives

$$P(k, k + \log_q k) \geq 1 - \frac{1}{k}.$$

□

Proof of Theorem 3.1

We start with a general observation. Consider any $\mathbf{r} = \mathbf{c}+\mathbf{e}$ where $\mathbf{c} \in C$ and $w_H(\mathbf{e}) = s \leq t$. If we change the ith coordinate of $\mathbf{r}$, which can be described by adding a vector $\mathbf{u}$ of weight 1 and with support $\{i\}$ to $\mathbf{r}$, then there are three possibilities (only two if $q = 2$) for the resulting $\mathbf{r}' = \mathbf{r} + \mathbf{u} = \mathbf{c} + \mathbf{e}'$.

1. $w_H(\mathbf{e}') = s - 1$ iff $e_i \neq 0$ and $u_i = -e_i$;

2. $w_H(\mathbf{e}') = s$ iff $e_i \neq 0$ and $u_i \neq -e_i$
   This is impossible if $q = 2$, because both are also non–zero in this case)

3. $w_H(\mathbf{e}') = s + 1$ iff $e_i = 0$.

Consider step 1 of Algorithm 3.1. For the range $0 \leq i \leq 2t + 1$, let $\mathbf{e}^{(i)} = \mathbf{r}^{(i)} - \mathbf{c}$, that is, $\mathbf{r}^{(i)} = \mathbf{c} + \mathbf{e}^{(i)}$. Of course each $\mathbf{e}^{(i)}$ is unknown to Eve and $\mathbf{e}^{(0)} = \mathbf{e}_A$. As $w_H(\mathbf{e}^{(0)}) = w_H(\mathbf{e}_M) \leq t$ it follows that there exists a first $0 < i \leq 2t + 1$ in step 1, such that $w_H(\mathbf{e}^{(i)}) = t$ and $w_H(\mathbf{e}^{(i+1)}) = t+1$. So, for $0 \leq j \leq i$ execution of the decoding algorithm $\mathcal{A}_t$ applied by Bob to $\mathbf{r}^{(j)}$ does not result in an error–message.

Note that i can only reach the value $2t+1$ in the (extremely unlikely) case that the $2t + 1$ errors introduced by Eve in this step of the SA algorithm include all the errors that Alice originally has added to $\mathbf{c}$. In this case, we can immediately proceed to step 3 of the algorithm since all other coordinates will be error–free, and contain enough independent columns of the generator matrix G.

Next, we claim that the $\mathcal{A}_t$ applied to $\mathbf{r}^{(i+1)} = \mathbf{c} + \mathbf{e}^{(i+1)}$ with $\mathbf{e}^{(i+1)}$ of weight $t + 1$, will result in an error–message (and we go to step 2). Indeed, if $\mathcal{A}_t$ applied to $\mathbf{r}^{(i+1)}$ does not issue an error–message then $\mathbf{r}^{(i+1)}$ lies at distance at most t from C by MEP. This means that there exist a codeword $\mathbf{c}'$ in C and a vector $\mathbf{e}'$ in $V_n(q)$ of weight at most t such that $\mathbf{r}^{i+1} = \mathbf{c}' + \mathbf{e}'$. Hence, we are in the situation of Lemma 3.1, from which it follows that this situation has a negligible probability of occurring.

In step 2, write $\mathbf{r}' = \mathbf{c}+\mathbf{e}'$, with $w_H(\mathbf{e}') = t$. We follow the same reasoning as above. When the f–th coordinate in $\mathbf{r}$ is changed, the obtained vector $\hat{\mathbf{r}}$ will not be accepted by $\mathcal{A}_t$ (i.e. without error–messages) if and only if (with a negligible probability of failure) the f–th coordinate of $\mathbf{e}'$ is zero.

The number of loops in this part of the SA algorithm follows directly from Lemma 3.2. Since we only take coordinates in step 2 different from those selected in step 1, we have to add an extra tern $+X$ that reflects the (marginal) possibility that X times a change made in this step canceled out an error introduced by Alice.

Because $S_1 + X \leq 2t + 1$ with large probability the algorithm takes at most $k + \log_q(k) + 2t + 1$ rounds.

□

4 Countermeasures to SA attacks on the McEliece cryptosystem

Countermeasures to the SA attack on the McEliece cryptosystem, should – at least – aim to achieve that there is no correlation between deciphering problems and the number of errors applied to the plaintext.

First, in the case that Bob is Eve's oracle, Bob could come up with the idea of checking for repeated messages. This would detect a SA–attack as described above, but nothing prevents the attacker Eve from adding a random codeword from $\mathcal{C}$ to her probe each round. This preserves the error–vector e, and will allow Eve to conduct her attack as usual with little additional effort.

As a second idea one might consider to fix the weight of the error vector to *exactly* t, (or any $t' \leq t$). (cf. equation (1)) and to return an error–message when in the deciphering process an error vector is encountered of weight unequal to t (or the chosen t'), that is, irrespective of whether successful decoding is still possible.

Let us illustrate how also in this setting effective SA attacks are still possible. First of all, suppose that the used code is non–binary. If one 'probes' a coordinate in step 2, say the i–th coordinate, twice but with different values, then Bob will always return an error–message if that coordinate is error–free (since there are $t + 1$ errors in both probes). However, if there's an error on that coordinate, at most one probe will return no error message (since Eve only alters the value of $\mathbf{e}_i$ and not the weight of the error–vector).

So we have the following situation:

- if both probes give an error–message, then $e_i = 0$;
- if only one probe gives an error–message, then $e_i \neq 0$ and e_i is in fact determined;
- if none of the probes gives an error–message, then $e_i \neq 0$.

It easily follows that the plaintext can be found with around $k+t$ probes, i.e. in around $2(k+t)$ rounds.

If the used code is binary, then one starts by changing any coordinate, say the i–th, of $\mathbf{r}'$. In the setting here, this will always lead to an error–message from the oracle. Now we distinguish two possibilities.

With probability $(n-t)/n$ one has that $e_i = 0$. If one changes an additional coordinate in all possible ways, then $n-t-1$ times another error will have been introduced, resulting in an error–message from the oracle. Further, t times an error (introduced originally by the original sender) will be eliminated and so in this case one is back at t errors, leading to a correct decryption. In this way, all errors introduced by Alice can be found.

If $e_i = 1$ (with probability t/n), then additionally introduced single errors will lead $n-t$ times to correct decryptions (and coordinates with $e_i = 0$) and $t-1$ times to an error message.

In the case that Bob is in fact serving as Eve's oracle, a better countermeasure to the SA attack technique may be to introduce further redundancy in the system to enable Bob to check if an active eavesdropper is altering a proper ciphertext.

For instance, let Alice choose her plaintext $\mathbf{m}$ from $V_{k-64}(q)$ instead of $V_k(q)$. As before, she choose a random error vector of weight $\leq t$. Bob has published as part of his public key a cryptographically secure hash–function $h(.)$ that maps elements in $V_n(q)$ of weight $\leq t$ to words in $V_{64}(q)$ (the hash function can also be a system parameter). To encrypt $\mathbf{m}$ Eve computes (cf. equality (1)):

$$\mathbf{r} = (\mathbf{m}||h(\mathbf{m}||\mathbf{e}))G' + \mathbf{e} \quad (= (\mathbf{m}||h(\mathbf{e}))SGP + \mathbf{e}), \tag{4}$$

where || stands for concatenation. When Bob receives a vector $\mathbf{r}'$, he will attempt to decode it. If this works, he will will find an error vector $\mathbf{e}'$ and an element $\mathbf{m}' \in V_k(q)$ satisfying

$$\mathbf{r}' = \mathbf{m}'G' + \mathbf{e}'.$$

Finally, he checks if $h(\mathbf{e}')$ equals the 64 right most coordinates of $\mathbf{m}'$. If this verification fails, an error–message is issued. This error–message will not give any information to Eve about the original choice of $\mathbf{e}$ by Alice. It follows that the SA attack as described in Section 3 fails.

However, if $h(.)$ is additive, i.e. $h(e_1 + e_2) = h(e_1) + h(e_2)$, then one can easily extend the described SA attack. So it follows that is $h(.)$ should be reasonable "non–linear".

Note that the choice of 64 bits in the above example is arbitrary: this is a security parameter of the system which indicates how many message bits are used for increased security. We refer to [8] for a more general description of this construction. Also observe that as a side effect of this variation to the McEliece cryptosystem it loses its inherent error–correcting capabilities. This seems to be inevitable.

5 Conclusion

We have introduced the Sloppy Alice attack, an adaptive chosen cipher-text attack, which is based on the idea that (ordinary) users see no problem in overtly revealing whether or not an encrypted message deciphers correctly. We have described such an attack on the McEliece Public–Key cryptosystem.

The general conclusion is that such error–messages can be used to efficiently break the McEliece Public–Key cryptosystem. Therefore, at the very least, error–messages should be as non–descriptive as possible and users should be alerted when many encrypted messages do not decrypt properly. We have also proposed a variant of the McEliece system which is immune to SA attacks.

A Proof of Proposition 2.1

Before proving Proposition 2.1, we recall the definition and some properties of Goppa codes (see [14]). Let $g(x)$ be a monic polynomial of degree u over the finite field $GF(p^m)$ Also, let $\gamma_1, \gamma_2, \ldots, \gamma_n$ be n distinct members of $GF(p^m)$ such that $g(\gamma_i) \neq 0$ for all $1 \leq i \leq n$. The *Goppa code* C generated by the Goppa polynomial $g(x)$ consists of all words $\mathbf{c} = (c_1, c_2, \ldots, c_n) \in V_n(q)$ satisfying

$$\sum_{i=1}^{n} \frac{c_i}{x - \gamma_i} = 0 \quad (\mathrm{mod} g(x)).$$

The code C is linear of dimension $\geq n - m \cdot u$. Its error–correcting capability t is greater than or equal to $\lfloor u/2 \rfloor$. Moreover, if $p = 2$ and $g(x)$ does not have multiple zeros, then $t \geq \lfloor u \rfloor$.

Actually, in his original proposal, McEliece let $g(x)$ be an irreducible polynomial of degree 50 over $GF(2^{10})$.

An important features of Goppa codes is the existence of an efficient decoding algorithm $A_{t'}$ for any t' less than or equal to the designed error–correcting capability t. In practice, one therefore only corrects up–to the designed error–correcting capability.

Decoding goes as follows: suppose that the vector $\mathbf{r} = (r_1, r_2, \ldots, r_n)$ is received. The *syndrome* polynomial $S_{\mathbf{r}}(x)$ of the $\mathbf{r}$ is defined by

$$S_{\mathbf{r}}(x) \equiv \sum_{i=1}^{n} \frac{r_i}{x - \gamma_i} \quad (\mathrm{mod}\ g(x)).$$

Observe that the inverse of each polynomial $x - \gamma_i$ exists modulo $g(x)$, as $g(x)$ is non–zero on the γ_i's. It is given by the solution of $u(x)(x - \gamma_i) \equiv 1$ (mod $g(x)$). Also, the syndrome polynomial $S_{\mathbf{r}}(x)$ is zero if and only if $\mathbf{r} \in C$. Now suppose that $\mathbf{r}$ is of the following form

$$\mathbf{r} = \mathbf{c} + \mathbf{e} \ ; \quad \mathbf{c} \in C, \ w_H(\mathbf{e}) \leq t, \tag{5}$$

Let $\mathcal{E}$ be the set of non–zero coordinates of $\mathbf{e} = (e_1, e_2, \ldots, e_n)$. Then the *error–locator* polynomial $\sigma(x)$ and the *error evaluator* polynomial $\omega_{\mathbf{r}}(x)$ are defined by

$$\sigma(x) = \prod_{i \in \mathcal{E}} (x - \gamma_i) \quad ; \quad \omega(x) = \sum_{i \in \mathcal{E}} e_i \prod_{j \in \mathcal{E} \setminus \{i\}} (x - \gamma_j).$$

Then $\deg(\omega(x)) < \deg(\sigma(x)) \leq t$, $\gcd(\sigma(x), \omega(x)) = 1$, and the following, so–called *key equation* holds:

$$S_{\mathbf{r}}(x)\sigma(x) \equiv \omega(x) \pmod{g(x)}.$$

The next (general) result is crucial for the decoding process. The proof is a direct consequence of [17, Theorem 8.5].

Lemma A.1 *Let $g(x)$, $u = \deg g(x)$ be defined as above. Let $S(x), \Sigma(x)$, and $\Omega(x)$ be q–ary polynomials and let the latter two be monic. Then the following two properties are equivalent:*

P–1 $S(x)\Sigma(x) \equiv \Omega(x) \pmod{g(x)}$, $\deg(\Omega(x)) < \deg(\Sigma(x)) \leq \lfloor u/2 \rfloor$, $\deg \Sigma(x)$ *is minimal among all such* $\Sigma(x), \Omega(x)$*;*

P–2 $S(x)\Sigma(x) \equiv \Omega(x) \pmod{g(x)}$, $\deg(\Omega(x)) < \deg(\Sigma(x)) \leq \lfloor u/2 \rfloor$, $\gcd(\Sigma(x), \Omega(x)) = 1$.

Moreover, a pair $(\Sigma(x)\ ,\ \Omega(x))$ *satisfying P1 (or P2) will be unique.*

In the remainder of this discussion, we shall first focus on the non–binary case (so $t = \lfloor u/2 \rfloor$). If the number of errors in $\mathbf{r}$ is not more than t, then the functions $\sigma(x)$ and $\omega(x)$ defined above form precisely the pair mentioned in Lemma A.1 w.r.t. $S(x) = S_{\mathbf{r}}(x)$. Clearly, from $\sigma(x)$ and $\omega(x)$ one can easily determine $\mathbf{c}$ and $\mathbf{e}$. Indeed, the error locations set $\mathcal{E}$ is completely determined by the roots of $\sigma(x)$. Further, for each i in $\mathcal{E}$ one can compute the error value e_i (the i–th coordinate of $\mathbf{e}$) from the relation $e_i = \omega(\gamma_i)/\sigma(\gamma_i)$. Finally, the original codeword $\mathbf{c}$ can be computed from as $\mathbf{c} = \mathbf{r} - \mathbf{e}$.

Proposition 2.1 states a property of the two main algorithms to determine the pair $\sigma(x), \omega(x)$ satisfying property P1 (or P2). The Berlekamp–Massey algorithm tries to find the pair $(\sigma(x), \omega(x))$ satisfying the first condition (w.r.t. $S(x) = S_{\mathbf{r}}(x)$), whereas Euclid's algorithm tries to find a pair satisfying the second condition. It is important to note that the Berlekamp–Massey algorithm is successful if and only if Euclid's algorithm is; they also lead to the same result. This is irrespective of whether $S(x)$ is a syndrome polynomial $S_{\mathbf{r}}(x)$. As is clear from Lemma A.1, the behavior of the Berlekamp–Massey algorithm and Euclid's algorithm with "bad" input, i.e. when inputting a syndrome polynomial of a vector $\mathbf{r}$ at distance more than t from the code, is the same.

Proof of Proposition 2.1 (non–binary case):
Suppose that $\mathbf{r} \in V_n(q)$ is the input to A_t and suppose that A_t outputs a vector $\mathbf{v}$ in $V_n(q)$ (so without error–message). As we do not assume that the implementation explicitly checks that the output is actually a codeword we do not know beforehand that $\mathbf{v} \in \mathcal{C}$. Certainly, if $\mathbf{r}$ is of the form (5) then A_t will output $\mathbf{c}$. However, to prove the proposition we have to show the converse, i.e. that equality (5) holds with $\mathbf{c} = \mathbf{v}$.

To this end, let's analyze the decoding process. First the decoding process tries – using either Euclid's or Berlekamp–Massey's algorithm – to find $\sigma_{\mathbf{r}}(x)$ and $\omega_{\mathbf{r}}(x)$ satisfying property P1 (P2). If this fails we assume that an error–message will be returned. After that step the decoding process determines the presumed set of error locations $\mathcal{E}' = \{1 \leq i \leq n \mid \sigma_{\mathbf{r}}(\gamma_i) = 0\}$. If $\mathcal{E}'$ has cardinality strictly less than $\deg(\sigma_{\mathbf{r}}(x))$ we assume that an error–message will be given. Next, the decoding process determines a vector $\mathbf{e}'$ of weight $\leq t$ by

$$e'_i = \begin{cases} \omega_{\mathbf{r}}(\gamma_i)/\sigma'_{\mathbf{r}}(\gamma_i) & \text{if } i \in \mathcal{B}, \\ 0 & \text{otherwise.} \end{cases}$$

Note that all e'_i with $i \in \mathcal{E}'$ are necessarily non–zero, as otherwise the $\sigma_{\mathbf{r}}$ and $\omega_{\mathbf{r}}$ would not be relatively prime, as dictated by Lemma A.1(2). Finally, the decoding process determines $\mathbf{v} = \mathbf{r} - \mathbf{e}'$, which is returned by A_t.

Now to see that $\underline{v}$ is a member of $\mathcal{C}$, we consider the polynomials $\sigma_{\mathbf{e}'}(x)$ and $\omega_{\mathbf{e}'}(x)$ (corresponding to $\mathbf{e}'$ written as $\mathbf{e}' = \mathbf{0} + \mathbf{e}'$), i.e. $S_{\mathbf{e}'}(x)\sigma_{\mathbf{e}'}(x) \equiv \omega_{\mathbf{e}'}(x) \pmod{g(x)}$. First observe that the pair $(\sigma_{\mathbf{e}'}(x), \omega_{\mathbf{e}'}(x))$ is equal to the pair $(\sigma_{\mathbf{r}}(x), \omega_{\mathbf{r}}(x))$. That $\sigma_{\mathbf{r}}(x) = \sigma_{\mathbf{e}'}(x)$ is trivially true, because the error vectors in $\mathbf{r} = \mathbf{v} + \mathbf{e}'$ and $\mathbf{e}' = \mathbf{0} + \mathbf{e}'$ are the same. That $\omega_{\mathbf{r}}(x) = \omega_{\mathbf{e}'}(x)$ follows from the fact that both are polynomials of degree $\leq \deg(\sigma_{\mathbf{r}}(x)) - 1$ that coincide on $\deg(\sigma_{\mathbf{r}}(x))$ distinct points.

So

$$S_{\mathbf{r}}(x)\sigma_{\mathbf{r}}(x) \equiv \omega_{\mathbf{r}}(x) = \omega_{\mathbf{e}'}(x) \equiv S_{\mathbf{e}'}(x)\sigma_{\mathbf{e}'}(x) \equiv S_{\mathbf{e}'}(x)\sigma_{\mathbf{r}}(x) \pmod{g(x)},$$

that is,

$$(S_{\mathbf{r}}(x) - S_{\mathbf{e}'}(x))\sigma_{\mathbf{r}}(x) \equiv 0 \pmod{g(x)}.$$

The polynomials $g(x)$ and $\sigma_{\mathbf{r}}(x)$ are relatively prime, because a common factor would also divide $\omega_{\mathbf{r}}(x)$ by the key equation and this contradicts property P2. Hence, it follows that $S_{\mathbf{v}}(x) = S_{\mathbf{r}}(x) - S_{\mathbf{e}'}(x) = 0$, i.e. $\mathbf{v} \in C$. This concludes the proof that $\mathbf{r}$ is of the form as described in equality (5). It also concludes the proof of the proposition.

Proof of Proposition 2.1 (binary case):
In this case t is equal to the degree of $g(x)$. Then the crucial equation $S_{\mathbf{r}}(x)\sigma_{\mathbf{r}}(x) \equiv \omega_{\mathbf{r}}(x) \pmod{g(x)}$ reduces to

$$S_{\mathbf{r}}(x)\sigma_{br}(x) = \sigma_{\mathbf{r}}(x). \tag{6}$$

Separating the squares from the non–squares in the expansion of the unknown $\sigma_{\mathbf{r}}(x)$ one can write $\sigma_{\mathbf{r}}(x) = (\alpha_{\mathbf{r}}(x))^2 + x(eta_{\mathbf{r}}(x))^2)$, one can use either Euclid's algorithm or the Berlekamp–Massey algorithm to find $\alpha_{\mathbf{r}}(x)$ and $eta_{\mathbf{r}}(x)$ and thus a $\sigma_{\mathbf{r}}(x)$ satisfying equation (6) (see for instance [18, Section 4.5]). As before, both algorithms lead to the same result (a codeword and an error vector) or they both lead to an error–message. The proof that the output $\mathbf{v}$ of $\mathcal{A}_t$ is of the form as in equation (5), with $\mathbf{v} \in C$ is similar to the non–binary case.

□

References

[1] A. Barg, E. Krouk, and H.C.A. van Tilborg, *On the Complexity of Minimum Distance Decoding of Long Linear Codes*, preprint.

[2] T.A. Berson, *Failure of the McEliece Public–Key Cryptosystem Under Message–Resend and Related–Message Attack*, Advances in Cryptology – Crypto '97 Proceedings, Springer–Verlag, 1997, pp. 213–220.

[3] E.R. Berlekamp, R.J. McEliece, H.C.A. van Tilborg, *On the inherent intractability of certain coding problems*, IEEE Transactions on Information Theory, 24 (1978), pp. 384–386.

[4] E. Biham, A. Shamir *Differential Cryptanalysis of the Data Encryption Standard*, Springer–Verlag, 1993.

[5] E. Biham, A. Shamir, *Differential Fault Analysis of Secret Key Cryptosystems*, Advances in Cryptology - CRYPTO '97 Proceedings, Springer–Verlag, 1997, pp. 513–525.

[6] D. Boneh, R. DeMillo, R. Lipton, *On the importance of checking cryptographic Protocols for faults*, Advances in Cryptology - EUROCRYPT '97 Proceedings, Springer–Verlag, 1997, pp. 37–51.

[7] I. Dumer, *Suboptimal Decoding of Linear Codes: partition Technique*, IEEE Trans. on Inform. Theory, IT–42, 1996, pp. 1971–1986.

[8] Fujisaki, Okamoto, *How to Enhance the Security of Public–Key Encryption at a Minimum Cost*, PKC 1999, LNCS 1560, pp.53-68, Springer-Verlag, 1999.

[9] S. Goldwasser, S. Micali and P. Tong, *Why and How to Establish a Private Code On a Public Network*, 23rd symposium on FOCS 1982, pp 134–144

[10] C. Hall, I. Goldberg, B. Schneier, *Reaction Attacks Against Several Public–Key Cryptosystems*, Proceedings of Information and Communication Security, ICICS'99, Springer-Verlag, 1999, pp. 2-12.

[11] T. Kasami, T. Fujiwara, and Shu Lin, *An Approximation to the Weight Distribution of Binary Linear Codes*, IEEE Trans. on Inform. Theory, Vol.IT–31, No. 6, 1985, pp.769–780.

[12] I. Krasikov and S. Litsyn, *On the Accuracy of the Binomial Approximation to the Distance Distance Distribution of Codes*, IEEE Trans. on Inform. Theory Vol.IT–41, 1995, pp. 1472–1475.

[13] P.C. Kocher, *Timing Attacks on Implementations of Diffie–Hellman, RSA, DSS, and Other Systems*, Advances in Cryptology – Crypto '96 Proceedings, Springer–Verlag, 1996, pp. 104–113.

[14] J.H. van Lint, *Introduction to Coding Theory*, Springer–Verlag, 1982.

[15] M. Matsui, *Linear Cryptanalysis Method for DES Cipher*, Advances in Cryptology - EUROCRYPT '93 Proceedings, Springer–Verlag, 1994, pp. 74–87.

[16] R.J. McEliece, *A public–key cryptosystem based on algebraic coding theory*, DSN Progress Report 42–44, Jet Propulsion Laboratory, Pasadena, 1978.

[17] R.J. McEliece, *The theory of information and coding,* Encyclopedia of Math. and its Applications, Vol. 3, Addison–Wesley Publishing Company, Reading, Mass., 1977.

[18] H.C.A van Tilborg, *Error–correcting Codes – a first introduction*, Chartwell Bratt Ltd, 1993.

[19] E.R. Verheul, *Sloppy Alice Attacks! Differential Fault Attacks on ElGamal, Chor–Rivest and Mandatory LEAF–reconstruction Recovery Systems*, in preparation.

Reducible Rank Codes and Applications to Cryptography

Ernst M. Gabidulin
Alexei Ourivski
Moscow Institute of Physics and Technology
Institutskii per., 9
Dolgoprudny, 141700, Russia

gab@pop3.mipt.ru
ourivski@mail.ru

Bassem Ammar
Bahram Honary
DCS
Lancaster University,
Lancaster LA1 4YR, UK

b.ammar@lancaster.ac.uk
b.honary@lancaster.ac.uk

Abstract

A new family of so called reducible rank codes, which are a generalization of rank product codes [8], is introduced. A McEliece type [1] public key cryptosystem based on these codes and on the idea of *a column scrambler* is proposed. The column scrambler 'mixes' columns of a generator matrix of a code. It makes the system more resistant to structural attacks such as Gibson's attacks [5, 6]. Possible attacks are studied. The possibility of using the system for combined error protection is investigated. The system is found to be secure against known attacks for public keys of about 10 kbits

1. Introduction

Rank codes were introduced in [2] to correct array errors. They also were used in the McEliece type [1] public-key cryptosystem (PKC), known as GPT [3]. It was shown by Gibson in [5,6] that this cryptosystem was insecure for practical values of parameters (lengths of codes $n \le 30$, the alphabet size $q = 2^{30}$). Recently PKC's based on *subcodes* of rank codes have been proposed to avoid Gibson's attack [4].

The original McEliece cryptosystem is based on Goppa codes and is much faster than the RSA system but its key size is too large (500 000 bits) to be practical. In this paper we present another approach to prevent Gibson's attacks and reduce the size of public key by using a new family of rank codes, the so-called *reducible* rank codes. These codes are a generalization of the family of rank *product* codes introduced and investigated in [8]. A notion of *column scrambler* is also employed. A column scrambler plays for the rank metric a role similar to a permutation matrix for a Hamming metric. A permutation of columns of a generator matrix of a code *does not change* the code Hamming weight distribution. A column scrambler forms linear combinations of columns of code's generator matrix with coefficients in the base field and *does not change* the code rank weight distribution.

2. Rank codes

Let $\mathbf{F}_N$ be a finite field with q^N elements, and let $\mathbf{F}_1$ be the base field of q elements; q is a power of a prime. Let $\mathbf{x} = (x_1, x_2, \ldots, x_n) \in \mathbf{F}_N^n$ be an n-vector with coordinates in $\mathbf{F}_N$. The *rank* $r(\mathbf{x}|\mathbf{F}_1)$ of $\mathbf{x}$ is defined to be the maximal number of x_i that are linearly independent over the base field $\mathbf{F}_1$. It is clear that $r(\mathbf{x} \mid \mathbf{F}_1) \le \min(n, N)$ for any $\mathbf{x} \in \mathbf{F}_N^n$

For a matrix $\mathbf{A}$ the *column rank* $r(\mathbf{A}|\mathbf{F})$ over a field $\mathbf{F}$ is the maximal number of columns that are linearly independent over $\mathbf{F}$. The column rank of a matrix depends on the field. In particular, $r(\mathbf{A} \mid \mathbf{F}_1) \ge r(\mathbf{A} \mid \mathbf{F}_N)$.

The *rank distance* $d_r(\mathbf{x}, \mathbf{y})$ between two vectors $\mathbf{x}$ and $\mathbf{y}$, $\mathbf{x}, \mathbf{y} \in \mathbf{F}_N^n$, is the rank of their difference: $d_r(\mathbf{x}, \mathbf{y}) = r(\mathbf{x} - \mathbf{y} \mid \mathbf{F}_1)$. For any code C the *rank distance* $d(C) = d$ is defined by $d = \min\{d_r(\mathbf{x}, \mathbf{y}) \mid \mathbf{x} \in C, \mathbf{y} \in C, \mathbf{x} \ne \mathbf{y}\}$.

If a code C has the rank distance d then it can correct all errors $\mathbf{e}$ with $r(\mathbf{e}|\mathbf{F}_1) \le t = \lfloor (d-1)/2 \rfloor$.

Any linear (*n*, *k*, *d*) code $C \subset \mathbf{F}_N^n$ satisfies the Singleton-style bound [2] for the rank distance

$$\begin{array}{ll} k \le n-d+1 & \text{if } n \le N, \\ Nk \le n(N-d+1) & \text{if } n > N \end{array} \qquad (1)$$

A code *C* that reaches this bound is called a *maximal rank distance* (MRD) code. The theory of MRD codes is given in [2].

For $n \le N$ a linear (*n*, *k*, *d*) MRD-code is defined by a $k \times n$ matrix:

$$\mathbf{G} = \left[g_j^{[i]}\right] \qquad (2)$$

where $g_j \in \mathbf{F}_N$, $j = 1, \ldots, n$; $g_1, g_2, \ldots, g_n$ are linearly independent over $\mathbf{F}_1$, and $g^{[i]} = g^{q^i}$ means the *i*-th Frobenius power of *g*. For such a code, there exists a *fast decoding algorithm* up to the rank error-correcting capacity $t = \lfloor (d-1)/2 \rfloor$.

3. Optimal rank codes for lengths $n > N$

For $n > N$ the construction (2) cannot be used since the first row of the generator matrix always contains coordinates that are linearly dependent over the base field $\mathbf{F}_1$. It is known that an $\mathbf{F}_1$-linear (or group) MRD code always exists [2, 8]. Constructions of $\mathbf{F}_N$-linear MRD codes are given in [8]. They are called the *direct product rank* codes.

In this paper, we present a new family of rank codes, which we call *reducible rank* codes.

Definition 1. *A code C is called a reducible rank code if its generator matrix can be represented as*

$$\mathbf{G} = \begin{bmatrix} \mathbf{G}_1 & \mathbf{0} \\ \mathbf{X} & \mathbf{G}_2 \end{bmatrix} \qquad (3)$$

for some matrix **X**.

If $\mathbf{X} = \mathbf{0}$, then *C* is a direct product code.

Lemma 1. *Let* $\mathbf{G}_1$ *and* $\mathbf{G}_2$ *be generator matrices of linear MRD* $(N, m, d = N - m + 1)$ *codes over the field* $\mathbf{F}_N$. *Let* **X** *be an arbitrary* $m \times N$ *matrix over* $\mathbf{F}_N$. *Then a reducible code C with the generator matrix of the form (3) is an* $\mathbf{F}_N$-*linear (n, k, d) MRD code with length* $n = 2N$ *and dimension* $k = 2m$.

Proof. Each non-zero codeword **g** may be represented as $\mathbf{g} = (\mathbf{g}_1, \mathbf{g}_2)$ and at least one of $\mathbf{g}_1$, $\mathbf{g}_2$ is of rank greater than or equal to $d = N - m + 1$. On the other hand, the Singleton inequality (1) turns into equality in this case: $Nk = n(N - d + 1)$.

Corollary 1. *Let* $\mathbf{G}_1, \mathbf{G}_2, \ldots, \mathbf{G}_r$ *be generator matrices of linear* $(N, m, d = N - m + 1)$ *MRD codes over the field* $\mathbf{F}_N$. *Let* $\mathbf{G}_{ij}$, $i = 2, \ldots, r; j = 1, \ldots, r-1;$ *be arbitrary* $m \times N$ *matrices over* $\mathbf{F}_N$. *Then a reducible code C with the generator matrix*

$$\mathbf{G} = \begin{bmatrix} \mathbf{G}_1 & \mathbf{0} & \mathbf{0} & \cdots & \mathbf{0} & \mathbf{0} \\ \mathbf{G}_{21} & \mathbf{G}_2 & \mathbf{0} & \cdots & \mathbf{0} & \mathbf{0} \\ \vdots & \vdots & \vdots & \cdots & \vdots & \vdots \\ \mathbf{G}_{r-1,1} & \mathbf{G}_{r-1,2} & \mathbf{G}_{r-1,3} & \cdots & \mathbf{G}_{r-1} & \mathbf{0} \\ \mathbf{G}_{r,1} & \mathbf{G}_{r,2} & \mathbf{G}_{r,3} & \cdots & \mathbf{G}_{r,r-1} & \mathbf{G}_r \end{bmatrix} \quad (4)$$

is an $\mathbf{F}_N$*-linear* (n, k, d) *MRD code with length* $n = rN$ *and dimension* $k = rm$. If all the matrices $\mathbf{G}_{ij} = \mathbf{0}$, then the code *C becomes a direct product* MRD code.

Encoding of reducible rank codes. Let $\mathbf{b} = (b_1, b_2, \ldots, b_k)$ be an information sequence with entries in $\mathbf{F}_N$. We will also represent this sequence as a concatenation of r sub-blocks of length m each: $\mathbf{b} = (\mathbf{b}_1, \mathbf{b}_2, \ldots, \mathbf{b}_r)$. Then the codeword is defined by

$\mathbf{g} = \mathbf{bG} = (\mathbf{g}_1, \mathbf{g}_2, \ldots, \mathbf{g}_r)$, where $\mathbf{g}_i$, $i = 1, 2, \ldots, r$, are sub-blocks of length N each and

$$\begin{aligned} \mathbf{g}_1 &= \mathbf{b}_1\mathbf{G}_1 + \mathbf{b}_2\mathbf{G}_{21} + \ldots + \mathbf{b}_r\mathbf{G}_{r1} \\ \mathbf{g}_2 &= \mathbf{b}_2\mathbf{G}_2 + \mathbf{b}_3\mathbf{G}_{32} + \ldots + \mathbf{b}_r\mathbf{G}_{r2} \\ &\vdots \\ \mathbf{g}_{r-1} &= \mathbf{b}_{r-1}\mathbf{G}_{r-1} + \mathbf{b}_r\mathbf{G}_{r,r-1} \\ \mathbf{g}_r &= \mathbf{b}_r\mathbf{G}_r \end{aligned}$$

Decoding of reducible rank codes. To decode, start with the last sub-block $\mathbf{g}_r$ and apply the usual fast decoding algorithm for rank codes to obtain the last information sub-block $\mathbf{b}_r$. Then proceed with $\mathbf{g}_{r-1} - \mathbf{b}_r\mathbf{G}_{r-1} = \mathbf{b}_{r-1}\mathbf{G}_{r-1}$ and obtain $\mathbf{b}_{r-1}$, and so on until $\mathbf{b}_1$.

3.1 Equivalence of rank codes

Definition 2. *Codes C and* $\widetilde{C}$ *are said to be equivalent if and only if their rank weight enumerators are identical.*

Let C be a code with code vectors $\mathbf{c}$. Let $\widetilde{C}$ be a code with code vectors $\{\mathbf{cP}\}$, where $\mathbf{P}$ is a non-singular $n \times n$ over $\mathbf{F}_1$

Lemma 2. *Codes C and* $\widetilde{C}$ *are equivalent.*

Proof. For an $\mathbf{x} \in \mathbf{F}_N^n$ and a non-singular $n \times n$ matrix **P** over the field $\mathbf{F}_1$, we have: $r(\mathbf{x}|\mathbf{F}_1) = r(\mathbf{xP}|\mathbf{F}_1)$.
The matrix **P** is called a *column scramble.* A column scramble 'mixes' columns of a generator matrix of a code.

4. Applications to Cryptography

We propose a new public key cryptosystem that is based on reducible rank codes. Choose at random r linear MRD $(N, N-d+1, d)$ codes of rank distance d over $\mathbf{F}_N$ with canonical generator matrices $\mathbf{G}_1, \mathbf{G}_2, \ldots, \mathbf{G}_r$. Denote $m := N - d + 1, k := rm, n := rN$.
Choose at random $m \times N$ matrices $\mathbf{G}_{i,j}$, $i = 2, \ldots, r; j = 1, \ldots, r-1$, over $\mathbf{F}_N$.
A *reducible rank MRD code* is defined by a canonical $k \times n$ generator matrix of the form (4). Choose at random a non-singular *row scrambling* $k \times k$ matrix **Q** over $\mathbf{F}_N$.
Choose at random a *distortion* matrix of the form

$$\mathbf{X} = \begin{bmatrix} \mathbf{X}_{11} & \mathbf{X}_{12} & \cdots & \mathbf{X}_{1r} \\ \mathbf{0} & \mathbf{X}_{22} & \cdots & \mathbf{X}_{2r} \\ \vdots & \vdots & \ddots & \vdots \\ \mathbf{0} & \mathbf{0} & \cdots & \mathbf{X}_{rr} \end{bmatrix},$$

where matrices

$$\mathbf{X}_{11}, \begin{bmatrix} \mathbf{X}_{12} \\ \mathbf{X}_{22} \end{bmatrix}, \ldots, \begin{bmatrix} \mathbf{X}_{1r} \\ \mathbf{X}_{2r} \\ \vdots \\ \mathbf{X}_{rr} \end{bmatrix}$$

are *distortion* sub-matrices each of column rank t_1 over $\mathbf{F}_1$. Here $t_1 < t = \lfloor (d-1)/2 \rfloor$ is a *design* parameter.
Choose at random a non-singular *column scrambling* $n \times n$ matrix **P** over the base field $\mathbf{F}_1$.

- **Secret** keys are four matrices $\{\mathbf{Q}, \mathbf{G}, \mathbf{X}, \mathbf{P}\}$.
- The **public** key is defined by $\mathbf{G}_{pub} = \mathbf{Q}(\mathbf{G} + \mathbf{X})\mathbf{P}$.
- **Plaintexts** are vectors $\mathbf{b} = (b_1, b_2, \ldots, b_k)$ of length $k = rm$ over the field $\mathbf{F}_N$. We also represent a plaintext as a concatenation of r sub-blocks of length m each: $\mathbf{b} = (\mathbf{b}_1, \mathbf{b}_2, \ldots, \mathbf{b}_r)$.

- **Encryption** given a plaintext **b** is fulfilled as **c** = (c_1, c_2, ..., c_n) = $\mathbf{b}\mathbf{G}_{pub}$ + **e**, where **e** is an *artificial* pattern of errors of rank $t_2 = t - t_1$ or less. Note that **e** must be chosen randomly.
- **Decryption.** Upon receiving **c**, the legitimate user calculates $\widetilde{\mathbf{c}} = \mathbf{c}\mathbf{P}^{-1}$, cuts $\widetilde{\mathbf{c}}$ into r successive sub-blocks of length N each, applies the fast decoding algorithm to each sub-block starting with the last one and obtains as a result **mQ**. Then he multiplies it by $\mathbf{Q}^{-1}$ to get the plaintext **m**.

In this system, the size of the public key is $V = knN\log_2 q$ $= r^2N^2(N-d+1)\log_2 q$ bits, and the information rate is $R = k/n$ $= (N-d+1)/N$.

4.1 Cryptanalysis

4.1.1 Direct attacks

The security of this new system against direct attacks, which try to get the plaintext from the ciphertext not knowing the private keys, is at least as high as that of the GPT system for a given N, k and t_2. All known direct attacks are exponential in t_2, and N or k. One of the fastest attacks is the decoding technique described in [7] and requires $O((N+nt_2)^3 q^{(t_2-1)(N-t_2)})$. For instance, for $N = 20$, $m = 10$, $r = 2$, $q = 2$, $t_2 = 3$, it takes about 2^{55} operations. Direct attacks are apparently ineffective for $N > 20$, $t_2 > 2$.

4.1.2 Structural attacks

A cryptanalyst faces the problem:

Given $\mathbf{G}_{pub}$, what is the complexity to find **Q, G, P** and **X**?

Note that for $r = 1$ we get the original GPT cryptosystem, so the column scrambling matrix **P** has no significance and Gibson's attacks can be applied. At $r \geq 2$ the situation changes and Gibson's attack cannot be applied anymore. Moreover, we can choose a distortion matrix **X** with total rank beyond error-correcting capacity $t = \lfloor (d-1)/2 \rfloor$.

1. **Special case X = 0: the system is resistant to known attacks.**

In this case, $\mathbf{G}_{pub} = \mathbf{QGP}$. For brevity, we consider the case $r = 2$.

A cryptanalyst may want to find a representation $\mathbf{G}_{pub} = \widetilde{\mathbf{Q}}\widetilde{\mathbf{G}}\widetilde{\mathbf{P}}$, where

$$\widetilde{\mathbf{G}} = \begin{bmatrix} \widetilde{\mathbf{G}}_1 & \mathbf{0} \\ \widetilde{\mathbf{G}}_{21} & \widetilde{\mathbf{G}}_2 \end{bmatrix} \tag{5}$$

and $\widetilde{\mathbf{G}}_j, j = 1,2$, are generator matrices of MRD $(N, N-d+1, d)$ codes, not necessarily coinciding with original codes; $\widetilde{\mathbf{P}}$ is a column scrambler, $\widetilde{\mathbf{Q}}$ is a row scrambler. It is clear that it is enough to break the system.

1.1 An easy case: no row scrambler, no distortion, column scrambler only. In this case the system is insecure and can be broken in $O(N^3)$ operations. The public key is the product of two matrices $\mathbf{G}_{pub} = \mathbf{GP}$.

Represent the matrix $\mathbf{P}$ as a block matrix $\mathbf{P} = \begin{bmatrix} \mathbf{P}_{11} & \mathbf{P}_{12} \\ \mathbf{P}_{21} & \mathbf{P}_{22} \end{bmatrix}$, where $\mathbf{P}_{ij}$, $i = 1,2, j = 1,2$ are $N \times N$ matrices. Then

$$\mathbf{G}_{pub} = \begin{bmatrix} \mathbf{G}_1\mathbf{P}_{11} & \mathbf{G}_1\mathbf{P}_{12} \\ \mathbf{G}_{21}\mathbf{P}_{11} + \mathbf{G}_2\mathbf{P}_{21} & \mathbf{G}_{21}\mathbf{P}_{12} + \mathbf{G}_2\mathbf{P}_{22} \end{bmatrix}.$$

We may consider $\mathbf{P}_{11}$ to be non-singular, and thus $\widetilde{\mathbf{G}}_1 = \mathbf{G}_1\mathbf{P}_{11}$ is a generator matrix of some $(N, N - \mathrm{d} + 1, d)$ MRD code. If it is not the case just permute the columns of $\mathbf{G}_{pub}$. From $\widetilde{\mathbf{G}}_1$ and $\mathbf{G}_1\mathbf{P}_{12} = \widetilde{\mathbf{G}}_1\mathbf{P}_{11}^{-1}\mathbf{P}_{12}$ it is now easy to compute $\mathbf{P}_{11}^{-1}\mathbf{P}_{12}$

The public key can be decomposed as

$$\mathbf{G}_{pub} = \begin{bmatrix} \widetilde{\mathbf{G}}_1 & \mathbf{0} \\ \mathbf{G}_{21}\mathbf{P}_{11} + \mathbf{G}_2\mathbf{P}_{21} & \mathbf{G}_2(\mathbf{P}_{22} - \mathbf{P}_{21}\mathbf{P}_{11}^{-1}\mathbf{P}_{12}) \end{bmatrix} \cdot \begin{bmatrix} \mathbf{E}_N & -\mathbf{P}_{11}^{-1}\mathbf{P}_{12} \\ \mathbf{0} & \mathbf{E}_N \end{bmatrix},$$

where $\mathbf{E}_N$ is an identity matrix of dimension N.

It can easily be shown that if $\mathbf{P}$ and $\mathbf{P}_{11}$ are non-singular matrices, then $(\mathbf{P}_{22} - \mathbf{P}_{21}\mathbf{P}_{22}^{-1}\mathbf{P}_{12})$ is also non-singular. From this we see that the matrix $\mathbf{G}_2(\mathbf{P}_{22} - \mathbf{P}_{21}\mathbf{P}_{22}^{-1}\mathbf{P}_{12})$ is a generator matrix of some $(N, N - \mathrm{d} + 1, d)$ MRD code.

Thus, we obtain decomposition $\mathbf{G}_{pub} = \widetilde{\mathbf{G}}\widetilde{\mathbf{P}}$ as in (5) with

$$\widetilde{\mathbf{G}}_1 = \mathbf{G}_1\mathbf{P}_{11}; \qquad \widetilde{\mathbf{G}}_2 = \mathbf{G}_2(\mathbf{P}_{22} - \mathbf{P}_{21}\mathbf{P}_{11}^{-1}\mathbf{P}_{12});$$

$$\widetilde{\mathbf{G}}_{21} = \mathbf{G}_{21}\mathbf{P}_{11} + \mathbf{G}_2\mathbf{P}_{21} \qquad \widetilde{\mathbf{P}} = \begin{bmatrix} \mathbf{E}_N & -\mathbf{P}_{11}^{-1}\mathbf{P}_{12} \\ \mathbf{0} & \mathbf{E}_N \end{bmatrix}.$$

1.2 General case: row scrambler, column scrambler, no distortion. Consider the matrix $\mathbf{G}_{pub} = \mathbf{QGP}$. For r = 2, using block representation of matrices

$$\mathbf{Q} = \begin{bmatrix} \mathbf{Q}_{11} & \mathbf{Q}_{12} \\ \mathbf{Q}_{21} & \mathbf{Q}_{22} \end{bmatrix}, \qquad \mathbf{P} = \begin{bmatrix} \mathbf{P}_1 \\ \mathbf{P}_2 \end{bmatrix},$$

we obtain

$$\mathbf{G}_{pub} = \begin{bmatrix} \mathbf{Q}_{11}\mathbf{G}_1\mathbf{P}_1 + \mathbf{Q}_{12}\mathbf{G}_2\mathbf{P}_2 + \mathbf{Q}_{12}\mathbf{G}_{21}\mathbf{P}_1 \\ \mathbf{Q}_{21}\mathbf{G}_1\mathbf{P}_1 + \mathbf{Q}_{22}\mathbf{G}_2\mathbf{P}_2 + \mathbf{Q}_{22}\mathbf{G}_{21}\mathbf{P}_1 \end{bmatrix}$$

One can see that in the first (respectively, second) row block the matrix $\mathbf{Q}_{11}\mathbf{G}_1\mathbf{P}_1 + \mathbf{Q}_{12}\mathbf{G}_2\mathbf{P}_2$ (respectively, $\mathbf{Q}_{21}\mathbf{G}_1\mathbf{P}_1 + \mathbf{Q}_{22}\mathbf{G}_2\mathbf{P}_2$) is a sum of two scrambled Frobenius matrices and the matrix $\mathbf{Q}_{12}\mathbf{G}_{21}\mathbf{P}_1$ (respectively, $\mathbf{Q}_{22}\mathbf{G}_{21}\mathbf{P}_1$) may be chosen with a large column rank. Thus, no attacks are known to recover (5) and break the system.

2. **General case: row scrambler, column scrambler and distortion.**

In the most general case, the public key is $\mathbf{G}_{pub} = \mathbf{Q}(\mathbf{G} + \mathbf{X})\mathbf{P} = \mathbf{QGP} + \mathbf{Y}$, where $\mathbf{Y} = \mathbf{QXP}$ is a distortion matrix. Using representation

$$\mathbf{X} = \begin{bmatrix} \mathbf{X}_{11} & \mathbf{X}_{12} \\ \mathbf{0} & \mathbf{X}_{22} \end{bmatrix}, \text{ one obtains } \mathbf{Y} = \begin{bmatrix} \mathbf{Q}_{11}\mathbf{X}_{11}\mathbf{P}_1 + \mathbf{Q}_{12}\mathbf{X}_{12}\mathbf{P}_2 + \mathbf{Q}_{12}\mathbf{X}_{22}\mathbf{P}_1 \\ \mathbf{Q}_{21}\mathbf{X}_{11}\mathbf{P}_1 + \mathbf{Q}_{21}\mathbf{X}_{12}\mathbf{P}_2 + \mathbf{Q}_{22}\mathbf{X}_{22}\mathbf{P}_1 \end{bmatrix}$$

.

Hence, one has extra degrees of freedom to destroy a visible structure of reducible codes by a proper choice of **X**.

5 Combined security and error protection

Based on the above section, one can have a reasonably secure system in $GF(2^{20})$ of Public key size 10 kbits and information rate (*plaintext/ciphertext* = *k/n*) 0.5. Though this represents a considerable improve to McEliece original system where the size of the key is about 500 kbits [1], the size is still too large compared to RSA (1 kbits). Moreover, the information rate is 1 in RSA. On the other hand, it is possible to have an added capability for error protection in McEliece type cryptosystems especially if we do without the distortion matrix **X**.

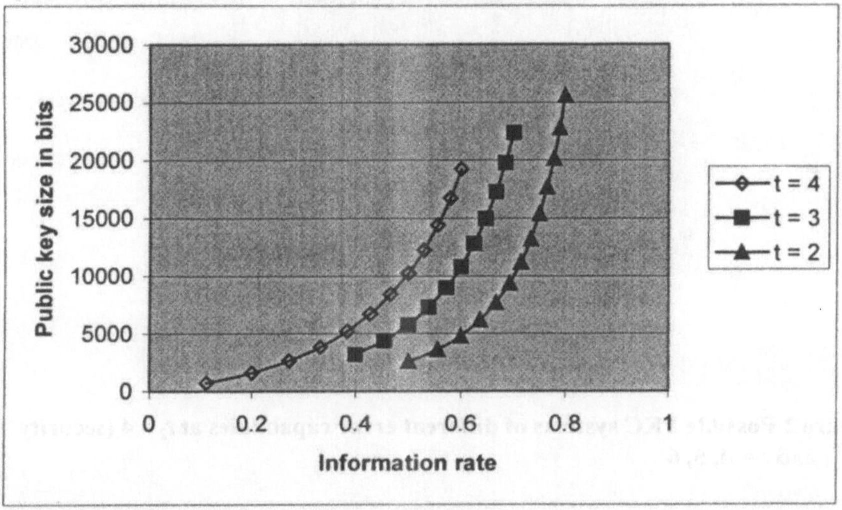

Figure 1 relation between key size and information rate for PKC of different security levels, $t = t_2$.

Figure 1 shows the relation between the public key size and the information rate for PKC at different values of t. We assume $GF(2^{20})$, $t = t_2$ and $r = 2$ to have the key size as small as possible. There is a trade-off between rate and key size. We can for instance have a secure PKC ($t = 4$) starting from a system with maximum rate of 0.6 and key size of 19200 bits, or we can reduce the key size to 10240 bits and lower the information throughput to 0.5. Changing t_2 would affect the degree of security of the system against direct attacks [7]. The security of the system is calculated to be about 2^{70} at $t = t_2 = 4$, 2^{55} at $t = t_2 = 3$ and 2^{37} at $t = t_2 = 2$ which is completely insecure.

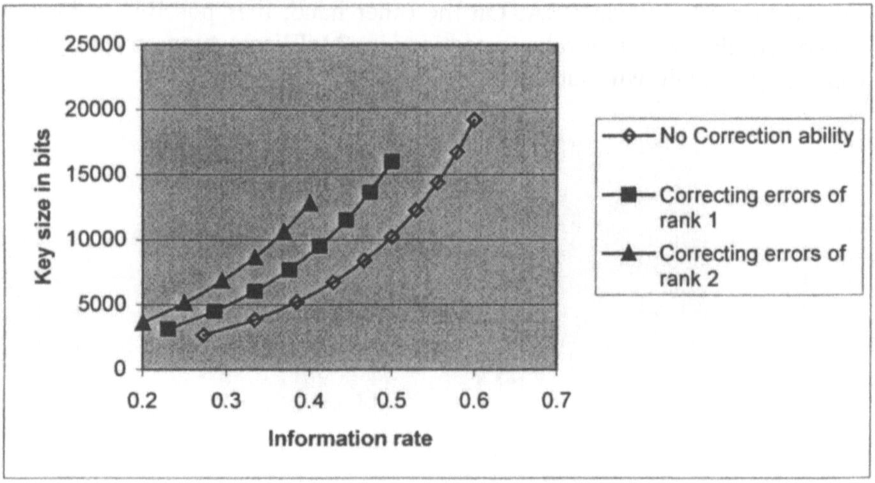

Figure 2 Possible PKC systems of different error capabilities at $t_2 = 4$ (security ~2^{70}) and $t = 4, 5, 6$

Figure 2 shows the relation between information rate and the key size for PKC's with different error correcting abilities. The systems in the figure are again all in $GF(2^{20})$ and have $t = t_2 + t_{ch}$ and $t_2 = 4$ always, thus we keep the same level of security (2^{70}) against direct attacks and we change t.

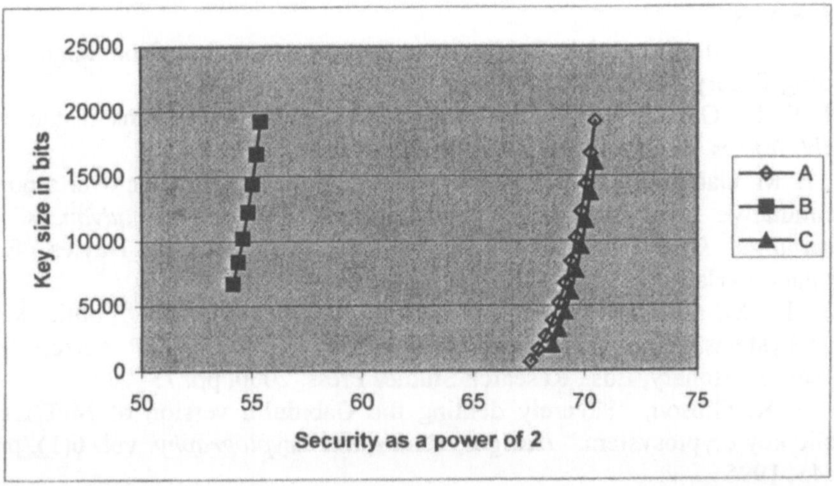

Figure 3 Relation between public key size and security for different systems. Systems of curve A has $t = t_2 = 4$ (No Error Correction), curve B has $t = 4$, $t_2 = 3$ (rank 1 errors correction) and curve C has $t = 5$, $t_2 = 4$ and rank 1 error correction.

In figure 3, we show the relation between public key size and the security of the system for different PKC in $GF(2^{20})$ with $r = 2$. Curve A shows this relation for system with $t = t_2 = 4$, so that we have no error protection. Curves B and C compare two different methods of adding error protection. In Curve B $t = 4$, $t_2 = 3$ and we have added protection against errors of rank $t - t_2 = 1$ at the expense of reducing the security (from $\sim 2^{70}$ to $\sim 2^{55}$). In curve C, we have $t = 5$ whereas the security is maintained by keeping $t_2 = 4$. Here we have again error protection against errors of rank 1, yet we have to increase the complexity of implementation and decrease the data throughput.

Conclusion

A new family of reducible rank codes were introduced. It has been shown that they could be used in McEliece type PKC. The key size is much reduced from McEliece original one. It has been shown that the system can provide error protection. There are different tradeoffs between key size/ error protection/ information throughput/ complexity that were discussed.

References:

[1] R. J. McEliece, "A Public Key Cryptosystem Based on Algebraic Coding Theory," *JPL DSN Progress Rep. 42-44*

[2] E. M. Gabidulin, "Theory of Codes with Maximum Rank distance," *Probl. Inform. Transm.*, vol. 21(1), pp.1–12, July 1985.

[3] E. M. Gabidulin, A.V. Paramonov, O.V. Tretjakov, "Ideals over a non-commutative ring and their application in cryptology," *advances in Cryptology, Proc. EUROCRYPT' 91, LNCS 547*, D. W. Davies, Ed. Springer-Verlag, 1991, pp. 482–489

[4] E. M. Gabidulin, A. V. Ourivski, "Improved GPT public key cryptosystems," *Coding, Communications and Broadcasting*, P. Farrell, M. Darnell, B. Honary, Eds., Research Studies Press, 2000, pp. 73–102.

[5] J. K. Gibson, "Severely denting the Gabidulin version of McEliece public key cryptosystem," *Designs, Codes and Cryptography,* vol. 6(1), pp. 37–45, 1995

[6] J. K. Gibson, "Algebraic coded cryptosystems, " *PhD. Thesis*, University of London, Royal Holloway and Bedford New College, 1995.

[7] F. Chabaud, J. Stern, "The cryptographic security of the syndrome decoding problem for rank distance codes," *Advances in Cryptology, Proc. ASIACRYPT'96, LNCS,* K. Kim, T. Matsumoto, Eds., Springer-Verlag, 1996, pp. 368–381.

[8] E. M. Gabidulin, P. Loidreau, "Subfield subcodes of rank codes and applications," *Proc. 7-th Int. Workshop on Algebraic and Combinatorial Coding Theory ACCT' 2000,* June 2000.

On a Boolean Maximization Problem

Solomon W. Golomb
University of Southern California
Department of Electrical Engineering
EEB 504
Los Angeles, CA 90089-2565
milly@mizar.usc.edu
Wensong Chu
University of Southern California
Department of Natural Sciences & Mathematics
Los Angeles, CA 90089-2565
wensong@scf.usc.edu

Abstract

Among all boolean functions f_j of n variables, $n = 2m+1$, the $\max_j \min_i (f_j \cdot x_i)$ value of correlation between f_j and the n independent binary variables x_i, $1 \leq i \leq n$, is $\binom{2m}{m} \cdot 2^{-2m}$, which is uniquely achieved by $f^* = \text{maj}\,(x_1, \ldots, x_n)$, the majority decision function on the x_i's. This value is asymptotically less, by a factor of $\sqrt{\frac{\pi}{2}} \approx 1.2533$, than is achieved by the real vector α^* which achieves $\max_{|\alpha|=1} \min_i(\alpha \cdot x_i) = \frac{1}{\sqrt{n}}$.

Keywords: Boolean functions, correlation, majority decision function.

1 Introduction

Any device which takes n binary inputs to produce one binary output is a realization of a *boolean function of n variables.* Such a device may be used sequentially, to produce a stream of successive output bits from n parallel streams of input bits. One application [1] is in the feedback computation for a (in general, nonlinear) feedback shift register. Another application may be to combine relatively short periodic binary input sequences with relatively prime periods to produce one long-period output sequence, as in the JPL ranging system [2]. Yet another reason for combining several sequences in this way may be to obtain a random-looking *key sequence* for cryptographic application [3].

For cryptographic purposes, it is generally desired that the "combined sequence" be uncorrelated, to the maximum extent possible, with the individual component sequences, as well as with modulo 2 sums of the component sequences taken two or more at a time. However, for applications like the JPL ranging system, the idea is to determine, by correlating a portion of the long-period sequence with individual short-period sequences, what the "phase" of the short sequence is. Knowing the "phase" of each short sequence makes it possible to determine (using the "Chinese Remainder Theorem" of elementary number theory) precisely where one is in the long-period combined sequence.

With the latter application in mind, it is natural to ask: Which of the 2^{2^n} possible boolean functions of n variables will produce a combined sequence with the largest possible values for its correlations with the individual component sequences? The "obvious" answer for odd n, which the first author conjectured in 1959, is that $\mathrm{maj}(x_1, x_2, \ldots, x_n)$, the function which is 0 or 1 according to whether the majority of the x_i's are 0 or 1, achieves this "maximum simultaneous correlation". In this paper, this conjecture is stated precisely and proved. The "simultaneous maximum value of correlation" between the combined sequence and each of its n components, when $n = 2m + 1$, is shown to be exactly $\frac{\binom{2m}{m}}{2^{2m}}$, which by Stirling's formula is asymptotic to $\frac{1}{\sqrt{\pi m}}$ for large m. However, a vector space model is described in which the n input variables correspond to orthogonal basis vectors, and the normalized vector (not restricted to binary values) with the largest simultaneous correlation with each of these achieves a correlation value of $\frac{1}{\sqrt{n}} = \frac{1}{\sqrt{2m+1}}$, which is asymptotically larger than $\frac{1}{\sqrt{\pi m}}$ by a factor of $\sqrt{\frac{\pi}{2}} = 1.253314137\ldots$. In view of the theorems proved about the function $f(x_1, x_2, \ldots, x_n) = \mathrm{maj}(x_1, x_2, \ldots, x_n)$, the reduction in simultaneous correlation value is necessitated by the binary output format of boolean functions of n variables.

2 Achieving the Maximum Simultaneous Correlation

Let $\alpha = (a_1, a_2, \ldots, a_L)$ and $\beta = (b_1, b_2, \ldots, b_L)$ be two binary vectors of length L. Let $D = D(\alpha, \beta)$ be the Hamming distance between α and β, i.e. the number of components where they *disagree*; and let $A = A(\alpha, \beta) = L - D$ be the number of components where they *agree*. We define the (normalized) *correlation* $C(\alpha, \beta)$ by

$$C(\alpha, \beta) = \frac{A - D}{A + D} = \frac{L - 2D}{L}. \tag{1}$$

Note that this definition is independent of which two binary symbols are used for the components of α and β. If specifically these two binary symbols are +1 and -1, then

$$C(\alpha, \beta) = \frac{1}{L} \sum_{k=1}^{L} a_k b_k = \frac{1}{L}(\alpha \cdot \beta). \tag{2}$$

By a *boolean function* $f_j(x_1, x_2, \ldots, x_n)$ *of* n *variables*, we mean a mapping from $B \times B \times \cdots \times B = B^n$ into B, where $B = \{0, 1\}$. There are 2^{2^n} such functions $f_j(x_1, x_2, \ldots, x_n)$. The standard "truth table" of f_j has 2^n rows and $n + 1$ columns. The first n columns correspond to the n "input variables" $x_1, x_2, \ldots, x_n$, and the $(n + 1)^{st}$ column corresponds to the "output variable" f_j. The rows of the truth table show the 2^n "input states" from $00 \cdots 0$ to $11 \cdots 1$, each followed by the value of $f(x_1, x_2, \ldots, x_n)$ for that state. We will regard these $n + 1$ columns as binary vectors of length $L = 2^n$, and we are interested in the correlation coefficient (the "normalized dot product") between pairs of these vectors.

We first observe that $C(x_i, x_j) = 0$ for $i \neq j$, for any two columns x_i and x_j, because clearly *agreements* and *disagreements* occur equally often; and $C(x_i, x_i) = 1$, because our definition of correlation is normalized.

For later use, we now observe that the set of n vectors, $\mathcal{B} = \{x_1, x_2, \ldots, x_n\}$, form an *orthonormal basis* for the n-dimensional subspace W which they generate in the vector space $V = R^{2^n}$ (of all real vectors with 2^n components).

We are interested in finding the boolean function $f^*(x_1, x_2, \ldots, x_n)$ which "maximizes the simultaneous correlation" with each of $x_1, x_2, \ldots, x_n$. Specifically, we seek the function f^* which achieves

$$\max_{1 \le j \le 2^{2^n}} \min_{1 \le i \le n} C(f_j, x_i) \tag{3}$$

for each odd $n = 2m + 1, m \geq 0$. In the next section, we will further show that this max-min value is $\frac{\binom{2m}{m}}{2^{2m}}$.

Theorem 1 *The unique maximizing function f^* in (3) is $f^*(x_1, x_2, \ldots, x_n) = \text{maj}\ (x_1, x_2, \ldots, x_n) = 1$ if at least $m+1$ of the x_is have the value 1,*
$f^*(x_1, x_2, \ldots, x_n) = \text{maj}\ (x_1, x_2, \ldots, x_n) = 0$ if at most m of the x_is have the value 1.

Proof. To simplify computation, we will use +1's and −1's (instead of 0's and 1's) as our two binary symbols, which has no effect on $C(\alpha, \beta)$ as defined in (1), but enables us to calculate $C(\alpha, \beta)$ by (2). We now observe:

$$\max_{1 \le j \le 2^{2^n}} \min_{1 \le i \le n} C(f_j, x_i) \le \max_{1 \le j \le 2^{2^n}} \operatorname*{average}_{1 \le i \le n} C(f_j, x_i) \tag{4}$$

$$= \max_{1 \le j \le 2^{2^n}} \left(\frac{1}{n} \sum_{i=1}^{n} C(f_j, x_i) \right) = \frac{1}{n} \max_{1 \le j \le 2^{2^n}} (f_j \cdot (x_1 + x_2 + \cdots + x_n))/2^n,$$

by the linearity of the dot product. Since the x_i's take only the values +1 and −1, the sum $S = x_1 + x_2 + \cdots + x_n$ is an odd integer (since n is odd) on the range $-n \leq S \leq +n$. Since $S \neq 0$, S is either negative or positive. Thus the f_j which achieves the final maximum in (4) takes the value −1 when $S < 0$ and the value +1 when $S > 0$. Strictly speaking, this f_j, whose output value agrees with the majority of its input values, has been shown to maximize the *average* $C(f_j, x_i)$, which upper-bounds the maximization of the *minimum* $C(f_j, x_i)$. However, since this f_j is symmetric in all n of its inputs, its minimum $C(f_j, x_i)$ *equals* its average $C(f_j, x_i)$. By an obvious symmetry argument, this maximizing function f_j gives the same value of $C(f_j, x_i)$ for all i, $1 \leq i \leq n$, which must therefore be not only the max average $C(f_j, x_i)$ but also the max min $C(f_j, x_i)$. Translating back to the binary values 0 and 1, we have f^* as stated in this Theorem as the unique "maximizing function". □

3 Evaluating the Max-Min Correlation Value

Since $f^*(x_1, x_2, \ldots, x_n) = \text{maj}(x_1, x_2, \ldots, x_n)$ is obviously a symmetric function of its n input variables (odd $n = 2m+1$), we can calculate $C(f^*, x_i)$ using any one of the x_i's, since all n values are the same.

Looking at the truth table of f^* as a binary table with 2^n rows and $n+1$ columns, as described in the proof of Theorem 1, we let the rows (except for the right-most column) look like the binary representations of the integers from 0 to $2^n - 1$. The left-most column (i.e. x_1) has the value 0 in the "top half" of the truth table, and the value 1 in the "bottom half" of the truth table. We look at $C(f^*, x_1) = (f^* \cdot x_1)/(2^{2m})$ in the top half of the table, where $x_1 = 0$. We have *agreements* when $wt(N) \leq m$ for integers $N, 0 \leq N \leq 2^m - 1$, and *disagreements* when $wt(N) > m$ for integers $N, 0 \leq N \leq 2^m - 1$. (By a symmetry argument, the bottom half of the truth table, where x_1 has the value 1, will give the same value for $C(f^*, x_1)$, so it suffices to compute this for the top half of the table only.)

In the top half of the table, letting $wt(x_1, x_2, \ldots, x_n) = wt(x_2, \ldots, x_n)$ be the "Hamming weight", or number of 1's, in the binary vector for N, $0 \leq N \leq 2^m - 1$, $x_1 = 0$ *agrees* with $f^* = 0$ if and only if $0 \leq wt(x_2, x_3, \ldots, x_n) \leq m = \frac{n-1}{2}$. The number of *agreements* is thus the number of ways to insert at most m 1's in a binary vector of length $n - 1 = 2m$. Thus, $A = \binom{2m}{0} + \binom{2m}{1} + \binom{2m}{2} + \cdots + \binom{2m}{m}$. The remaining cases are the *disagreements*: $D = \binom{2m}{m+1} + \binom{2m}{m+2} + \cdots + \binom{2m}{2m}$. Using $\binom{2m}{k} = \binom{2m}{2m-k}$, we see that $A - D = \binom{2m}{m}$, while of course $A + D = 2^{2m}$. Thus we have proved:

Theorem 2 *For each* $i, 1 \leq i \leq n$, *we have* $C(f^*, x_i) = \frac{\binom{2m}{m}}{2^{2m}}$. □

(The only effect of considering the bottom half, as well as the top half, of the truth table would be to double both the numerator and the denominator of $\frac{\binom{2m}{m}}{2^{2m}}$*, with no effect on their ratio.)*

Note. *Using Stirling's formula:* $t! \sim \sqrt{2\pi t}\left(\frac{t}{e}\right)^t$ *as* $t \to \infty$*, we see that*

$$\left\{\frac{\binom{2m}{m}}{2^{2m}} = \frac{(2m)!}{2^{2m}(m!)(m!)} \sim \frac{2\sqrt{\pi m}(\frac{2m}{e})^{2m}}{(2\pi m)\left(\frac{m}{e}\right)^{2m} \cdot 2^{2m}} = \frac{\sqrt{\pi m}}{\pi m} = \frac{1}{\sqrt{\pi m}}\right\} \tag{5}$$

as $m \to \infty$.

4 The Vector Space Model

As noted in Section 2 above, the n independent boolean variables $x_1, x_2, \ldots, x_n$, regarded as column vectors in the truth table for any f_j, are "orthonormal" relative to our definition of $C(x_i, x_j) = \frac{A-D}{A+D} = \delta_{ij} = \left\{\begin{array}{ll} 1 \text{ if } & i = j \\ 0 \text{ if } & i \neq j \end{array}\right\}$, and their analogues as $+1, -1$ vectors, which are orthonormal by formula (2), form an orthonormal basis for the subspace W, of dimension n, in the space $V = R^{2^n}$ consisting of all vectors with 2^n real components. Hence there exists an "orthogonal transformation" from W to W', which rotates these basis vectors, from $\mathcal{B} = \{x_1, x_2, \ldots, x_n\}$, to a new orthonormal basis $\mathcal{B}' = \{\epsilon_1, \epsilon_2, \ldots, \epsilon_n\}$ of W', where $\epsilon_1 = (1, 0, \ldots, 0), \epsilon_2 = (0, 1, \ldots, 0), \cdots, \epsilon_n = (0, 0, \ldots, 1)$. (In a manner of speaking, the other $2^n - n$ formal dimensions in V shrivel to 0 in W', somewhat reminiscent of superstring theory.)

Analogous to the relationship of f^* to the vectors $x_1, x_2, \ldots, x_n$, we can look for the n-component real vector α^* which achieves

$$\max_{\alpha \neq \vec{0}, \alpha \in W'} \min_{1 \leq i \leq n} C(\alpha, \epsilon_i). \tag{6}$$

The normalized maximizing vector α^* is readily seen to be $\alpha^* = (\frac{1}{\sqrt{n}}, \frac{1}{\sqrt{n}}, \ldots, \frac{1}{\sqrt{n}})$, where $(\alpha^* \cdot \alpha^*) = 1$, and $(\alpha^* \cdot \epsilon_i) = \frac{1}{\sqrt{n}}$ for each $i, 1 \leq i \leq n$.

Since $n = 2m + 1$, this value $\frac{1}{\sqrt{2m+1}}$ for α^* versus the coordinate vectors is asymptotically larger than the corresponding $\frac{\binom{2m}{m}}{2^{2m}} \sim \frac{1}{\sqrt{\pi m}}$, as

$m \to \infty$, by a factor of $\sqrt{\frac{\pi}{2}} = 1.253314317\ldots$. In view of Theorems 1 and 2, this factor of $\sqrt{\frac{\pi}{2}}$ is the price paid for having f^* constrained to have 2^n binary components, when correlated against each x_i, also with 2^n binary components.

In Table 1, we see a comparison of $\frac{\binom{2m}{m}}{2^{2m}}$, $\frac{1}{\sqrt{\pi m}}$, and $\frac{1}{\sqrt{2m+1}}$, for selected values of m.

m	$\frac{\binom{2m}{m}}{2^{2m}}$	$\frac{1}{\sqrt{\pi m}}$	$\frac{1}{\sqrt{2m+1}}$
1	0.50000000	0.56418958	0.57735027
2	0.37500000	0.39894228	0.44721360
3	0.31250000	0.32573501	0.37796447
4	0.27343750	0.28209479	0.33333333
5	0.24609375	0.25231325	0.30151134
6	0.22558594	0.23032943	0.27735010
7	0.20947266	0.21324362	0.25819889
8	0.19638062	0.19947114	0.24253563
9	0.18547058	0.18806319	0.22941573
10	0.17619705	0.17841241	0.21821789
20	0.12537069	0.12615663	0.15617376
30	0.10257817	0.10300645	0.12803688
40	0.08892788	0.08920621	0.11111111
50	0.07958924	0.07978846	0.09950372

Table 1: Maximum-Minimum Correlation Values With $n = 2m + 1$ Boolean Variables

References

[1] *Shift Register Sequences*, S. Golomb, Holden-Day, Inc. 1967. Second edition, Aegean Park Press, 1982.

[2] *Digital Communications – With Space Applications*, S. Golomb (ed.), Prentice-Hall, 1964. Reprint edition, Peninsula Publishers,

1982.

[3] "On the Cryptanalysis of Nonlinear Shift Register Sequences", S. Golomb, in *Proceedings of the Seventh IMA Conference on Coding and Cryptography*, Cirencester, U.K., December, 1999.

On the Security of the McEliece Public-Key Cryptosystem

Nicolas Sendrier
INRIA Rocquencourt, projet CODES
B.P. 105, 78153 Le Chesnay Cedex, FRANCE

Nicolas.Sendrier@inria.fr

Abstract

As RSA, the McEliece public-key cryptosystem has successfully resisted more than 20 years of cryptanalysis effort. However, despite the fact that it is faster, it was not as successful as RSA as far as applications are concerned. This is certainly due to its very large public key and probably also to the belief that the system could not be used for the design of a digital signature scheme. We present here the state of art of the implementation and the security of the two main variants of code-based public-key encryption schemes (McEliece's and Niedereitter's) as well as the more recent signature scheme derived from them. We also show how it is possible to formally reduce the security of these systems to two well identified algorithmic problems. The decoding attack (aimed on one particular ciphertext) is connected to the NP-complete syndrome decoding problem. The structural attack (aimed on the public key) is connected to the problem of distinguishing binary Goppa codes from random codes. We conjecture that both these problems are difficult and present some arguments to support this claim.

Keywords: Public-key cryptography, McEliece cryptosystem, Niedereitter cryptosystem, error correcting codes, Goppa codes, syndrome decoding, proof of security, digital signature.

1 Introduction

Public key cryptography was introduced by Diffie and Hellman in 1976. Systems based on the error correcting codes arrived short after that. The first encryption scheme was proposed by Bob McEliece in 1978 [18] but we had to wait until 2001 for the first practical code-based signature scheme [8].

Designing a cryptosystem is a tradeoff between its security, its speed and its implementation aspects. By implementation, we mean everything connected with the key size, block length, information rate, algorithm design, ...

McEliece system achieves an excellent tradeoff between security and speed. Unfortunately some of the implementation aspects may prove problematic in certain contexts. The worst problems are an information rate smaller than one (the best security is obtained for a rate ≈ 0.63), and a very large key size (hundreds of kbits instead of 1 kbit for the same security with RSA).

We present in this paper the main aspects of the McEliece cryptosystem and of Niederreiter's variant [19]. We will assume that these systems use binary Goppa codes, which is probably the best choice today. In §2 we will present the encryption schemes and we describe in §3 how they can be implemented. The §4 is devoted to the cryptanalysis which are divided into two categories: the decoding attacks which are directed on one particular ciphertext, and the structural attacks aimed on the public key. We present in §5 the main features of the digital signature scheme, and finally in §6 we formally prove that breaking code-based public-key cryptosystem is at least as hard as solving one out of two supposedly difficult problems: the distinguishability of Goppa codes and the decoding of a linear code with Goppa parameters.

2 Public-key cryptography using error correcting codes

A public-key encryption scheme consists of two functions $E()$ and $D()$ such that [9]

- $D()$ and $E()$ are the inverse of one another,

- both $E()$ and $D()$ are easy to compute,
- it is computationally unfeasible to derive $D()$ from $E()$.

The function $E()$ is called one-way trapdoor and is made public so that anybody can encrypt a message x into the ciphertext $y = E(x)$ but only those knowing the trapdoor $D()$ can recover x from y.

The most famous example is the RSA encryption scheme [21], where

$$E(x) = x^e \bmod n, \quad D(y) = y^d \bmod n, \quad de = 1 \bmod \varphi(n).$$

When n is composite, finding d from e and n requires the knowledge of the Euler's indicator $\varphi(n)$ (*i.e.* the number of positive integers smaller than n and prime to n) and thus of the prime factors of n. In that case the public key is formed by the exponent e and the modulus n. The secret key is d, the inverse of e modulo $\varphi(n)$.

2.1 Definition of McEliece and Niederreiter cryptosystems

A solution using coding theory and due to Bob McEliece [18] appeared at approximatively the same time as RSA. A dual variant was proposed a few years later by Harald Niederreiter [19]. The finite field with q elements is denoted by $\mathbf{F}_q$, the Hamming weight is denoted by $w_H()$ and the set of all words of $\mathbf{F}_q^n$ of Hamming weight t is denoted by $W_{n,t}$.

	McEliece	Niederreiter
	A linear code $C[q;n,k]$	
public key	generator matrix G	parity check matrix H
plaintext	$x \in \mathbf{F}_q^k$	$x \in W_{n,t}$
ciphertext	$y = xG + e,\ w_H(e) = t$	$y = Hx^T$
trapdoor	A t-error correcting procedure for C	

In these systems, the ciphertext is a corrupted codeword (or its syndrome) and the decryption consists in applying a decoding procedure.

A code C_0, for which a fast algebraic decoding procedure is known, is chosen by the user. The algebraic structure of C_0 (and thus its decoder) constitutes the trapdoor of the system and is hidden by randomly permuting the coordinates. A random generating (or parity check) matrix of the permuted code C is used as the public key. In practice, if we denote G_0 and H_0 respectively a generator and a parity check matrix of C_0, the public keys for McEliece and Niederreiter systems are respectively

$$G = UG_0P \text{ and } H = VH_0P,$$

where U and V are non-singular matrices and P is a permutation matrix. We will refer to C_0 as the *public code* and to C as the *secret code*.

The security of the system requires two assumptions

1. no efficient general purpose decoder exists,
2. no (relevant) information on the algebraic structure of C_0 can be easily deduced from the public key.

The first assumption has been widely studied, and is connected with the NP-complete syndrome decoding problem [2]. The second assumption depends on how the secret code is chosen, so far the binary Goppa codes proposed by McEliece seem safe, while other choices as generalized Reed-Solomon codes [28] or concatenated codes [25] are not.

2.2 Goppa codes

Definition 2.1 *Let $L = (\alpha_1, \ldots, \alpha_n)$ be a sequence of n distinct elements of $\mathbf{F}_{q^m}$ and let $g(z) \in \mathbf{F}_{q^m}[z]$ be a monic polynomial of degree t without roots in L. The q-ary Goppa code of support L and generator $g(z)$ is denoted $\Gamma(L, g)$ and is formed by the words $a = (a_1, \ldots, a_n) \in \mathbf{F}_q^n$ such that*

$$R_a(z) = \sum_{i=1}^{n} \frac{a_i}{z - \alpha_i} = 0 \bmod g(z). \tag{1}$$

Goppa codes are subfield subcodes of some particular generalized Reed-Solomon codes, characterized by the following parity check matrix,

$$a \in \Gamma(L, g) \Leftrightarrow Ha^T = 0 \tag{2}$$

with

$$H = \begin{pmatrix} 1 & \cdots & 1 \\ \alpha_1 & \cdots & \alpha_n \\ \vdots & & \vdots \\ \alpha_1^{t-1} & \cdots & \alpha_n^{t-1} \end{pmatrix} \begin{pmatrix} g(\alpha_1)^{-1} & & \\ & \ddots & \\ & & g(\alpha_n)^{-1} \end{pmatrix}.$$

This implies in particular that Goppa codes have a minimum distance at least equal to $\deg(g) + 1 = t + 1$.

2.2.1 Binary Goppa codes

In the binary case there is another convenient characterization:

$$a \in \Gamma(L, g) \Leftrightarrow g(z) \mid \frac{d\sigma_a(z)}{dz}. \tag{3}$$

Because the derivative of any polynomial in characteristic 2 is a square, we can deduce from (3) that $\Gamma(L, g) = \Gamma(L, g^2)$ when $g(z)$ is square-free. This implies that each time $g(z)$ is square-free the minimum distance of $\Gamma(L, g)$ is at least $2t + 1$.

2.2.2 Number of binary Goppa codes

We refer to the notation of Definition 2.1. Irreducible binary Goppa codes ($g(z)$ is irreducible in $\mathbf{F}_{2^m}[z]$) of length $n = 2^m$ were originally proposed by McEliece. They are well suited for cryptography because they are numerous. The public code is obtained by choosing arbitrarily a support L and a monic generator polynomial $g(z)$. There are $n!$ choices for L and approximatively $2^{tm}/t$ choices for $g(z)$ (the number of monic irreducible polynomials, see [15, p. 93] for instance). Some choices however lead to the same public code, for each code the number of duplicates is, in general, $m2^m$ corresponding to the permutations $\sigma : z \mapsto z^{2^i} + a$, $0 \leq i < m$ and $a \in \mathbf{F}_{2^m}$ (see [17, Ch. 12]).

If we consider extended Goppa codes $\widehat{\Gamma}(L, g)$ by adding an overall parity bit, the number of duplicates is about $m2^{3m}$ corresponding to the permutations of the form $\sigma : z \mapsto (az^{2^i} + b)/(cz^{2^i} + d)$ with $0 \leq i < m$ and $a, b, c, d \in \mathbf{F}_{2^m}$ such that $ad - bc = 0$ (see [17, Ch. 12]).

2.2.3 Decoding

The mapping $a \mapsto R_a(z)$ is linear and thus for any $a \in \Gamma(L, g)$ and any $e \in \mathbf{F}_q^n$ we have a key equation

$$R_{a+e}(z)\sigma_e(z) = \omega_e(z) \mod g(z) \tag{4}$$

where $\sigma_e(z) = \prod_{e_i \neq 0}(z - \alpha_i)$ is the locator polynomial of the error vector e and $\omega_e(z) = \sum_{i=1}^{n} e_i \sigma_e(z)/(z - \alpha_i)$ its derivative. With the extended Euclidean algorithm we can recover $\sigma_e(z)$ (and $\omega_e(z)$) if $R_{a+e}(z)$ is known and $w_H(e) \leq t/2$. In the binary case, when $g(z)$ is square-free, we can replace $g(z)$ by $g^2(z)$ and up to t errors can be corrected.

The best and earliest description of an algebraic decoder for Goppa codes is due to Patterson [20]. Decoding a binary Goppa code (provided $g(z)$ is square-free) is done in three steps

1. Compute the syndrome $R_b(z)$ mod $g^2(z)$ of the received vector $b \in \mathbf{F}_2^n$.

2. Find the locator polynomial of the error by solving the key equation (4) with $g^2(z)$.

3. Compute the roots of the locator polynomial.

Provided some pre-computations, the number of elementary operations in $\mathbf{F}_{2^m}$ are approximatively of nt for step 1, $8t^2$ for step 2 and $2mt^2$ for step 3.

The actual solution proposed by Patterson is slightly different, as the key equation is transformed to be solved with the Berlekamp-Massey algorithm. The total cost is not significantly different.

3 Implementation

We briefly describe here the encryption and decryption procedures for McEliece and Niederreiter systems. We give some numbers which are summarized in Table 1. These values have to be handled with great care, their purpose is not to give an accurate estimate of running time, but simply to indicate which part of the implementation are expensive and how the cost will evolve with the parameters.

3.1 Encryption

We consider a binary linear code of length n, dimension k correcting t errors.

McEliece Encryption consists mainly in multiplying a vector x of length k by a generating matrix.

$$y = xG + e, \quad w_H(e) = t$$

If G is in systematic form, this will cost $k(n-k)/2$ binary operations in average. Compared to that generating and adding an error vector e of weight t will be negligible in terms of complexity, though it requires a random number generator, which might be difficult to implement.

Niederreiter We have to add t columns of the parity check matrix.

$$y = Hx^T, \quad w_H(x) = t$$

If H is in systematic form, only tk/n columns in average have to be added, and the cost will be $tk(n-k)/n$ operations in $\mathbf{F}_2$. In addition to that, the information has to be transformed from a binary string into words of $W_{n,t}$, this can be achieved efficiently in time proportional to $\log_2 \binom{n}{t} \approx nH_2(t/n)$ [23, 27], which can be neglected in practice.

3.2 Decryption

We assume that binary t-error correcting Goppa codes of length $n = 2^m$ dimension $k = n - tm$ are used. We denote $R = k/n$ the transmission rate of the code.

McEliece The ciphertext is a word at distance t of the code. If the public generating matrix is in systematic form, the plaintext is formed by the first k bits of the decoded codeword. From §2.2.3, the whole decryption procedure will take $tn(3-2R)$ operations in $\mathbf{F}_{2^m}$.

Niederreiter The ciphertext is a syndrome y (but not suited to the decoding procedure) of the plaintext $x \in W_{n,t}$. If the public parity-check matrix is in systematic form, the proper syndrome is equal to $R_b(z)$, where $b = (0, y) \in \mathbf{F}_2^n$ is formed by k '0' followed by y. The cost for computing this syndrome is mt^2 operations in $\mathbf{F}_{2^m}$. The other two steps are as described in §2.2.3, and the total cost is $3mt^2$ operations in $\mathbf{F}_{2^m}$.

	McEliece	Niederreiter
Block size	2^m	tm
Message size	$2^m - tm$	$\log_2 \binom{n}{t}$
Information rate	$R = 1 - tm/2^m$	$\log_2 \binom{n}{t}/tm$
Encryption cost	$(R/2)tmn$	Rt^2m
Decryption cost	$(3 - 2R)tmn$	$3t^2m^2$
Public-key size	$tm(2^m - tm)$	$tm(2^m - tm)$

Table 1: Code-based encryption systems with t-error correcting binary Goppa codes of length $n = 2^m$. *All sizes are in bits, all costs are in binary operations (one field operation in* $\mathbf{F}_{2^m}$ *is counted for* m *binary operations).*

4 Cryptanalysis

4.1 Decoding attacks

All the best known techniques for general purpose decoders lead to an exponential complexity. The best known implementation with a polynomial space complexity is a variant of the information set decoding [5]. The information set decoder takes repeatedly and randomly a set of k information positions and it is successful if this set contains no errors. Its cost is the inverse of the probability of success $\binom{n}{t}/\binom{n-k}{t}$ multiplied by the cost of a Gaussian elimination on the generator matrix. Using approximates for the binomial coefficients we obtain, when t is small compare to n and R is not too close to 1, a cost of

$$P(n)2^{-t\log_2(1-R)}, \tag{5}$$

where $R = n/k$ and $P(n)$ is asymptotically bounded by a polynomial of degree at most 3. With the Canteaut-Chabaud algorithm, the degree of the polynomial is reduced to 1, maybe 0. The actual evaluation uses a Markov chain and thus no easy closed formula is available.

4.1.1 Binary Goppa codes

If we apply (5) to Goppa codes ($n = 2^m$, $k = n - tm$), we have a general formula

$$2^{cn/m(1+o(1))},$$

where $c = -(1-R)\log_2(1-R)$ (see Figure 1). If one allows exponential

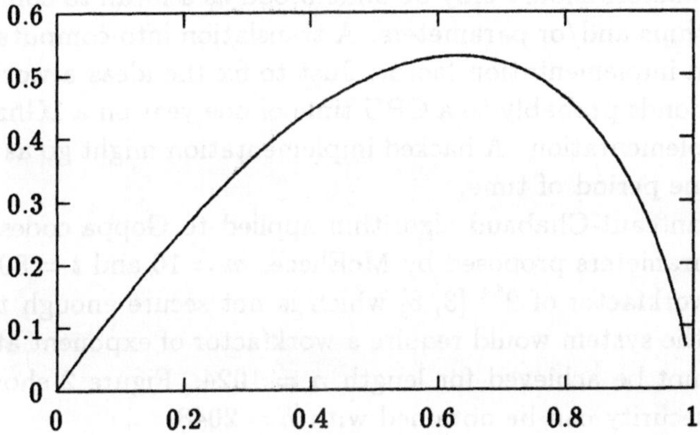

Figure 1: Plot of $R \mapsto -(1-R)\log_2(1-R)$

space complexity, then for a given rate R the constant c can be made smaller, and the shape of the curve may vary slightly.

When the transmission rate tends to 1 (a case of interest for the digital signature scheme), there are simple and effective attacks of the divide and conquer type. It is possible to obtain both time and space complexities equal to

$$2^{\frac{1}{2}tm(1+o(1))}.$$

More generally, with the *split syndrome algorithm* we can obtain for rates close to one any time and space complexities such that their exponents add to tm.

4.1.2 Workfactors

The best known algorithm with small space complexity (*i.e.* an implementation which can run on today's computers, even if it takes a lot of time) is due to Canteaut and Chabaud. It is in fact the last of many works on the subject [12, 13, 29, 30, 7, 4, 3, 5]. The time complexity is expressed in terms of a binary workfactor, which is the theoretical average number of binary operations to achieve the decoding. Note that these workfactors should only be understood as a mean to compare various algorithms and/or parameters. A translation into computation time requires an implementation factor. Just to fix the ideas a workfactor of 2^{52} corresponds probably to a CPU time of one year on a 1Ghz PC with a good implementation. A hacked implementation might go as far as 2^{55} for the same period of time.

The Canteaut-Chabaud algorithm applied to Goppa codes with the original parameters proposed by McEliece, $m = 10$ and $t = 50$, leads to a binary workfactor of 2^{64} [3, 6] which is not secure enough today. To be secure the system would require a workfactor of exponent at least 80, which cannot be achieved for length $n = 1024$. Figure 2 shows that a sufficient security can be obtained with $n = 2048$.

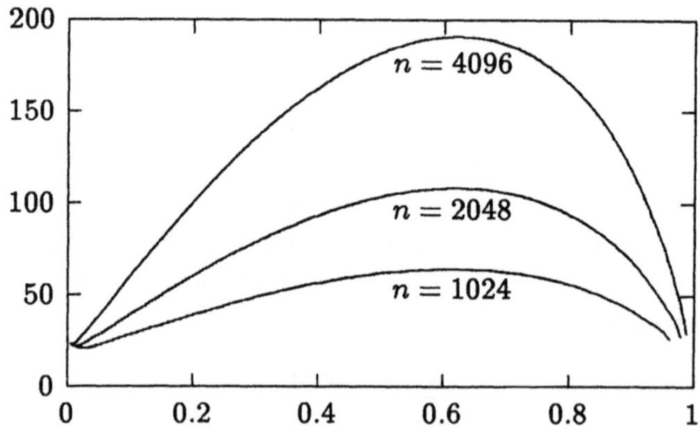

Figure 2: $\log_2(WF)$ versus the code rate for $n = 1024, 2048, 4096$ with the Canteaut-Chabaud decoder

Scalability – Moore's law The maximum workfactor for all n is obtained for a rate close to $R = 1 - 1/e$. For a fixed rate, the security of the system grows exponentially with $n/\log_2 n$. Thus, if the length of the code doubles the security exponent doubles too (almost).

If we apply Moore's law (a factor 2 in speed and memory every 18 months) using codes of length 2048 should be enough for forty years. Note that we are not limited to powers of two for the code length, we may build t-error correcting Goppa codes of any length $n \leq 2^m$ and dimension $n - tm$.

		$n = 2048$ $t = 26$	$n = 2048$ $t = 71$	$n = 1472$ $t = 51$
McEliece	block size (bits)	2048	2048	1472
	message size (bits)	1772	1267	911
	information rate	0.86	0.62	0.62
Niederreiter	block size (bits)	276	781	561
	message size (bits)	197.4	440.6	315.5
	information rate	0.69	0.56	0.56
	public key size (bits)	503932	989527	511071
Security exponent	decoding attack	81.0	108.4	80.6
	structural attack	≈ 268	≈ 771	≈ 552

Table 2: Some secure parameters for McEliece and Niederreiter systems

4.2 Structural attacks on Goppa codes

Structural attacks take as input the public key and try to deduce some information on the secret code. It has been shown that permuting the columns of a generating matrix of a generalized Reed-Solomon code cannot efficiently hide its structure [22, 28]. Also, using the small dual distance, it is possible to recover the first order concatenated structure of a code from any permuted generating matrix [25]. For binary Goppa codes no efficient (i.e. polynomial) structural attack is known, though some weak keys have been exhibited [16].

Two general attacks are known and both involve some kind of exhaustive search, the first is due to Gibson [10] and searches through all possible permutations. The second uses the support splitting algorithm [26] and searches through all binary Goppa codes of given support.

4.2.1 Gibson attack

We consider binary Goppa codes of length $n = 2^m$. If we have a generator matrix G of $\Gamma(L, g)$ and if L is known, the generator polynomial can be easily deduced from (3). All we have to do then is to try all possible L until we find a suitable one. In practice we pick two linear combinations a and b of the rows of G and compute the gcd of $d\sigma_a(z)/dz$ and $d\sigma_b(z)/dz$ which will very likely be equal to $g^2(z)$.

When $n = 2^m$ the support L contains all the elements of $\mathbf{F}_{2^m}$ in any order. In general, there are mn mutually equivalent binary Goppa codes (see §2.2.2). The average number of supports L to try will then be in average $n!/mn$. For each possible L a small number of gcd (usually one) will produce $g(z)$. If we assume a cost of mn^2 binary operations for a gcd, the average cost of the attack will be $n \cdot n!$.

The cost can be divided by mn^3 instead of nm by considering extended binary Goppa codes (see §2.2.2). The average number of binary operations required for the attack becomes $(n-1)!$. In any case the order of magnitude is $2^{nm(1+o(1))}$.

4.2.2 Support splitting algorithm

The support splitting algorithm recovers the permutation between two equivalent linear codes of length n for a cost of $O(n^3 + 2^h n^2)$ binary operations, where h is the dimension of the hull (intersection of the code with its dual). Thus if we know a code $C = \Gamma(L, g)$ with L and g unknown, we may try to compute the permutation between C and $\Gamma(L_0, g)$, for a fixed L_0 and for all possible irreducible polynomials $g(z)$, that is $\approx 2^{tm}/t$. Considering extended Goppa codes and the size of the orbits under the action of all permutation of the form $z \mapsto (az^{2^i} + b)/(cz^{2^i} + d)$, the average number of calls to the support splitting algorithm is divided by mn^3, that is $2^{(t-3)m}/tm$. The cost of the support splitting algorithm in this case cannot be smaller than $n(tm)^2$ binary operations (Goppa codes behave like random linear codes as far as the dimension of the hull is concerned, thus h is almost always a small number [24]). We finally get a binary workfactor lower bounded by $tm2^{(t-2)m}$. Asymptotically, we get $2^{tm(1+o(1))}$, which is substantially better than Gibson's attack but still not a real threat to the system.

5 Digital signature

A digital signature is a small piece of information which, appended to a document, ascertains at the same time its integrity and its authenticity. A signature scheme can be derived from a public-key encryption scheme in the following manner:

- the document M to be signed is transformed into a ciphertext y by a public one-way collision-free hash function $h()$ (*i.e.* finding (x, y) such that $h(x) = h(y)$ is computationally difficult).
- the ciphertext is deciphered into a plaintext s which will be used as the signature.

Using the notation of §2, we have $s = D(h(M))$. The verifier can use the hash function $h()$ and the encryption function $E()$ to check that $h(M) = E(s)$. Only the legal user should be able to compute s from $h(M)$.

5.1 Digital signature using McEliece encryption scheme

The function $h()$ is a public way to produce ciphertexts. Moreover, when we run through all possible messages M the distribution of $h(M)$ should be close to the uniform distribution over the ciphertext space. If we consider Niederreiter's variant, a ciphertext is a syndrome of a word of given weight t according to the public parity check matrix H. This implies that the set of valid ciphertexts depends of H. We are not aware of any easy way to produce valid syndromes without explicitly multiplying H by a vector of weight t, which would lead to an invalid protocol as it would not prove that the signer knows the secret decoder.

In practice the hash function $h()$ behaves like a random generator producing uniformly distributed bits. What we should be able to do to design a digital signature scheme is to find an error pattern corresponding to any syndrome, or equivalently to find the closest codeword to any element of the Hamming space.

In all this section we will consider an encryption/decryption pair $(E(), D())$ which is an instance of the Niederreiter cryptosystem. The following procedure takes as input a message and produces a quantity which can be used as its signature:

1. for all $i \geq 0$, let $y_i = h(h(M) \oplus i) \in \mathbf{F}_2^{n-k}$ (where $\oplus$ denotes the bitwise exclusive or),
2. let i_0 be the smallest integer such that y_{i_0} can be deciphered, $x_{i_0} = D(y_{i_0})$,
3. the signature of M is defined as $s = (x_{i_0}, i_0)$.

The decryption function $D()$ is needed to produce the signature, but the signature can be publicly checked by verifying that $h(M, i_0)$ is equal to $E(x_{i_0})$. In the case of McEliece's original parameters ($m = 10$, $n = 1024$, $k = 524$ and $t = 50$) the probability for a syndrome chosen at random to correspond to an error pattern of weight t is $\binom{n}{t}/2^{n-k} \approx 2^{-216}$. Obviously, the cost for signing is not satisfactory in that case.

5.2 Finding the proper parameters

The idea of the digital signature scheme presented in [8] is to find parameters for which the proportion of correctable syndromes is not too small. For a t error correcting binary Goppa codes of length $n = 2^m$ and dimension $k = n - tm$, this ratio is equal to

$$\frac{1}{2^{n-k}}\binom{n}{t} \approx \frac{1}{t!}, \quad \text{if } t \ll n.$$

By keeping t very small, we may obtain a reasonable signing cost and still remain safe against both decoding and structural attack.

$n = 2^m$	2^{11}	2^{12}	2^{13}	2^{14}	2^{15}	2^{16}	2^{17}
$t = 9$	$2^{54.6}$	$2^{59.9}$	$2^{69.3}$	$2^{74.0}$	$2^{78.8}$	$2^{83.7}$	$2^{88.2}$
$t = 10$	$2^{60.9}$	$2^{66.8}$	$2^{72.3}$	$2^{77.4}$	$2^{87.4}$	$2^{90.9}$	$2^{94.6}$

Table 3: Binary workfactor for complete decoding in a $(n, n-tm)$ binary code

The numbers of Table 3 are for the Canteaut-Chabaud algorithm applied to complete decoding. The best structural attack has a binary workfactor of at least $tm2^{(t-2)m}$ which is always greater than the workfactor for decoding.

5.3 Implementation

It can be seen in Table 3 that for $t = 9$ and $m = 16$ we obtain a reasonable level of security. For these parameters we have to generate in average $t! \approx 3.6\,10^5$ syndromes before we can find one corresponding to an error pattern of weight t. With a non optimized software implementation we were able to sign a message in about two minutes in average.

Another big advantage of this signature scheme is that the signature produced is very short, it consists of an error pattern of weight t and the index i_0 for which the decoding was achieved. This produces signatures of less than 150 bits which are extremely fast to verify. Many tradeoffs between signature length and verification time are available (see [8] for details), and it is possible to produce signatures of 80 bits, with about one minute of verification time (with a non optimized software implementation).

6 Security arguments

6.1 Distinguishability of Goppa codes

We will denote by $\mathcal{G}_{n,t}$, with $n = 2^m$, the set of all parity check matrices of Goppa codes of support $\mathbf{F}_{2^m}$ in any order and of generator polynomial of degree t. We will denote by $\mathcal{M}_{r,n}$ the set of all $r \times n$ binary matrices. Note that $\mathcal{G}_{n,t}$ is a subset of $\mathcal{M}_{tm,n}$.

Definition 6.1 *A Goppa code distinguisher $\mathcal{D}$ is a probabilistic Turing machine taking as input a binary matrix and whose output is 0 or 1. For any positive integers t, m, $n = 2^m$ and $r = tm$, we denote $T_{\mathcal{D}}(n,t)$ the maximum running time for an input in $\mathcal{M}_{r,n}$ and the* advantage *of $\mathcal{D}$ is defined to be*

$$Adv_{\mathcal{D}}(n,t) = \left| Prob\left[\mathcal{D}(H) = 1 \mid H \in \mathcal{M}_{r,n}\right] - Prob\left[\mathcal{D}(H) = 1 \mid H \in \mathcal{G}_{n,t}\right] \right|.$$

The advantage of a Goppa code distinguisher is thus a measure of how well it is able to make a difference when its input is a Goppa code rather than a random linear code. The advantage has to be compared to the running time, it does not matter if the advantage is significant if the running time is large (exponential). Neither is it a problem if the running time is small (polynomial), as long as the advantage is negligible.

The best known Goppa code distinguisher of advantage equal to 1 is based on the support splitting algorithm and has an exponential running time (proportional to 2^{tm}). On the other hand, we are not aware of any distinguisher which runs in polynomial time and distinguishes more than an exponentially small proportion of Goppa codes.

In practice we are interested in the average difficulty of the following problem

Problem 6.1 (Goppa Code Distinguishing - GD)

Instance: *An $r \times n$ binary matrix H.*

Question: *Does H belong to $\mathcal{G}_{n,t}$?*

It is obviously NP, as checking that H is a parity check matrix of a given $\Gamma(L, g)$ can be done in polynomial time. Any instance of GD can be written as a system of polynomial equations in $\mathbf{F}_{2^m}$. Existence of solution to such a system is NP-complete, however we are dealing with very particular instances, whose solutions are sometimes easy to find (for instance for GRS codes [28]).

6.2 Difficulty of decoding a linear code

For all integers $r = tm$ and $n = 2^m$, a matrix/syndrome pair (H, s) of parameter (r, n) is constituted by an $r \times n$ binary matrix H and a word s of $\mathbf{F}_2^r$. We will denote $W_{n,t}$ the set of words of $\mathbf{F}_2^n$ of weight t.

Definition 6.2 *A Goppa-Bounded decoder (GB-decoder) φ is a probabilistic Turing machine which takes as input a matrix/syndrome pair and outputs 0 and 1. For any positive integers t, m, $n = 2^m$ and $r = tm$, we denote $T_\varphi(n, t)$ the maximum running time for an input of parameter*

(r, n) *and the* advantage *of* φ *is defined to be*

$$Adv_{\varphi}(n,t) = \big|Prob\left[\varphi(H, He^T) = 1 \mid H \in \mathcal{M}_{r,n}, e \in W_{n,t}\right] - Prob\left[\varphi(H, s) = 1 \mid H \in \mathcal{M}_{r,n}, s \in \mathbf{F}_2^r\right]\big| .$$

The best known GB-decoders for codes of fixed rate $R = (n-r)/n$ have a running time proportional to $2^{cn/m(1+o(1))}$, with $c = -(1-R)\log_2(1-R)$. For codes with a rate close to 1 (used in the signature scheme), the running time is proportional to $2^{tm/2(1+o(1))}$.

6.2.1 NP problems connected with decoding

Instances of McEliece's or Niederreiter's encryption schemes can be reduced to instances of the Syndrome Decoding problem, which is known to be NP-complete [2].

Problem 6.2 (Syndrome Decoding - SD)

Instance: *An* $r \times n$ *binary matrix* H, *a word* s *of* $\mathbf{F}_2^r$ *and an integer* $w > 0$.

Question: *Is there a word* e *in* $\mathbf{F}_2^n$ *of weight* $\leq w$ *such that* $He^T = s$*?*

Actually, to guaranty the invertibility of the encryption, the instances must be particular and belong to the Bounded-Distance Decoding problem.

Problem 6.3 (Bounded-Distance Decoding - BDD)

Instance: *An* $r \times n$ *binary matrix* H *and a word* s *of* $\mathbf{F}_2^r$.

Promise: *Any* $d-1$ *columns of* H *are linearly independent.*

Question: *Is there a word* e *in* $\mathbf{F}_2^n$ *of weight* $< d/2$ *such that* $He^T = s$*?*

Because the predicate of the promise is NP-complete, BDD might not be in NP [31], however it is conjectured to be NP-hard [31, 1]. Still, this problem is too general for our purpose. In practice an instance of the encryption scheme is built with a particular family of codes, and the number of errors involved in the encryption process depends on the designed distance and not on the actual minimum distance. For Goppa codes we will state the following problem

Problem 6.4 (Goppa Bounded Decoding - GBD)

Instance: *An $r \times n$ binary matrix H and a word s of $\mathbf{F}_2^r$.*

Question: *Is there a word e in $\mathbf{F}_2^n$ of weight $\leq r/\log_2 n$ such that $He^T = s$?*

This problem is NP, and we have no clue to decide whether it is NP-complete or not. Building an efficient GB-decoder is very close to solving random (with uniform distribution) instances of GBD. We are therefore interested in knowing if code-based NP problems, and in particular GBD, are NP-complete *in average* (see [14, 11]) rather than just difficult in the worst case. This is an open problem [1, p. 747].

6.3 Security reduction

We will state that if an algorithm is able to efficiently break the McEliece cryptosystem, then it can be used to build either an efficient Goppa code distinguisher or an efficient GB-decoder.

Definition 6.3 *An attacker $\mathcal{A}$ is a probabilistic Turing machine running in time bounded by $T(n,t)$ when its input is a matrix/syndrome pair (H, s) of parameter (tm, n). Its probability of success is defined as*

$$P_{\mathcal{A}}(n,t) = Prob\left[\mathcal{A}(H, He^T) = e \mid H \in \mathcal{G}_{n,t}, e \in W_{n,t}\right].$$

Proposition 6.1 *Let $t \geq 2$, m and $n = 2^m$ be positive integers. If there exists an attacker $\mathcal{A}$ running in time T for all inputs in $\mathcal{M}_{tm,n} \times \mathbf{F}_2^{tm}$ which a probability of success ε, then there exists either a Goppa code distinguisher or a GB-decoder running in time bounded by $T + \delta$ (where δ is bounded by a polynomial in n) for all inputs in $\mathcal{M}_{tm,n} \times \mathbf{F}_2^{tm}$ and $\mathcal{M}_{tm,n}$ respectively, with an advantage of at least $\varepsilon/3$.*

Proof: From $\mathcal{A}$ we build the GB-decoder

$$\varphi(H,s) = \begin{cases} 1 & \text{if } He^T = s \text{ and } w_H(e) = t, \text{ where } e = \mathcal{A}(H,s), \\ 0 & \text{else,} \end{cases}$$

and the Goppa code distinguisher

$\mathcal{D}$: input H
$e \leftarrow random(W_{n,t})$
return $\varphi(H, He^T)$

where $random(W_{n,t})$ uniformly picks an element of $W_{n,t}$. Obviously both run in time with exceeds T by only a small quantity, bounded by a polynomial in n. We will denote

$$p = Prob\left[\mathcal{A}(H, He^T) = e \mid H \in \mathcal{M}_{tm,n}, e \in W_{n,t}\right].$$

1. Advantage of $\mathcal{D}$. It is equal to the difference of the two following numbers

$$Prob\left[\mathcal{D}(H) = 1 \mid H \in \mathcal{G}_{n,t}\right] = Prob\left[\mathcal{A}(H, He^T) = e \mid H \in \mathcal{G}_{n,t}, e \in W_{n,t}\right]$$
$$Prob\left[\mathcal{D}(H) = 1 \mid H \in \mathcal{M}_{tm,n}\right] = Prob\left[\mathcal{A}(H, He^T) = e \mid H \in \mathcal{M}_{tm,n}, e \in W_{n,t}\right]$$

and thus $Adv_{\mathcal{D}} = |p - \varepsilon|$.

2. Advantage of φ. It is equal to the difference of the two following numbers

$$Prob\left[\varphi(H, He^T) = 1 \mid H \in \mathcal{M}_{tm,n}, e \in W_{n,t}\right] = p,$$
$$Prob\left[\varphi(H, s) = 1 \mid H \in \mathcal{M}_{tm,n}, s \in \mathbf{F}_2^{tm}\right] = \frac{\binom{n}{t}}{n^t} p,$$

There are $\binom{n}{t}$ syndromes s corresponding to error patterns of weight t. A proportion p of them verify $\varphi(H, s) = 1$. Among the $2^{tm} = n^t$ words of $\mathbf{F}_2^{tm}$ only those corresponding to error patterns of weight t can verify $\varphi(H, s) = 1$. Thus we obtain the equalities above and $Adv_{\varphi} = p\left(1 - \binom{n}{t}/n^t\right)$.

3. Finally, since for $t \geq 2$ we have $1 - \binom{n}{t}/n^t > 1/2$, if $p \geq 2\varepsilon/3$ then $Adv_{\varphi} > \varepsilon/3$. Else if $p < 2\varepsilon/3$ then $Adv_{\mathcal{D}} = |\varepsilon - p| > \varepsilon/3$. ◇

The most important consequence of this result, is that, if both Problem 6.1 and Problem 6.4 are difficult in average, whatever that means, then code-based cryptosystems are arguably secure.

7 Conclusion

We saw that the security of McEliece (and Niederreiter) cryptosystem is good in all respects. First practically, all existing attacks have an asymptotic exponential cost. Also, we were able to prove that the security can be reduced to two well identified algorithmic problems: a subclass of the syndrome decoding problem and the distinguishability of Goppa codes. We conjecture that these problem are difficult in average and cannot be solved by a polynomial time algorithm.

In addition, encryption and decryption have a very low algorithmic complexity, at most quadratic in the block length (and the block length in Niederreiter's variant in small!). Furthermore, the algorithmic involvess polynomial over finite field and/or linear algebra over $\mathbf{F}_2$ whose implementation has been thoroughly studied in the past few years.

An important handicap was the belief that code-based public-key cryptography could not be used for digital signature. This is not true anymore, though the signing cost is relatively high (more than one minute). We believe however that an optimized implementation can sign in ten to twenty seconds on a fast PC, and that an hardware implementation could do even better.

Still two important drawbacks exist, the first is the fact that the encryption scheme has an information rate smaller than one. This disadvantage can be reduced by putting information in the error pattern in McEliece's scheme, which could lead to information rates greater than 90%. The other drawback is the key size. Unfortunately, apart from taking the public-key in systematic form (which should be done with care, see [27]), it does not seem possible to reduce it.

Finally, if we compare with the most important existing cryptosystems, code-based systems compare well with RSA who has a sub-exponential security and a very high decryption cost. Elliptic curves however seem to produce better systems except for the encryption/decryption speed which is still one order of magnitude lower with error correcting codes.

References

[1] A. Barg. Complexity issues in coding theory. In V. S. Pless and W. C. Huffman, editors, *Handbook of Coding theory*, volume I, chapter 7, pages 649–754. North-Holland, 1998.

[2] E. R. Berlekamp, R. J. McEliece, and H. C. van Tilborg. On the inherent intractability of certain coding problems. *IEEE Transactions*

on Information Theory, 24(3), May 1978.

[3] A. Canteaut. A new algorithm for finding minimum-weight words in large linear codes. In Colin Boyd, editor, *Cryptography and Coding ; proceedings of the 5th IMA conference*, number 1025 in LNCS, pages 205–212. Springer-Verlag, December 1995.

[4] A. Canteaut and H. Chabanne. A further improvement of the work factor in an attempt at breaking McEliece cryptosystem. In P. Charpin, editor, *Livre des résumé - EUROCODE 94*, pages 163–167, Abbaye de la Bussière sur Ouche, France, October 1994. INRIA.

[5] A. Canteaut and F. Chabaud. A new algorithm for finding minimum-weight words in a linear code: Application to McEliece's cryptosystem and to narrow-sense BCH codes of length 511.. *IEEE Transactions on Information Theory*, 44(1):367–378, January 1998.

[6] A. Canteaut and N. Sendrier. Cryptanalysis of the original McEliece cryptosystem. In *Advances in Cryptology - ASIACRYPT'98*, number 1514 in LNCS, pages 187–199. Springer-Verlag, 1998.

[7] H. Chabanne and B. Courteau. Application de la méthode de décodage itérative d' omura à la cryptanalyse du système de McEliece. Rapport de recherche 122, Université de Sherbrooke, Canada, October 1993.

[8] N. Courtois, M. Finiasz, and N. Sendrier. How to achieve a McEliece-based digital signature scheme. In *Asiacrypt 2001*, number 2248 in LNCS, pages 157–174. Springer-Verlag, 2001.

[9] W. Diffie and M. Hellman. New directions in cryptography. *IEEE Transactions on Information Theory*, 22(6):644–654, November 1976.

[10] J. K. Gibson. Equivalent Goppa codes and trapdoors to McEliece's public key cryptosystem. In D. W. Davies, editor, *Advances in Cryptology - EUROCRYPT'91*, number 547 in LNCS, pages 517–521. Springer-Verlag, 1991.

[11] Y. Gurevich. Average case completeness. *Journal of Computer and System Sciences*, 42(3):346–398, 1991.

[12] P. J. Lee and E. F. Brickell. An observation on the security of McEliece's public-key cryptosystem. In C. G. Günther, editor, *Advances in Cryptology - EUROCRYPT'88*, number 330 in LNCS, pages 275–280. Springer-Verlag, 1988.

[13] J. S. Leon. A probabilistic algorithm for computing minimum weights of large error-correcting codes. *IEEE Transactions on Information Theory*, 34(5):1354–1359, September 1988.

[14] L. Levin. Average case complete problems. *SIAM Journal on Computing*, 15(1):285–286, 1986.

[15] R. Lidl and H. Niederreiter. *Finite Fields*. Cambridge University Press, 1983.

[16] P. Loidreau and N. Sendrier. Weak keys in McEliece public-key cryptosystem. *IEEE Transactions on Information Theory*, 47(3):1207–1212, April 2001.

[17] F. J. MacWilliams and N. J. A. Sloane. *The Theory of Error-Correcting Codes*. North-Holland, 1977.

[18] R. J. McEliece. A public-key cryptosystem based on algebraic coding theory. *DSN Prog. Rep.*, Jet Prop. Lab., California Inst. Technol., Pasadena, CA, pages 114–116, January 1978.

[19] H. Niederreiter. Knapsack-type crytosystems and algebraic coding theory. *Prob. Contr. Inform. Theory*, 15(2):157–166, 1986.

[20] N. J. Patterson. The algebraic decoding of Goppa codes. *IEEE Transactions on Information Theory*, 21(2):203–207, March 1975.

[21] R. L. Rivest, A. Shamir, and L. M. Adleman. A method for obtaining digital signatures and public-key cryptosystems. *Communications of the ACM*, 21(2):120–126, February 1978.

[22] R. M. Roth and G. Seroussi. On generator matrices of MDS codes. *IEEE Transactions on Information Theory*, 31(6):826–830, May 1985.

[23] N. Sendrier. Efficient generation of binary words of given weight. In Colin Boyd, editor, *Cryptography and Coding ; proceedings of the 5th IMA conference*, number 1025 in LNCS, pages 184–187. Springer-Verlag, December 1995.

[24] N. Sendrier. On the dimension of the hull. *SIAM Journal on Discrete Mathematics*, 10(2):282–293, May 1997.

[25] N. Sendrier. On the concatenated structure of a linear code. *AAECC*, 9(3):221–242, 1998.

[26] N. Sendrier. Finding the permutation between equivalent codes: the support splitting algorithm. *IEEE Transactions on Information Theory*, 46(4):1193–1203, July 2000.

[27] N. Sendrier. *Cryptosystmes cl publique bass sur les codes correcteurs d'erreurs.* Mmoire d'habilitation diriger des recherches, Universit Paris 6, 2002. `http://www-rocq.inria.fr/codes/Nicolas.Sendrier/` or `Nicolas.Sendrier@inria.fr`.

[28] V. M. Sidel'nikov and S. O. Shestakov. On cryptosystem based on generalized Reed-Solomon codes. *Discrete mathematics (in russian)*, 4(3):57–63, 1992.

[29] J. Stern. A method for finding codewords of small weight. In G. Cohen and J. Wolfmann, editors, *Coding theory and applications*, number 388 in LNCS, pages 106–113. Springer-Verlag, 1989.

[30] J. van Tilburg. On the McEliece cryptosystem. In S. Goldwasser, editor, *Advances in Cryptology - CRYPTO'88*, number 403 in LNCS, pages 119–131. Springer-Verlag, 1990.

[31] A. Vardy. The intractability of computing the minimum distance of a code. *IEEE Transactions on Information Theory*, 43(6):1757–1766, November 1997.

Performance of MIMO Space-Time Coding with Discrete Modulations

Jung-Fu (Thomas) Cheng

Ericsson Research
511 Davis Dr., P.O. Box 13905, Research Triangle Park, NC 27709, USA
`Thomas.Cheng@aml.ericsson.se`

Abstract

We consider the maximum-likelihood (ML) decoding performance of concatenated space-time (ST) random coded communication utilizing multiple input and multiple output (MIMO) antennas on quasi-static fading channels. The ensemble-averaged frame error rates of such systems can be upper bounded by the outage probabilities of the corresponding conditional cutoff rates. We refer to these as the information cutoff probabilities and evaluate them numerically for a wide range of ST coding schemes. The slope behaviors of these cutoff probabilities are found to be consistent with the maximum achievable diversity orders given by a new unified bound. With simulations of a class of ST coding schemes based on bit-interleaved coded modulations and iterative spatial/temporal decoders, we further show these cutoff probabilities can provide accurate predictions of total performance gains. We then present numerical results of the *p*% cutoff capacities, which are defined as the highest supportable rates for target cutoff probabilities at *p*%. Our discussion show how this new notion can provide a unified view of all the three main theoretic results of MIMO space-time coding: rate enhancements, achievable diversity gains, and performance of specific codes. As a result, this analysis of cutoff capacities reveals the relative strengths and weaknesses of several space-time coding schemes in interesting and surprising ways.

Keywords: Space-time codes, random coding bounds, iterative decoding, fading channels.

1 Introduction

THE study of efficient communication over fading wireless environments utilizing multiple transmit and receive antennas has recently attracted much interest of researchers from diverse fields. Novel theoretical as well as practical results of such MIMO diversity communication systems have thus been derived from different perspectives ranging from information theory, coding theory to signal processing. Each of these results is well developed and understood in isolation, but there is a need of an analytic framework that ties, and reconciles the somewhat conflicting, conclusions together to provide guidelines for practical engineering deployment.

For instance, the information-theoretic formulations based on Gaussian signaling have shown that significant gains in capacity can be achieved in a flat Rayleigh-fading environments [5, 14]. This analysis demonstrates the ultimate potentials of coding across the spatial dimension as well as time that give rise to both coding and diversity gains. However, implicit in the Gaussian signaling assumption is that full diversity orders are always achieved. For a practical system with a specific discrete modulation of size Q, this cannot be the case from a coding perspective. Indeed, for ST trellis codes, it was shown that full diversity is attainable only when the information rate is smaller than $\log_2 Q$ bits per channel use [12]. For frequency hopped systems, a more elaborate diversity bound has also been reported in [8, 10]. Another approach to tackle this problem is the search of ST codes that guarantee full diversity orders using various coding and signal processing techniques. In particular, a set of short block spatial encoders is obtained from the elegant theory of orthogonal designs [1, 13]. However, the research has also concluded that, for cases with more than two transmitters, full-rate spatial encoders are not possible based on such designs.

Both the results from diversity bounding and encoder design analysis hence show that there are limits on the temporal and/or spatial coding rates for achieving certain diversity orders. Since system complexity scales generally with the modulation sizes, it is a desirable engineering practice to use the smallest modulation given a specified set of rate and performance requirements. The above discussion points out that, for a given modulation size, information rate enhancements and performance gains promised by the information-theoretic analysis have to be traded off with each other. It is not clear how this tradeoff should be optimized. In particular, it is not clear how the rates should be distributed in the temporal and spatial dimensions. Equivalently, it is a question of how to, or even if one should, trade off diversity gains with coding gains.

In this paper, we propose a new information-theoretic analytical approach to provide a unified view of the rate and diversity behaviors of a wide range of ST encoding techniques described in Section 2. This is accomplished by analyzing the outage events of conditional cutoff rates of the specific ST coded modulations. In Section 3, we show that these outage probabilities, referred to as information cutoff probabilities, are upper bounds on the ensemble-averaged frame error rates (FER) of the MIMO ST random coded systems with ML decoding. The cutoff probabilities for a wide range of MIMO ST coded schemes are evaluated numerically. The slope behaviors of these probabilities are compared to the maximum achievable diversity orders given by a new unified bound derived here. To further confirm the accuracy of the FER bounds, we conducted extensive simulations of a class of ST coding schemes based on bit-interleaved coded modulations and iterative spatial/temporal decoders in Section 4. In Section 5, we introduce the $p\%$ information cutoff capacities which are defined defined as the highest supportable rates for target cutoff probabilities at $p\%$. While the cutoff probabilities provide a snapshot of the performance of these different schemes at a particular information rate, the cutoff capacities can provide a more comprehensive overview of their performance across a wide spectrum of rates. Since these two notions are closely related, the same factors, most importantly the diversity orders, that affect the behaviors of the cutoff probabilities will also play roles in the cutoff capacities and reveal themselves in interesting and rather surprising ways (Fig. 5).

Using cutoff rates as the basis of this analysis enables a unified treatment of both flat and frequency-selective fading channels. For instance, we are able to present and compare the performance of delay diversity coding with other pure flat fading results in the same context. More results for frequency-selective channel are presented in [3].

2 System Models of MIMO Space-Time Codes

We consider a ST coded communication systems utilizing K transmit and M receive diversity antennas. A block of information bits $\mathbf{u}$ is encoded into a codeword $\mathbf{c} = (c(1), c(2), \ldots, c(N_c))^t$ of length N_c by an error control code $\mathcal{C}$ with symbols chosen from the Q-ary constellation $\mathfrak{X}$. The symbols belong to the complex number field and are assumed to be normalized such that the average symbol energy is one. The number of codewords in $\mathcal{C}$ is denoted by

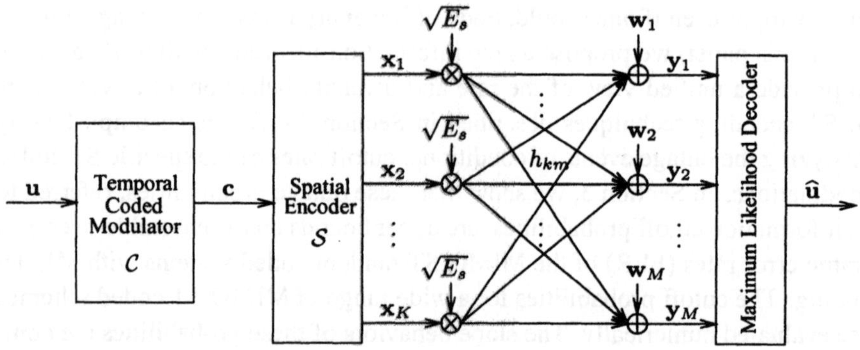

Figure 1: Space-time coded diversity communication system with K transmit and M receive antennas.

$|\mathcal{C}|$ and, hence, the code rate of $\mathcal{C}$ is given by $R_c \triangleq \frac{1}{N_c}\log_2|\mathcal{C}|$ bits per symbol. Furthermore, the equivalent binary coding rate of $\mathcal{C}$ can be defined as $r_c \triangleq R_c/\log_2 Q$. The temporal encoder can be based on any of the practical coded modulation techniques ranging from multilevel codes, trellis codes, to bit-interleaved coded modulations.

Before transmission, the codeword is first transformed by the spatial encoder $\mathcal{S}$ into an $N_s \times K$ matrix $\mathcal{S}(\mathbf{c}) \triangleq \begin{bmatrix}\mathbf{x}_1 & \mathbf{x}_2 & \cdots & \mathbf{x}_K\end{bmatrix}$, where $\mathbf{x}_k = (x_k(1), x_k(2), \ldots, x_k(N_s))^t$, for $k = 1, 2, \ldots, K$. Each row of the signal matrix is normalized such that $E\left[\Sigma_{k=1}^{K}|x_k(n)|^2\right] = 1, \forall\, n$. The columns of the signal matrix $\mathbf{X}$ are then scaled by $\sqrt{E_s}$ and transmitted simultaneously across the K antennas. Since the communication system uses the channel N_s times, the overall information rate is given by $R = N_c R_c/N_s = r_s R_c$ bits per channel use, where $r_s = N_c/N_s$ is the coding rate of the spatial encoder $\mathcal{S}$.

We adopt the standard quasi-static Rayleigh flat fading channel assumption [6, 12]: the fading coefficient h_{km} from the k-th transmit to the m-th receive antenna is taken to be a complex, circular symmetric, zero-mean Gaussian random variable with unit power which remains constant for the duration of a block of N_s channel uses but changes independently from block to block. Therefore, the signal at the m-th receive antenna is

$$\mathbf{y}_m = \sqrt{E_s}\sum_{k=1}^{K} h_{km}\mathbf{x}_k + \mathbf{w}_m = \sqrt{E_s}\mathcal{S}(\mathbf{c})\mathbf{h}_m + \mathbf{w}_m, \tag{1}$$

where $\mathbf{y}_m = (y_m(1), y_m(2), \ldots, y_m(N_s))^t$, $\mathbf{h}_m = (h_{1m}, h_{2m}, \ldots, h_{Km})^t$, and $\mathbf{w}_m = (w(1), w(2), \ldots, w(N_s))^t$. The additive white Gaussian noise (AWGN) samples $\mathbf{w}_m$ are assumed to be of zero mean and variance $N_0/2$ per dimension. The average received signal-to-noise ratio (SNR) at each antenna is therefore E_s/N_0. The noise is also assumed to be uncorrelated across the antennas. Throughout the paper, we assume the channel coefficients are available to the receiver but unknown to the transmitter. In the following, we give explicit definitions of several spatial encoders to be analyzed.

Unstructured Spatial Multiplexer Inspired by the results shown in [5, 14], this spatial encoder simply multiplexes a codeword across all the antennas to increase spectral efficiency.

$$\mathcal{S}_{\mathrm{USM}}(\mathbf{c}) \triangleq \frac{1}{\sqrt{K}}\left[\mathbf{c}_1 \quad \mathbf{c}_2 \quad \cdots \quad \mathbf{c}_K\right],$$

where $\mathbf{c}_k \triangleq \left(c((k-1)\frac{N_c}{K}+1), c((k-1)\frac{N_c}{K}+2), \ldots, c(k\frac{N_c}{K})\right)^t$ is the k-th sub-block of the codeword $\mathbf{c}$. Since $N_s = N_c/K$, the spatial coding rate is $r_s = K$. It is noted this abstract encoding model in fact includes many existing ST coding structures as special cases: BLAST [4], trellis ST codes [12], and natural ST codes for PSK [7].

Orthogonal Transmit Diversity Frequency and antenna hopping techniques are among the earliest known transmit diversity codes by transforming the MIMO channel into a set of independent parallel channels.

$$\mathcal{S}_{\mathrm{OTD}}(\mathbf{c}) \triangleq \begin{bmatrix} \mathbf{c}_1 & \mathbf{0} & \ldots & \mathbf{0} \\ \mathbf{0} & \mathbf{c}_2 & \ldots & \mathbf{0} \\ \vdots & \vdots & \ddots & \vdots \\ \mathbf{0} & \mathbf{0} & \ldots & \mathbf{c}_K \end{bmatrix},$$

Since $N_s = N_c$, the spatial coding rate $r_s = 1$.

Delay Diversity The delay diversity coding method creates an artificial dispersive channel by sending symbol-delayed versions of the same sequence on the all antennas simultaneously. The signal matrix of such K-antenna delay diversity (DD_K) is given by the $(N_c + K - 1) \times K$ matrix:

$$\mathcal{S}_{\mathrm{DD}_K}(\mathbf{c}) \triangleq \frac{1}{\sqrt{K}} \begin{bmatrix} c(1) & c(2) & \ldots & 0 & 0 \\ 0 & c(1) & \ldots & c(N_c) & 0 \\ \vdots & \vdots & \ddots & \vdots & \vdots \\ 0 & \ldots & \ldots & c(N_c-1) & c(N_c) \end{bmatrix}^t.$$

Since the spatial code length $N_s = N_c + K - 1$, the spatial coding rate $r_s = N_c/(N_c + K - 1)$ approaches one asymptotically with the code length N_c.

Orthogonal Design Codes A set of short ST block codes based on generalized complex orthogonal designs are discussed in [13]. For two transmit antennas, a full-rate design ($r_s = 1$), also known as the Alamouti code [1], is given by

$$\mathcal{S}_{\text{Ala}}(\mathbf{c}) \triangleq \frac{1}{\sqrt{2}} \begin{bmatrix} c(1) & c(2) \\ -c^*(2) & c^*(1) \\ \vdots & \vdots \\ c(N_c-1) & c(N_c) \\ -c^*(N_c) & c^*(N_c-1) \end{bmatrix}.$$

In this paper, we also consider the $r_s = 3/4$ code for four transmit antennas, which is denoted by $\mathcal{S}_{\text{QOC}}$. Noted that, for more than two transmit antennas, full-rate encoders based on orthogonal designs are not possible [13].

Hybrid Space-Time Codes Full-rate spatial encoders for more than two transmit antennas can be created by combining existing codes, though full diversity orders cannot be guaranteed in these cases. In this paper, we consider two hybrid techniques that combine orthogonal transmit diversity with either the Alamouti or delay diversity codes. Assuming K is even, let $\mathbf{c}_1, \mathbf{c}_2, \ldots, \mathbf{c}_{K/2}$, each of length $2N_c/K$, be the $K/2$ sub-blocks of codeword $\mathbf{c}$. An OTD-Alamouti encoder is defined by the following $N_c \times K$ matrix

$$\mathcal{S}_{\text{OTD-Ala}}(\mathbf{c}) \triangleq \begin{bmatrix} \mathcal{S}_{\text{Ala}}(\mathbf{c}_1) & \mathbf{0} & \ldots & \mathbf{0} \\ \mathbf{0} & \mathcal{S}_{\text{Ala}}(\mathbf{c}_2) & \ldots & \mathbf{0} \\ \vdots & \vdots & \ddots & \vdots \\ \mathbf{0} & \mathbf{0} & \ldots & \mathcal{S}_{\text{Ala}}(\mathbf{c}_{K/2}) \end{bmatrix}.$$

That is, the K antennas are divided into $K/2$ pairs, each of which is used to transmit a sub-block $\mathbf{c}_k$ employing Alamouti coding. The rate of this encoder is $r_s = 1$. Similarly, assuming K is divisible by t, the K antennas can then be divided into K/t sets of antennas. Each set of the t antennas can be used to transmit a sub-codeword employing t-antenna delay diversity coding:

$$\mathcal{S}_{\text{OTD-DD}}(\mathbf{c}) \triangleq \begin{bmatrix} \mathcal{S}_{\text{DD}_t}(\mathbf{c}_1) & \mathbf{0} & \ldots & \mathbf{0} \\ \mathbf{0} & \mathcal{S}_{\text{DD}_t}(\mathbf{c}_2) \ldots & & \mathbf{0} \\ \vdots & \vdots & \ddots & \vdots \\ \mathbf{0} & \mathbf{0} & \ldots & \mathcal{S}_{\text{DD}_t}(\mathbf{c}_{K/t}) \end{bmatrix},$$

where $\mathbf{c}_1, \mathbf{c}_2, \ldots, \mathbf{c}_{K/t}$, each of length tN_c/K, are the K/t sub-blocks of codeword $\mathbf{c}$. Since $N_s = N_c + K - \frac{K}{t}$, the spatial coding rate r_s is asymptotically one in this case.

Before leaving the section, we note that the encoders defined in the above satisfy the property that $\mathcal{S}(\mathbf{c}) - \mathcal{S}(\widehat{\mathbf{c}}) = \mathcal{S}(\mathbf{c} - \widehat{\mathbf{c}})$.

3 Random Coding Upper Bounds

With perfect channel state estimates at the receiver, the maximum-likelihood decoder searches a codeword $\mathbf{c} \in \mathcal{C}$ that maximize the log-likelihood $\log \Pr(\mathbf{Y}|\mathcal{S}(\mathbf{c}), \mathbf{H}) = \kappa - \frac{1}{N_0}\Sigma_{m=1}^{M} \left\|\mathbf{y}_m - \sqrt{E_s}\mathcal{S}(\mathbf{c})\mathbf{h}_m\right\|^2$, where $\mathbf{Y} = (\mathbf{y}_1, \mathbf{y}_2, \ldots, \mathbf{y}_M)$, $\mathbf{H} = (\mathbf{h}_1, \mathbf{h}_2, \ldots, \mathbf{h}_M)$, and κ is a constant independent of $\mathbf{c}$. Let $\mathbf{c}$ be the transmitted codeword and $\widehat{\mathbf{c}}$ be an alternative codeword preferred by this decoder. Given the channel coefficients $\mathbf{H}$, the probability of such pairwise codeword error can be upper-bound by the Bhattachayya bound:

$$\Pr(\mathbf{c} \to \widehat{\mathbf{c}}|\mathbf{H}) \leq \int \sqrt{\Pr(\mathbf{Y}|\mathcal{S}(\mathbf{c}), \mathbf{H}) \Pr(\mathbf{Y}|\mathcal{S}(\widehat{\mathbf{c}}), \mathbf{H})} d\mathbf{Y}. \quad (2)$$

Consider using a code $\mathcal{C}$ with its codewords chosen randomly from the space $\mathfrak{X}^{N_c}$. The conditional codeword error probability averaged over the ensemble of such random codes can be upper bounded by [15, Chapter 3]:

$$\overline{P_{N_c}(E|\mathbf{H})} \leq |\mathcal{C}| \cdot E_{\mathbf{c},\widehat{\mathbf{c}}}\left[\int \sqrt{\Pr(\mathbf{Y}|\mathcal{S}(\mathbf{c}), \mathbf{H}) \Pr(\mathbf{Y}|\mathcal{S}(\widehat{\mathbf{c}}), \mathbf{H})} d\mathbf{Y}\right]. \quad (3)$$

We show the following holds.

Theorem 1 *The unconditioned ensemble-averaged codeword error probability is upper bounded by*

$$\lim_{N_c \to \infty} \overline{P_{N_c}(E)} \leq \Pr(\mathcal{R}_0(\mathcal{S}|\mathbf{H}) \leq R), \quad (4)$$

where the conditional cutoff rate $\mathcal{R}_0(\mathcal{S}|\mathbf{H})$ *is defined as*

$$\lim_{N_c \to \infty} -\frac{r_s}{N_c} \log_2 E_{\mathbf{c},\widehat{\mathbf{c}}}\left[\int \sqrt{\Pr(\mathbf{Y}|\mathcal{S}(\mathbf{c}), \mathbf{H}) \Pr(\mathbf{Y}|\mathcal{S}(\widehat{\mathbf{c}}), \mathbf{H})} d\mathbf{Y}\right]. \quad (5)$$

In this paper, we limit our discussion to the so-called symmetric cutoff rates, which means that the above expectation is evaluated according to the following probability measure: $q_{N_c}(\mathbf{c}) = Q^{-N_c}$, where Q is the size of the coded symbol alphabet $\mathfrak{X}$. Consider the error symbol $\delta(n) = c(n) - \widehat{c}(n)$. It belongs to the error symbol alphabet $\Delta(\mathfrak{X})$ derived from $\mathfrak{X}$: $\Delta(\mathfrak{X}) \triangleq \{c - c' : \forall c, c' \in \mathfrak{X}\}$. The derived probability measure for $\delta(n)$ is given by

$$\widetilde{q}(\delta(n) = \delta) = \frac{1}{Q^2}\left|\{c - c' : c - c' = \delta, \forall c, c' \in \mathfrak{X}\}\right|. \quad (6)$$

For example, if the coded symbols are taken from BPSK, the error alphabet $\Delta(\mathfrak{X}) = \{-2, 0, 2\}$ and the derived probability measures for each symbol are $\{\frac{1}{4}, \frac{2}{4}, \frac{1}{4}\}$, respectively. By expanding the Bhattachayya bound and using the property that $\mathcal{S}(\mathbf{c}) - \mathcal{S}(\widehat{\mathbf{c}}) = \mathcal{S}(\mathbf{c} - \widehat{\mathbf{c}})$, we have

Theorem 2 *The conditional cutoff can be computed as*

$$\mathcal{R}_0(\mathcal{S}|\mathbf{H}) = \lim_{N_c\to\infty} -\frac{r_s}{N_c}\log_2 E_{\boldsymbol{\delta}}\left[\exp\left(-\frac{E_s}{4N_0}\sum_{m=1}^{M}\|\mathcal{S}(\boldsymbol{\delta})\mathbf{h}_m\|^2\right)\right], \quad (7)$$

where the error symbols $\delta(n)$ are treated as independent identically distributed random variables with the measure (6).

With this result, explicit cutoff rate formulae for memoryless spatial encoders can be readily derived as follows.

Unstructured Spatial Multiplexer The conditional cutoff rate of USM is given by $\mathcal{R}_0(\mathcal{S}_{\text{USM}}|\mathbf{H}) =$

$$-\log_2 \sum_{\delta_1,\ldots,\delta_K\in\Delta(\mathfrak{X})}\cdots\sum \widetilde{q}(\delta_1)\cdots\widetilde{q}(\delta_K)\exp\left(-\frac{E_s}{4KN_0}\sum_{m=1}^{M}\left|\sum_{k=1}^{K}h_{kf}\delta_k\right|^2\right) \quad (8)$$

For a conventional single-input-single-output (SISO) system, this reduces to

$$\mathcal{R}_0(\mathcal{S}_{\text{SISO}}|h_{11}) = -\log_2\sum_{\delta\in\Delta(\mathfrak{X})}\widetilde{q}(\delta)\exp\left(-\frac{E_s}{4N_0}|\delta|^2|h_{11}|^2\right), \quad (9)$$

which agrees with the conventional cutoff rate expression for AWGN channels [15]. For a single-input-multiple-output (SIMO) diversity system, the fading coefficient matrix $\mathbf{H}$ reduces to a row vector $\boldsymbol{\rho}_1 = (h_{11}, h_{12}, \ldots, h_{1M})$ of length M. The conditional cutoff rate is given by

$$\mathcal{R}_0(\mathcal{S}_{\text{SIMO}}|\boldsymbol{\rho}_1) = -\log_2\sum_{\delta\in\Delta(\mathfrak{X})}\widetilde{q}(\delta)\exp\left(-\frac{E_s}{4N_0}|\delta|^2\sum_{m=1}^{M}|h_{1m}|^2\right). \quad (10)$$

Comparing the last two equations, the SNR gain from diversity reception in the SIMO system is evident.

Orthogonal Transmit Diversity

$$\mathcal{R}_0(\mathcal{S}_{\text{OTD}}|\mathbf{H}) = \frac{1}{K}\sum_{k=1}^{K}\mathcal{R}_0(\mathcal{S}_{\text{SIMO}}|\boldsymbol{\rho}_k), \quad (11)$$

where $\boldsymbol{\rho}_k \triangleq (h_{k1}, h_{k2}, \ldots, h_{kM})$ is the k-th row of $\mathbf{H}$ and $\mathcal{R}_0(\mathcal{S}_{\text{SIMO}}|\boldsymbol{\rho}_k)$ is the conditional cutoff rates of the k-th equivalent SIMO channel.

Orthogonal Design Codes The general expression is given by

$$\mathcal{R}_0(\mathcal{S}_{\text{OD}}|\mathbf{H}) = -r_s\log_2\sum_{\delta\in\Delta(\mathfrak{X})}\widetilde{q}(\delta)\exp\left(-\frac{g_{od}E_s}{4KN_0}|\delta|^2\sum_{m=1}^{M}\sum_{k=1}^{K}|h_{km}|^2\right)$$

The cutoff rate $\mathcal{R}_0(\mathcal{S}_{\text{Ala}}|\mathbf{H})$ for the Alamouti code is obtained with with $r_s = 1$, $g_{od} = 1$, and $K = 2$. The cutoff rate $\mathcal{R}_0(\mathcal{S}_{\text{QOC}}|\mathbf{H})$ for the $r_s = 3/4$ orthogonal design code is obtained with $g_{od} = 4/3$, and $K = 4$.

OTD-Alamouti Coding

$$\mathcal{R}_0(\mathcal{S}_{\text{OTD-Ala}}|\mathbf{H}) = \frac{1}{K/2} \sum_{k=1}^{K/2} \mathcal{R}_0\left(\mathcal{S}_{\text{Ala}} \left| \begin{bmatrix} \rho_{2k-1} \\ \rho_{2k} \end{bmatrix} \right.\right), \tag{12}$$

where ρ_{2k-1} and ρ_{2k} are the $(2k-1)$- and $2k$-th rows of $\mathbf{H}$.

Delay Diversity For spatial encoders with memory, (7) can be evaluated with the help of the Perron-Frobenius theorem [15, Chapter 5]:

$$\mathcal{R}_0(\mathcal{S}_{\text{DD}_K}|\mathbf{H}) = -\log_2 \lambda_{\max}, \tag{13}$$

where $\lambda_{\max}$ is the largest eigenvalue of a $|\Delta(\mathfrak{X})|^{K-1} \times |\Delta(\mathfrak{X})|^{K-1}$ matrix whose elements represent transition gains from state $(\delta_2, \delta_3, \ldots, \delta_K)$ to state $(\delta_1, \delta_2, \ldots, \delta_{K-1})$:

$$\tilde{q}(\delta_1) \exp\left(-\frac{E_s}{4KN_0} \sum_{m=1}^{M} \left| \sum_{k=1}^{K} h_{km}\delta_k \right|^2\right).$$

OTD-Delay-Diversity This encoder creates K/t equivalent independent parallel t-tap frequency-selective channels. Hence, its conditional cutoff rate is the average of those of the K/t equivalent channels:

$$\mathcal{R}_0(\mathcal{S}_{\text{OTD-DD}}|\mathbf{H}) = \frac{1}{K/t} \sum_{k=1}^{K/t} \mathcal{R}_0\left(\mathcal{S}_{\text{DD}_t} \left| \begin{bmatrix} \rho_{tk-t+1} \\ \vdots \\ \rho_{tk} \end{bmatrix} \right.\right), \tag{14}$$

where $\rho_{tk-t+1}, \ldots, \rho_{tk}$ are the $(tk-t+1)$ through tk-th rows of $\mathbf{H}$.

4 Analysis of Frame Error Rates

In the now classic analysis presented in [6, 12], an upper bound on the unconditioned pairwise codeword error probability is obtained by averaging the Bhattachayya bound (2) over the distribution of $\mathbf{H}$:

$$\Pr(\mathbf{c} \to \hat{\mathbf{c}}) \triangleq E\left[\Pr(\mathbf{c} \to \hat{\mathbf{c}}\,|\mathbf{H})\right] \leq \left(\tfrac{E_s}{4N_0}\Lambda\right)^{-\Gamma M}, \tag{15}$$

where Γ is the rank of $\mathcal{S}(\mathbf{c}-\hat{\mathbf{c}})^{\dagger}\,\mathcal{S}(\mathbf{c}-\hat{\mathbf{c}})$, and Λ is the geometrical average of the corresponding Γ nonzero singular values. This analysis shows the importance of diversity gain $\mathcal{D} \triangleq \Gamma M$ which determines th asymptotic slope of the FER curve on a log scale. Maximizing this quantity has becomes an important criterion for designing spatial encoders. For example, the orthogonal design codes achieve full diversity ($\mathcal{D} = KM$) for all information rates R by construction. The delay diversity scheme has also been established for achieving full diversity. For the rest of the spatial encoders considered in the paper, however, full diversity orders can not be guaranteed for all information rates. We apply a bounding technique from [8, 10] to these encoders to arrive at the following unified diversity bound.

Theorem 3 *For K-transmit M-receive space-time coded communication systems on the flat Rayleigh fading channel, the achievable diversity orders are bounded by*
(a) for the USM and OTD encoders,

$$\mathcal{D} \leq M\big(1 + \lfloor(1 - r_c)K\rfloor\big),$$

(b) for the OTD-Alamouti encoder,

$$\mathcal{D} \leq 2M\big(1 + \lfloor(1 - r_c)K/2\rfloor\big),$$

(c) for the OTD-Delay-Diversity encoder,

$$\mathcal{D} \leq tM\big(1 + \lfloor(1 - r_c)K/t\rfloor\big),$$

where $\lfloor x \rfloor$ is the largest integer no greater than x.

As with the capacity outage probabilities [11], we define the probability that the realized conditional cutoff rate is lower than the actual information rate as the information cutoff probability of the system that transmits at R bits per channel use (pcu):

$$P_{\text{co}}(\mathcal{S}, R) \triangleq \Pr\left(\mathcal{R}_0(\mathcal{S}|\mathbf{H}) \leq R\right). \tag{16}$$

From (4), this quantity is thus an upper bound on the average codeword error probability of randomly chosen codes of very large length with ML decoding. In this section, we illustrated how this quantity can be used as a unified framework for comparing the average frame error rates (FER) of different spatial coding and modulation configurations. To this end, we consider two system design scenarios where an information rate of $R = 1$ bit pcu is desired in the first and $R = 1.2$ bits pcu in the second. The following alternatives using one, two or four transmit antennas (short-handed as 1T, 2T, or 4T systems, respectively) are capable of providing the required information rate for scenario I (and scenario II):

- 1T using QPSK with an equivalent binary coding rate of $r_c = 0.5$ (and $r_c = 0.6$).
- 2T using USM coded BPSK with $r_c = 0.5$ (and $r_c = 0.6$).

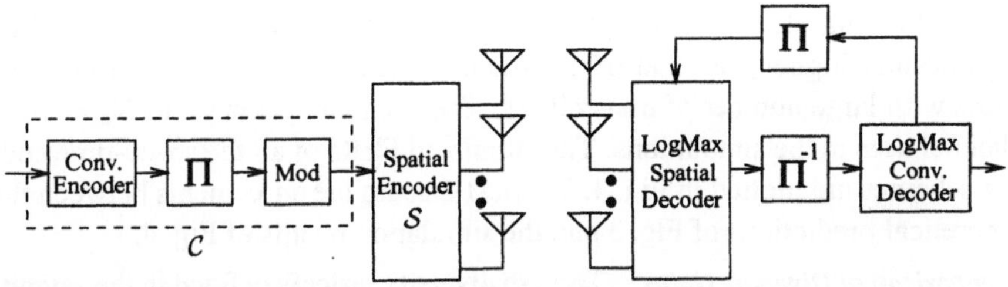

Figure 2: Practical MIMO system based on bit-interleaved coded modulation and iterative decoding.

- 2T using QPSK in conjunction with Alamouti, OTD, or DD encoders. The equivalent binary coding rate is set to $r_c = 0.5$ (and $r_c = 0.6$).
- 4T using USM coded BPSK with $r_c = 0.25$ (and $r_c = 0.3$).
- 4T using QPSK in conjunction with OTD, OTD-Alamouti, or OTD-DD ($t = 2$) encoders. The equivalent binary coding rate is set to $r_c = 0.5$ (and $r_c = 0.6$).

For 1T systems, we consider reception using only one antenna (short-handed as 1T1R systems). For the 2T systems, we consider the option of using only one or two antennas at the receiver (2T1R and 2T2R systems). For the 4T systems, reception with two antennas is considered (4T2R systems). Assuming independent Rayleigh fading, the information cutoff probabilities of these space-time coding schemes are evaluated using the Monte Carlo method and are plotted in Fig. 3.

To verify these predictions, we conducted simulations for these MIMO schemes with the setup illustrated in Fig. 2. We specialized the temporal code C to a class of bit-interleaved coded modulations in the simulations. For practicality, the receiver decodes the spatial and temporal codes separately. For QPSK with the Alamouti or OTD codes, a single-pass decoding using this structure is optimal. For the USM and DD codes, however, iterative spatial/temporal decoding is needed to approximate ML performance. These iterative decoding algorithms can be constructed by enabling exchange of extrinsic information between the two decoders using the guiding principle of belief propagation [9]. To make the comparison fairer, we use punctured convolutional codes of two different constraint lengths depending on whether iterative decoding is used. For the Alamouti or OTD codes, a convolutional code of 64 states is applied. For the USM and DD codes, we use a 16-state code with four iterations of spatial/temporal decoding. It is, however, noted that, even with these adjustments, the total decoding complexity of the USM and DD codes can still be significantly higher than that of the Alamouti and OTD codes, especially for cases with large number of transmitters. The frame length is set to $N_s = 1000$ channel uses in the simulations. The simulated FERs of all the space-time coding schemes are plotted in Fig. 4. We next discuss the agreements between the theoretical predictions of Fig. 3 and the simulation results of Fig. 4.

Comparison of Diversity Gains Though diversity order is defined in the asymptotic sense, some of its patterns can be readily observed even in the FER region between 10^{-1} and 10^{-3}. For example, it can be observed in figures 3(a) and 4(a) that the 1T1R system exhibits performance of diversity order one, i.e., the slope of the cutoff probability is 10 dB per decade. For Scenario I, all but the 4T2R OTD schemes should be able to achieve full diversity orders according to Theorem 3. It is easy to see in these figure that the 2T1R and 2T2R systems indeed exhibit performance consistent with diversity orders close to two and four, respectively. Furthermore, when the information rates are increased by 20% to $R = 1.2$ bits pcu for Scenario II, not all the space-time coding schemes can maintain full diversity performance by the Theorem. In particular, for 2T1R and 2T2R systems, those schemes with the OTD and USM encoders will see their diversity order reduced by half. These are markedly illustrated in figures 3(b) and 4(b). We assert that the behaviors of the other curves shown in these two figures are also consistent with Theorem 3.

Comparison of Performance Gain Predictions We observed that the cutoff probabilities give quite accurate predictions of the total coding gains in addition to the diversity gains just discussed. Using the 1T1R-QPSK scheme as baselines, we observed the gains at 1% FER range from as small as 1.2 dB for the 2T1R-USM-BPSK scheme at $R = 1.2$ bits pcu to as large as 18.2 dB for the 4T2R-USM-BPSK at $R = 1$ bit pcu. For 20 out of the 24 cases, the theoretical predictions based on cutoff probabilities differ by fractions of a dB from the observed gains. For the four cases where the predictions and observations differ by around 1.1 dB, iterative decoding approximation instead of true ML decoding is used.

It is noted that the cutoff probabilities are *upper bounds* by construction. Namely, unlike the capacity outage probabilities [11], these quantities should not be construed as unbeatable performance limits of the respective systems. Instead, as we have just demonstrated, these information cutoff probabilities can provide most utility when used as predictive tools against known baselines.

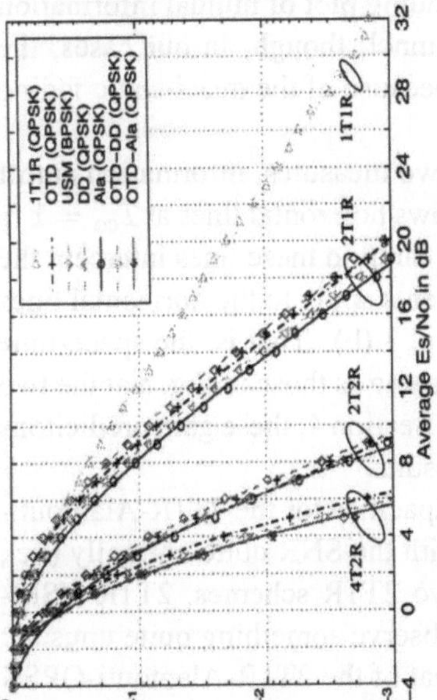

(a) Scenario I: $R = 1.0$ bit per channel use.

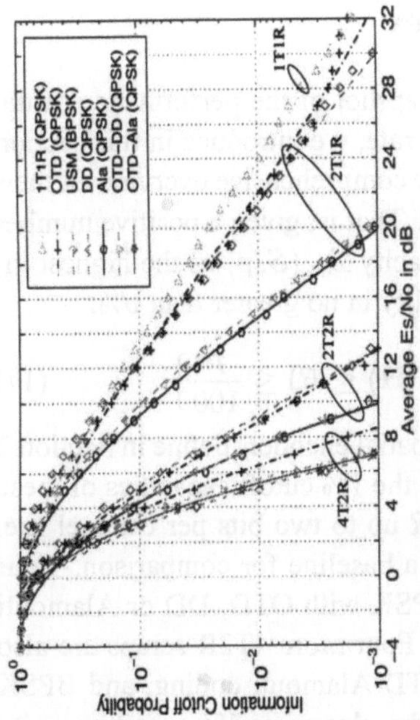

(b) Scenario II: $R = 1.2$ bits per channel use.

Figure 3: Information cutoff probabilities of space-time coded systems defined in Section 4 on flat fading channel.

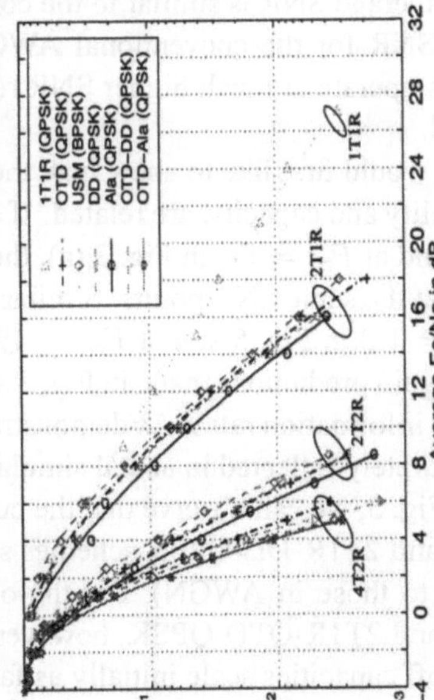

(a) Scenario I: $R = 1.0$ bit per channel use.

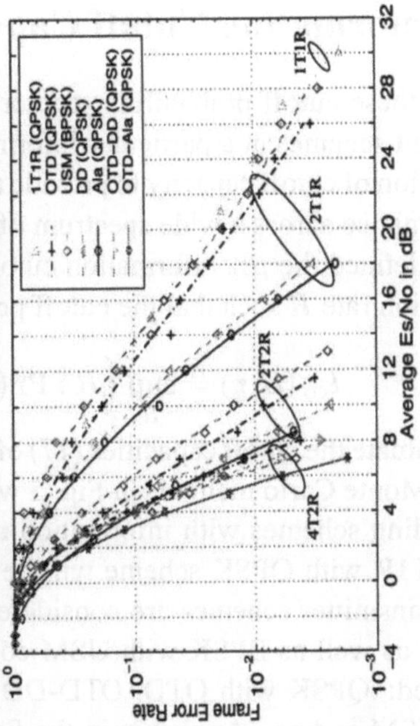

(b) Scenario II: $R = 1.2$ bits per channel use.

Figure 4: Simulated frame error rates of space-time coded systems using the same setups as in Fig. 3.

5 Information Cutoff Capacities

While these cutoff probabilities capture a snapshot of the performance of the different schemes at a particular information rate, we introduce in this section the notion of cutoff capacity to provide a more comprehensive overview of their performance across a wide spectrum of rates. That is, given a positive number p, we defined the $p\%$ information cutoff capacity $\mathcal{I}_{\text{co}}(\mathcal{S}, p)$ as the highest information rate R such that the cutoff probability in no greater than $p\%$:

$$\mathcal{I}_{\text{co}}(\mathcal{S}, p) \triangleq \sup_{R} \left\{ R : \Pr\left(\mathcal{R}_0(\mathcal{S}|\mathbf{H}) \le R\right) \le \frac{p}{100} \right\}. \quad (17)$$

We evaluate the cutoff capacities (17) of the spatial encoders define in Section 2 using Monte Carlo method. In Fig. 5 we plot the 1% cutoff capacities of these ST coding schemes with information rates R up to two bits per channel use. The 1T1R with QPSK scheme will be used a baseline for comparison. Four dual transmitter schemes are considered: QPSK with OTD, DD or Alamouti coding as well as BPSK with USM coding. Four more 4T2R setups are also included: QPSK with OTD, OTD-DD or OTD-Alamouti coding, and BPSK with USM coding. As shown in the figure, the plot of its 1% cutoff capacity versus average SNR is similar to the corresponding plot of mutual information versus SNR for the conventional AWGN channel, though, in our cases, the systems operate at much higher SNR region because of the quasi-static fading channel.

We would first like to show how the two two measures, information cutoff probability and capacity, are related. If one draws horizontal lines at $\mathcal{I}_{\text{co}} = 1$ in Fig. 5 and at $P_{\text{co}} = 1\%$ in Fig. 3 (a), then one can find these lines intercept the curves at the same SNR points. Similar comments apply to the horizontal lines at $\mathcal{I}_{\text{co}} = 1.2$ in Fig. 5 and at $P_{\text{co}} = 1\%$ in Fig. 3 (b). That is, the space-time coding gain predictions made in Fig. 3 are the same as those in Fig. 5 at the two specific information rates. As demonstrated in Section 4, these gain predicitons are accurately reflected in acutal simulation results.

In Fig. 5, we can observe that the cutoff capacities for the 2T1R-Alamouti-QPSK and 2T1R-DD-QPSK schemes scale with the SNR quite normally (i.e., similar to those in AWGN). For the other two 2T1R schemes, 2T1R-USM-BPSK and 2T1R-OTD-QPSK, however, we observe something quite unusual: the cutoff capacities scale initially as fast as that of the 2T1R-Alamouti-QPSK scheme but then, above one bit pcu, at much slower rate. These two different capacity scaling behaviors are the manifestations of the diversity order bounds

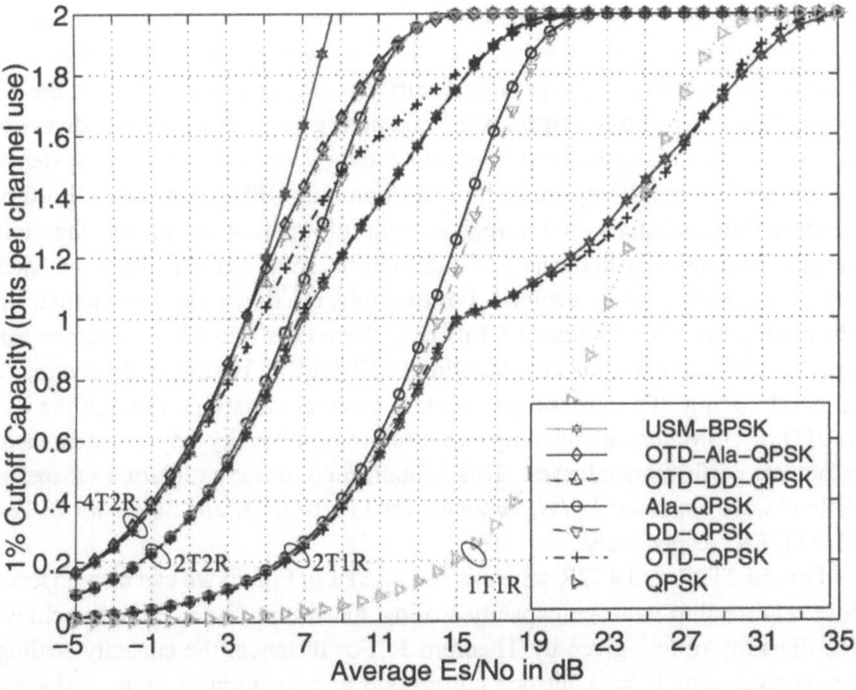

Figure 5: 1% information cutoff capacities on flat Rayleigh fading channel of ST coding configurations: (a) 1T1R system using QPSK; (b) 2T1R and 2T2R systems using QPSK with OTD, DD or Alamouti coding, and BPSK with USM coding; (c) 4T2R system using QPSK with OTD, OTD-DD or OTD-Alamouti coding, and BPSK with USM coding. Detailed defintions of the encoders are given in Section 2.

from Theorem 3. For all achievable information rates, the 2T1R-Alamouti-QPSK and 2T1R-DD-QPSK schemes have diversity orders of two by construction. For information rate below ones bits per channel use, the 2T1R-USM-BPSK and 2T1R-OTD-QPSK schemes can also achieve diversity orders up to two. For rates higher than that, however, the diversity orders of these two schemes are limited to one. Since the diversity order determines how fast the FER decay with SNR, a lower diversity order means a higher SNR requirement to achieve 1% FER. Therefore, for rates below ones bits pcu, the 1% cutoff capacities of all the four 2T1R schemes scale similarly because they can all achieve the same diversity orders. For rates higher than that, the cutoff capacity of the 2T1R-Alamouti-QPSK and 2T1R-DD-QPSK schemes will scale faster because their diversity orders are twice those of the other two schemes.

In addition to diversity gains, these curves, by their close connection with the cutoff probabilities, also provide accurate coding gain predictions of the ST codes. This is especially helpful when comparing codes with idential diversity orders. For example, since the Alamouti coding is optimal for 2T1R systems, we can see the cutoff capacity of 2T1R-Alamouti-QPSK dominates those of the other three schemes even in regions with information rates lower than one bits pcu. Moreover, with the help of this analysis, some system design mistakes based on intuition can be avoided. For example, the figure also shows that, for information rates higher than 1.5 bits pcu, these later two 2T1R schemes can actually perform worse than the baseline 1T1R-QPSK. That is, without enough temporal coding, the introduction of extra channel variations through the use of OTD of USM spatial encoding can result in no diversity improvements and in fact net performance losses. This is confirmed in our simulation where, at FER=1%, we observed $E_s/N_0 = 28$ dB for 1T1R-QPSK and 30 dB for 2T1R-OTD-QPSK, respectively.

For the 2T2R and 4T2R schemes considered in Fig. 5, we can also observe close relationship between capacity scaling speeds and the maximum achievable diversity orders given by Theorem 3. For instance, the capacity scaling speeds change at $R = 1$ bit pcu correspond to maximum diversity order reductions from 8 to 4 for the 4T2R-OTD-Ala-QPSK, and from 4 to 2 for the 2T2R-USM-BPSK and the 2T2R-OTD-QPSK schemes, respectively. For the 4T2R-OTD-QPSK, the diversity order predicts three scaling speed transition points at $R = 0.5, 1.0, 1.5$ bits per channel use, respectively.

References

[1] S. V. Alamouti, "A Simple Transmit Diversity Technique for Wireless Communications," *IEEE J. Sel. Areas Commun.*, vol. 16, no. 8, pp. 1451–1458, Oct. 1998.

[2] L. R. Bahl, J. Cocke, F. Jelinek, and J. Raviv, "Optimal Decoding of Linear Codes for Minimizing Symbol Error Rate," *IEEE Trans. Infor. Theory*, Mar. 1974.

[3] J.-F. Cheng, "Performance of MIMO Space-Time Coded Discrete Modulations — Part I: Introduction and Flat Fading Channels," and "Part II: Extensions and Frequency-Selective Fading Channels," preprint.

[4] G. J. Foschini, "Layered Space-Time Architecture for Wireless Communication in Fading Environments When Using Multiple Antennas," *Bell Labs. Technical Journal*, vol. 2, Autumn 1996.

[5] G. J. Foschini and M. J. Gans, "On Limits of Wireless Communications in a Fading Environment when Using Multiple Antennas," *Wireless Personal Communications*, vol. 6, no. 3, pp. 311-335, Mar. 1998.

[6] J.-C. Guey, M. P. Fitz, M. R. Bell, and W.-Y. Kuo, "Signal Design for Transmitter Diversity Wireless Communication Systems over Rayleigh Fading Channels," *IEEE Trans. Commun.,* vol. 47, no. 4, pp. 527–537, Apr. 1999.

[7] A. R. Hammons, Jr., and H. El Gammal, "On the Theory of Space-Time Codes for PSK Modulation," *IEEE Trans. Infor. Theory,* vol. 46, no. 2, pp. 524–542, Mar. 2000.

[8] R. Knopp and P. A. Humblet, "On Coding for Block Fading Channels," *IEEE Trans. Information Theory*, vol. 46, no. 1, pp. 189–205, Jan. 2000.

[9] D. J. C. MacKay, R. J. McEliece, and J.-F. Cheng, "Turbo decoding as an instance of Pearl's 'belief propagation' algorithm," *IEEE J. Sel. Areas Commun.,* vol. 16, pp. 140–152, Feb. 1998.

[10] E. Malkamäki and H. Leib, "Coded Diversity on Block-Fading Channels," *IEEE Trans. Information Theory*, vol. 45, no. 2, pp. 771–781, Mar. 1999.

[11] L. H. Ozarow, S. Shamai, and A. D. Wyner, "Information Theoretic Considerations for Cellular Mobile Radio," *IEEE Trans. on Vehicular Tech.,* vol. 43, no. 5, pp. 359–378, May 1994.

[12] V. Tarokh, N. Seshadri, and A. R. Calderbank, "Space-Time Codes for High Data Rate Wireless Communication: Performance Criterion and Code Construction," *IEEE Trans. Infor. Theory,* vol. 44, no. 2, Mar. 1998.

[13] V. Tarokh, H. Jafarkhani, and A. R. Calderbank, "Space-Time Block Codes from Orthogonal Designs," *IEEE Trans. Infor. Theory,* vol. 45, no. 5, July 1999.

[14] E. Teletar, "Capacity of Multi-antenna Gaussian Channels," *European Trans. on Telecom,* vol. 10, no. 6, pp. 585–596, Nov.–Dec. 1996.

[15] A. J. Viterbi and J. K. Omura, *Principles of Digital Communication and Coding*, McGraw-Hill, 1979.

Coding for Slow-Frequency-Hop Transmission: Variations on a Theme of McEliece

Thomas G. Macdonald
MIT Lincoln Laboratory
244 Wood St.
Lexington, MA 02420
tom_macdonald@ll.mit.edu

Michael B. Pursley
Clemson University
303 Fluor Daniel Bldg.
Clemson, SC 29634
pursley@ces.clemson.edu

Abstract

An important theme of McEliece's publications on slow-frequency-hop communications is that Reed-Solomon codes can protect against the effects of frequency-selective fading and partial-band interference, especially if side information is provided to the receiver. Two variations on this theme are described and the potential performance improvements are discussed. A new method is presented that permits the receiver to derive its own side information without requiring the transmitter to send side-information symbols, and Hermitian codes are evaluated as an alternative to Reed-Solomon codes for frequency-hop communications.

Keywords: Reed-Solomon codes, slow-frequency-hop communications, side information, Hermitian codes, interleaving, iterative decoding.

1 Prologue

In frequency-hop communications, a single message or packet is divided into segments and each segment is sent in a separate interval of time known as a *dwell interval.* During each dwell interval, the transmitted signal occupies a single frequency slot, but the frequency slot may be changed from one dwell interval to the next. Frequency hopping that occurs during the transmission of a packet or message provides diversity that is beneficial if the channel has frequency-selective fading or interference that is concentrated in part of the frequency band. Frequency-hop systems are typically classified according to the relationship between the hopping rate and the data rate. In *slow-frequency-hop* communications, the hopping rate does not exceed the data rate, so each dwell interval contains one or more data symbols. The hopping rate is greater than the rate at which the data symbols are transmitted in a *fast-frequency-hop* communication system.

There was tremendous interest in frequency-hop communications at the University of Illinois at the time Robert McEliece joined the faculty in the summer of 1978. Slow-frequency-hop transmission had been adopted for use in the new tactical communication system for the U.S. Army. This system, known as SINCGARS, was being developed and produced by ITT in Fort Wayne, Indiana. Faculty from the University of Illinois were beginning to collaborate with engineers from ITT and initiate research that would eventually find application in SINCGARS. Early literature on SINCGARS is scarce, but the interested reader may consult [13] for a general description of the initial SINCGARS system. At approximately the same time, Cooper and Nettleton [5] proposed a fast-frequency-hop modulation scheme for mobile cellular communications. The potential for applications to both military and civilian communications made it very attractive to pursue research in modulation, coding, and receiver processing for frequency-hop communications.

Much of the attention in frequency-hop communications research in the late 1970s and early 1980s was devoted to providing multiple-access capability to a number of asynchronous frequency-hop transmitters [8], but antijam techniques were also investigated for military applications. Repetition coding, which was employed in the first version of SINCGARS, was known to provide some protection against *hits* that occur when the frequency-hop signal uses one or more frequency slots that have interference. Diversity combining methods, such as a simple majority vote, can be employed in the receiver to mitigate the effects of interference due to hits, such as those that may arise when multiple signals are frequency hopped in a common set of frequency slots. Frequency-hop multiple-access interference is one example of *partial-band interference*, which is interference that is present in only a fraction of the frequency band used by the frequency-hop system.

If the receiver can determine which frequency slots have interference, the symbols transmitted in those frequency slots can be omitted from the combining process. Information about the presence of interference is often referred to as *side information*. For theoretical studies, the source of the side information is not important. For practical implementation, the side information should be derived in the receiver. A few methods were known in the late 1970s for deciding if a dwell interval is affected by interference, but many of those methods require energy measurements. For example, the energy in a frequency slot can be measured before, during, and after each dwell interval, and comparisons of the energy levels provide an estimate of the amount of interference in the frequency slot. An alternative method that does not require energy measurements relies on the demodulation of special side-information symbols that are included in each dwell interval prior to transmission. Because the symbols are known to the receiver, it can easily determine the number of erroneous side-information symbols. If the number of errors among the side-information symbols in a dwell interval is sufficiently large, the data symbols in that dwell interval might be omitted from the diversity combining process, or a request can be made for the retransmission of the affected data symbols.

Around 1980, McEliece initiated research that had a profound influence on the theory and practice of slow-frequency-hop communications. One theme of McEliece's research is that information theory and game theory can be used to obtain fundamental limits on the performance of frequency-hop communications for channels with frequency-selective fading and partial-band interference (e.g., [18], [19], and [37]). A second theme is that Reed-Solomon codes are natural choices for providing error-control in frequency-hop transmission over such channels (e.g., [17] and [37]) and the performance of such codes can be enhanced greatly if side information is available to permit the erasure of unreliable symbols (e.g., [21]). The effect of side information on the capacity is also explained in [21], and comparisons between Reed-Solomon and convolutional codes are given in [17] and [21].

Extensive research by McEliece and others followed these two themes and provided our current understanding of the theory of frequency-hop communications. Of greatest significance for implementation of slow-frequency-hop communication systems and networks is the role of the second theme in advancing the state-of-the art in transmission techniques, receiver processing methods, and even network protocols [29]. Methods for deriving and using side information, such as those described in [14], [21], and [25], have been applied in several system designs. In the SINCGARS Packet Overlay [33], Reed-Solomon codes are used for error control and the transmission and demodulation of test symbols [25] provide side information. The second-generation SINCGARS system [11] also employs Reed-Solomon codes, but parity-check symbols [14] are used to provide side information.

Two variations on McEliece's second theme are described in Sections 2 and 3. A method for obtaining side-information that does not require the transmission of side-information symbols is presented and evaluated in Section 2. In Section 3, Hermitian codes are evaluated as an alternative to Reed-Solomon codes for slow-frequency-hop communications.

2 Side Information from Staggered Interleaving and Iterative Decoding

Several methods have been developed for determining which symbols to erase among the demodulated symbols in slow-frequency-hop spread spectrum systems [28]. These methods can be divided into two classes. In the first class the side information is developed from the transmission and demodulation of side-information symbols that are included in each dwell interval [28]. One advantage of such methods is that they can be used with standard hard-decision demodulators in existing frequency-hop radio receivers. In fact, the SINCGARS frequency-hop radio has been enhanced by the use of side-information symbols and errors-and-erasures decoding without requiring a change to the demodulator [11]. One disadvantage of the use of side-information symbols is that the number of message symbols per dwell interval must be reduced in order to accommodate the inclusion of the side-information symbols. This reduces the information rate, and it may lead to a reduction in the throughput for the frequency-hop system [27].

In the second class of methods, the side information is developed from the demodulation of the message symbols themselves, and no additional symbols are transmitted as side-information symbols. The methods in the second class are more complex and cannot be used with the demodulators in existing frequency-hop radios. Bayesian methods are described in [1], and methods based on the ratio threshold technique are given in [1], [2], and [41]. Also included in the second class is Forney's generalized minimum-distance (GMD) decoding [6] and related soft-decision decoding methods for Reed-Solomon codes (e.g., see [4] and [39]). GMD and related methods require the demodulator to provide soft decisions consisting of likelihood ratios or some functions of likelihood ratios, which typically entails estimation of the strength of the signal and the power spectral density of the noise. These parameters are difficult to measure accurately in a mobile frequency-hop communication network.

An alternative method for the development and use of side information in frequency-hop systems is described in this section. No side-information symbols are used, and almost any form of demodulation is acceptable. Thus, the new method has the advantages of the first class without the need to transmit special symbols. Unlike the techniques in the second class, the proposed method requires no modification to existing hard-decision demodulators. Instead, the side information is obtained through a combination of staggered interleaving of the code words and an iterative errors-and-erasures decoding procedure similar to that employed in GMD decoding. As is true of the methods in the first two classes, the new method also provides side information for use in higher-layer protocols (e.g., as described in [7] and [29]).

Two frequency-hop systems are compared for channels with partial-band interference. In the first system, the transmitter sends binary test symbols along with the data symbols. Block interleaving is used, and the demodulation of the test symbols furnishes the side information for erasing symbols at the input to the errors-and-erasures decoder. The second system employs the proposed staggered interleaving and itera-

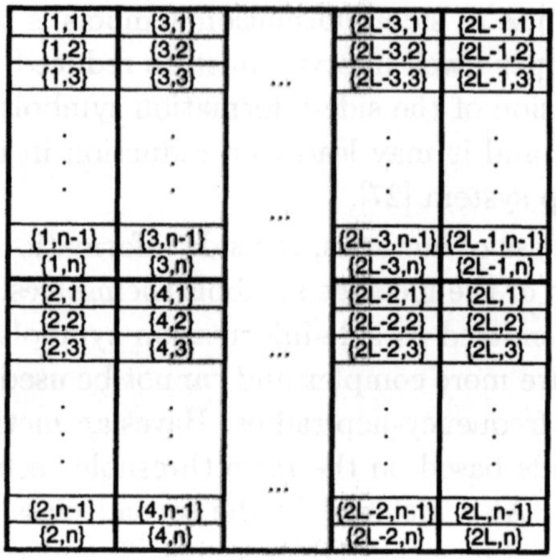

Figure 1: Block interleaving with two error-control blocks

tive decoding, thereby avoiding the need to transmit side-information symbols.

There are a number of features that are common to both systems, including the use of an (n, k) singly extended Reed-Solomon code to encode the message. The code words are interleaved in such a way that no code word has two code symbols in the same dwell interval. The code symbol alphabet size M is equal to the block length n. Transmission of the code symbols can be accomplished with binary modulation, in which case $m = \log_2 M$ binary symbols represents a code symbol. Alternatively, M-ary modulation may be employed with one transmitted symbol for each code symbol.

2.1 Block Interleaving

The packet format for a standard slow-frequency-hop system that employs block interleaving is shown in Figure 1. Each pair of integers $\{i, j\}$ represents one code symbol; specifically, $\{i, j\}$ refers to the jth symbol of the ith code word. The symbols in a given row are in the same dwell interval, and different rows correspond to different dwell intervals. Thus all symbols in a given dwell interval are transmitted in the same frequency slot.

There are two error-control blocks in each packet. Each error-control block has L code words of length n that are transmitted in n consecutive dwell intervals. An important feature of block interleaving is that if there is interference in one frequency slot, each code word in a given error-control block is affected in exactly the same way. Thus, for a given error-control block, errors among symbols that occupy the same position in different code words are highly correlated. One reason there are two error-control blocks per packet is to facilitate comparisons with staggered interleaving. However, the use of multiple error-control blocks per packet also has advantages for adaptive-rate coding in packet radio networks [7].

For the system that employs test symbols to provide side information, binary symbols are used for the data modulation, and a fixed number of binary test symbols are inserted in each dwell interval in the manner described in [28]. If the number of errors among the demodulated test symbols for a given dwell interval exceeds γ, the prescribed threshold, then all data symbols in that dwell interval are erased at the input to the decoder. The decoder is able to correct t errors and e erasures in a received word if $2t + e \leq n - k$. Cyclic redundancy check (CRC) codes or other error-detecting codes are employed to detect decoding errors, although such errors are rare unless $n - k - e$ is very small [22].

If a received word at the decoder input produces at the decoder output a code word in which no errors are detected, we say that the received word is *decoded*. If a received word at the decoder input does not produce a code word at the decoder output (i.e., if there is a decoding failure) or produces a code word in which errors are detected, we say that received word is *not decoded*. A received packet is said to be decoded if each of the received words in the packet is decoded. A packet error occurs if at least one received word in the packet does not decode.

2.2 Staggered Interleaving and Iterative Decoding

In staggered interleaving, which was introduced in [3], the code words in each packet are interleaved in the offset manner illustrated in Figure 2 for $L = n$. If $L \neq n$ the packet is constructed in a similar manner, but some offsets are not used if $L < n$ or some are used more than once if $L > n$.

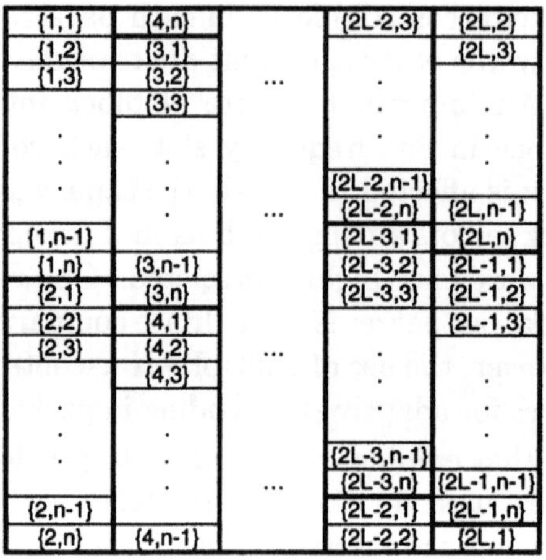

Figure 2: Staggered interleaving for $L = n$

The primary effect of staggering the code words is to reduce the correlation among the error patterns in different words at the receiver. For example, suppose the symbols in the nth dwell interval are transmitted in a frequency slot that contains very strong interference. Each odd-numbered received word is affected by this interference, and each such word is likely to have a symbol error in the position corresponding to the nth dwell interval. However, among the odd-numbered code words, only the first one uses all of the first n frequency slots. If many of these first n frequency slots have strong interference, it is likely that the first word does not decode correctly, but several other odd-numbered words may decode correctly. If they do, the symbol errors in the nth dwell interval can be detected, in which case the dwell interval is a strong candidate for erasure.

The iterative decoding procedure is described as follows. The received packet is deinterleaved and errors-only decoding is attempted. If at least one received word decodes and at least one received word does not decode, a second decoding step is employed. Side information is obtained from the received words that decoded correctly in the first step, and this side information is applied to the decoding of other received words. If,

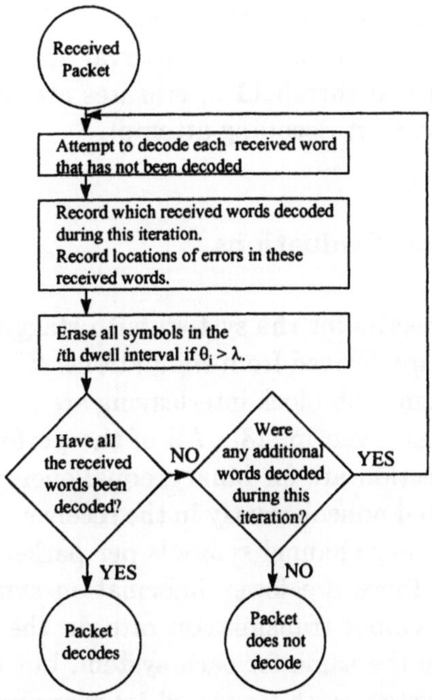

Figure 3: The decoding algorithm

after the second decoding attempt, at least one additional word has been decoded but at least one word remains undecoded, the decoding proceeds to the third step. This process continues until all words have been decoded or a decoding attempt produces no additional decoded words. In [16] it is shown that if the maximum number of iterations is limited to three, there is only a very small increase in the packet error probability compared to decoding with an unlimited number of iterations. Erasures are made without regard to the locations of the erasures in previous steps, so a symbol may be erased in one step but not the next. A flow diagram for the iterative decoding algorithm is shown in Figure 3.

In each step of the decoding procedure, the erasure decisions depend on the dwell-interval statistics θ_i, $1 \leq i \leq 2n$. Let μ_i denote the number of symbols from the ith dwell interval that are in words that decoded, and let ν_i denote the number of these symbols that are erroneous at the demodulator output. The ith dwell-interval statistic is $\theta_i = \nu_i/\mu_i$. If θ_i exceeds a prescribed threshold λ, erasures are made in the ith dwell interval prior to the next decoding attempt.

2.3 Performance Evaluations

The performance results for the system with staggered interleaving and iterative decoding are derived from simulations. The packet error probabilities for the system with block interleaving are determined analytically using the techniques given in [28]. All of the performance comparisons presented in this section are for binary, equal-energy, orthogonal signals that are demodulated noncoherently in the receiver. Each system has the same number of binary channel symbols per packet. If side-information symbols are used, there are fewer information symbols in the packet. Thus, for a fixed symbol transmission rate on the channel, the packet transmission rate is the same for each system, but the information rate is greater for the system with staggered interleaving.

The performance of each system is evaluated for transmission over a channel that has partial-band interference and thermal noise. The thermal noise is modeled as additive white Gaussian noise, which implies the effects of thermal noise at the receiver are statistically independent from one received symbol to the next. The conditional probability that a binary symbol is in error given that it is transmitted in a frequency slot that has no interference is referred to as the *quiescent probability of error.*

For one of the interference models we considered, the error probability is 1/2 for a binary symbol that is transmitted in a frequency slot with partial-band interference. In the terminology of [28], this is referred to as *catastrophic* partial-band interference. Very strong multiple-access interference is an example of catastrophic partial-band interference. The results for channels with catastrophic interference are given in [16]. In most situations with catastrophic interference, we found that the performance of the system that uses test symbols depends greatly on the erasure threshold γ, whereas the performance of the staggered-interleaving technique is fairly robust with respect to its threshold λ.

The results given here are for less severe partial-band interference. The partial-band interference is modeled as band-limited white Gaussian noise that occupies a fraction of the frequency slots used by the frequency-hop system. The one-sided power spectral density for the noise is $\rho^{-1}N_I$ in a fraction ρ of the frequency slots, and it is zero elsewhere. Thus, the average one-sided spectral density, averaged over all frequency slots, is N_I. The thermal noise in the receiver has one-sided power spectral density N_0. Therefore, the total power spectral density is $\eta_I = \rho^{-1}N_I + N_0$ in a fraction ρ of the frequency slots and $\eta_N = N_0$ in the remainder of the slots.

The partial-band-noise model has been used for more than 25 years in investigations of frequency-hop communications (e.g., [12] and [42]). It is a convenient model for analysis and simulation, and it provides a reasonably accurate representation for many forms of interference that occupy a portion of the frequency band. In the proper context, it is essentially the same as the Gaussian model for pulsed or partial-time interference.

For the results presented in this section each error-control block contains 32 code words ($L = 32$) if staggered interleaving is used. The corresponding error-control block for block interleaving has only 31 code words ($L = 31$). The reason there is one fewer code word per error-control block for block interleaving is that five binary symbols per dwell interval must be used as test symbols. These five binary symbols replace one code symbol in each dwell interval.

In Figure 4 the packet error probabilities for the two systems are shown as a function of E_b/N_I for several different values of the thresholds. Note that E_b/N_I is inversely proportional to the strength of the partial-band noise. The results in Figure 4 are for a (32,12) Reed-Solomon code, a quiescent probability of error of 10^{-2}, and $\rho = 1/4$. The fact that the performance does not always improve as E_b/N_I increases has been noted previously (e.g., [1]) for systems that derive and use side information to make erasures.

As shown in Figure 4, the selection of the threshold (γ or λ) for each system can can greatly affect the packet error probability. For $\rho = 1/4$ the staggered-interleaving scheme with $\lambda = 0.2$ results in the best performance for the entire range of E_b/N_I. For the test-symbol method, no single choice of γ is best for the range of E_b/N_I shown in Figure 4.

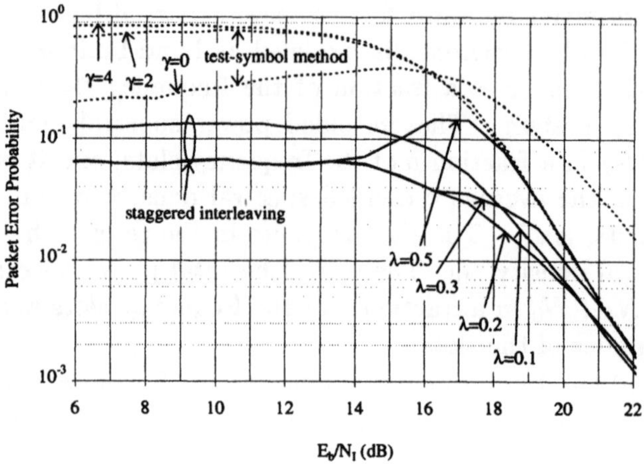

Figure 4: Packet error probabilities for a quiescent error probability of 10^{-2} and partial-band noise with $\rho = 1/4$

In addition to varying the values of E_b/N_I and ρ, we also varied the quiescent probability of error. We found that $\lambda = 0.25$ gives nearly optimal performance for most of the partial-band noise channels that we investigated.

Packet error probabilities for systems with a $(32, 12)$ Reed-Solomon code are shown in Figure 5 as a function of the quiescent error probability for partial-band noise channels with $\rho = 1/4$ and three values for E_b/N_I (4 dB, 16 dB, and 23 dB). The threshold is $\lambda = 0.25$ for the system with staggered interleaving and iterative decoding. Results for thresholds of $\gamma = 0$ and $\gamma = 2$ are shown for the test-symbol method. For $E_b/N_I = 19$ dB, the staggered interleaving scheme and the test-symbol method with $\gamma = 2$ perform almost identically, but the test-symbol method with $\gamma = 0$ performs slightly worse. For the smaller values of E_b/N_I, the performance of the staggered-interleaving scheme is significantly better than that of the test-symbol method. Because the

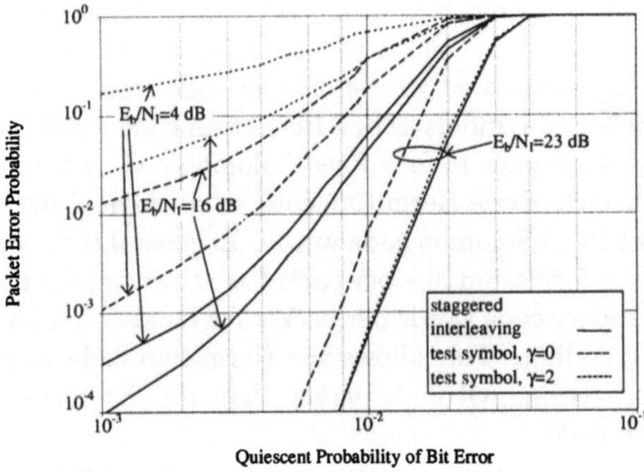

Figure 5: Packet error probabilities for partial-band noise with $\rho = 1/4$

staggered-interleaving method also has a higher information rate, it is clearly superior for channels with strong partial-band interference.

3 Hermitian Codes for Frequency-Hop Communications

Hermitian codes are an attractive alternative to Reed-Solomon codes for use in frequency-hop packet radio networks. Hermitian codes are not as well known as Reed-Solomon codes, but they provide many of the same benefits. In addition, they have the potential for extending the effective range of the frequency-hop transmissions beyond that attainable with Reed-Solomon codes. The potential improvement is due to the fact that for the same alphabet size M, the block length of a Hermitian code is longer by a factor of $\sqrt{M}$ than the block length of a singly extended Reed-Solomon code, which provides a performance advantage in combating Gaussian noise.

As an example consider Hermitian and Reed-Solomon codes that have an alphabet of size $M = 16$. The Hermitian code words are of length 64 and the singly extended Reed-Solomon code words are of length 16. Thus the number of symbols in one Hermitian code word is equal to the total number of symbols in four Reed-Solomon code words. Similarly, if $M = 64$ the length of one Hermitian code word is equal to the combined length of eight Reed-Solomon code words. In general if the code rate and the number of information bits per packet are the same for the two codes, there are $\sqrt{M}$ more code words per packet for Reed-Solomon coding than for Hermitian coding. This allows the Hermitian code to correct some symbol error patterns within a packet that are not correctable by the Reed-Solomon code.

Performance comparisons for Hermitian and Reed-Solomon are given in [43]. The comparisons are based on the bit error probability rather than the packet error probability, and they are for the additive white Gaussian noise channel only. Even for the additive white Gaussian noise channel, the bit errors at the output of the decoder are statistically dependent, so the packet error probability cannot be determined directly from the bit error probability. The performance comparisons provided in this section are more appropriate for the typical operating environment of a frequency-hop packet radio network, since the comparisons are based on the packet error probability and the channel model includes partial-band noise.

In this section a singly extended Reed-Solomon code is designated by (n_R, k_R) and its minimum distance is denoted by d_R. Because Reed-Solomon codes are maximum distance separable, $d_R = n_R - k_R + 1$. In a similar manner a Hermitian code is denoted by (n_H, k_H) and d_H denotes its minimum distance. As explained in [38], Hermitian codes do not exist for all values of M. We consider the class of Hermitian codes for which $M = 4^z$, where z is a positive integer. The block length for a Hermitian code with alphabet size M is $n_H = M\sqrt{M} = 8^z$. For the range of values of k_H that we consider, a lower bound on the minimum distance of a Hermitian code is given in [40]. In [44] it is shown that the lower bound is in fact the minimum distance, and this minimum distance is given by

$$d_H = n_H - k_H - \frac{1}{2}(M - \sqrt{M}) + 1.$$

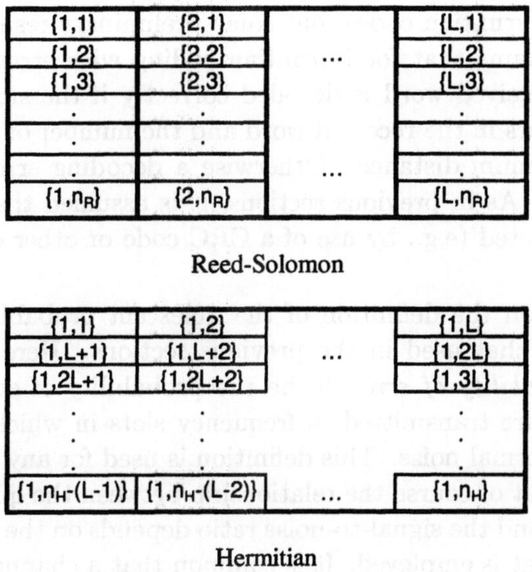

Figure 6: Packet formats

In order to ensure a fair performance comparison based on the packet error probability, the packet size must be the same for the two codes. The packet formats for the two codes are illustrated in Figure 6. For Hermitian coding there is one code word per packet, and for Reed-Solomon coding there are $L = \sqrt{M}$ code words per packet. All symbols in the each row of Figure 6 are transmitted in the same dwell interval, and different rows correspond to different dwell intervals. The block length for singly extended Reed-Solomon codes is $n_R = M$. For Hermitian codes, the block length is $n_H = M\sqrt{M}$, and there are $\sqrt{M}$ code symbols per dwell interval. Thus, for either code, the transmission of a packet requires M dwell intervals. The hopping pattern is modeled as a sequence of independent random variables, each of which is uniformly distributed over the set of allowable frequencies. The hopping patterns for different transmissions are independent.

An errors-only decoding algorithm for Hermitian codes is given in [32]. At the present time, an errors-and-erasures decoding algorithm is not available for Hermitian codes, but some preliminary results are encouraging. We assume that for Hermitian coding with errors-and-erasures decoding, a received word is decoded correctly if the sum of twice the number of errors in the received word and the number of erasures is less than the minimum distance. Otherwise a decoding error or decoding failure occurs. As in previous sections, it is assumed that all decoding errors are detected (e.g., by use of a CRC code or other error-detecting code).

In this section the definition of the quiescent probability of error is different from that used in the previous sections. Here we define the *quiescent probability of error* to be the probability of error for M-ary symbols that are transmitted in frequency slots in which the only disturbance is thermal noise. This definition is used for any type of M-ary modulation, but of course the relationship between the quiescent probability of error and the signal-to-noise ratio depends on the form of M-ary modulation that is employed. It is common that a channel in which the only disturbance is thermal noise is characterized by the ratio E_b/N_0, where E_b is the energy per information bit and N_0 is the one-sided power spectral density for the thermal noise. The quiescent probability of error is a continuous, strictly decreasing (and hence one-to-one) function of E_b/N_0, so the channel can also be characterized by its quiescent probability of error. For many channel models, an advantage of the latter characterization is that one set of performance curves is valid for several different forms of M-ary modulation.

The packet error probability for Hermitian coding with either errors-only decoding or errors-and-erasures decoding is given in [15] for three different channel models. The derivation of the packet error probability is given here for only one of the channel models, the partial-band noise channel defined in Section 2.3. The test-symbol method [25] is used to obtain side information for errors-and-erasures decoding. The packet error probabilities for Reed-Solomon coding are obtained from the analytical expressions given in [28].

The probability of error for a symbol that is transmitted in a dwell interval with interference is denoted by δ. Conditioned on either the presence or the absence of interference, the bit errors in a dwell interval are independent at the input to the decoder. Thus, if each M-ary code symbol is transmitted as m binary symbols, then $\delta = 1-(1-\delta_b)^m$, where

δ_b is the probability of error for a binary symbol. For binary orthogonal modulation and noncoherent demodulation, $\delta_b = \frac{1}{2}\exp\{-E_{bs}/2\eta_I\}$, where E_{bs} is the energy per binary symbol. If an (n, k) block code is used for error-control, then $E_{bs} = E_b(k/n)$.

Following [25] and [28], we say that if the side information correctly indicates that interference is present in a frequency slot, the interference is *detected*. If interference in a frequency slot used for transmission is not detected, a *miss* is said to occur. A *false alarm* occurs if the side information indicates interference is present in a frequency slot that has no interference. The false alarm probability and detection probability are denoted by α and β, respectively.

If there are N binary test symbols per dwell interval and the channel has partial-band noise, the detection probability is

$$\beta = \sum_{\ell=\gamma+1}^{N} \binom{N}{\ell} \delta_b^{\ell}(1-\delta_b)^{N-\ell}$$

and the false alarm probability is

$$\alpha = \sum_{\ell=\gamma+1}^{N} \binom{N}{\ell} p_0^{\ell}(1-p_0)^{N-\ell},$$

where p_0 is the probability of error for a binary symbol that sent in a frequency slot without interference (i.e., the only disturbance is thermal noise). Because the bit errors in a dwell interval are independent at the input to the decoder, the quiescent probability of error is $p = 1-(1-p_0)^m$.

For a partial-band noise channel, the packet error probability for Hermitian coding and errors-and-erasure decoding is $P_e = 1 - P_c$, where P_c is given by

$$P_c = \sum_{j=0}^{M} \binom{M}{j} \rho^j (1-\rho)^{M-j} P_3(j).$$

In this expression $P_3(j)$ is the conditional probability that the received word decodes given that partial-band interference is present in j of the M dwell intervals that are used by the transmission. The probability

$P_3(j)$ is given by

$$P_3(j) = \sum_{i=0}^{j} \binom{j}{i} \beta^i (1-\beta)^{j-i} P_2(i;j),$$

where $P_2(i;j)$ is the conditional probability that the received word decodes given there are j hits and i of them are detected. The probability $P_2(i;j)$ is given by

$$P_2(i;j) = \sum_{s=0}^{s^*} \binom{M-j}{s} \alpha^s (1-\alpha)^{M-j-s} P_1(s;i,j),$$

where $s^* = \min\{M-j, \lfloor (d_H-1)/L \rfloor - i\}$ and $P_1(s;i,j)$ is the conditional probability that the code word is correct given j hits, i detected hits, and s false alarms. The limits of the summation are chosen to ensure that the error-correcting capability of the code is not exceeded. Finally, the probability $P_1(s;i,j)$ is given by

$$\begin{aligned} P_1(s;i,j) &= \sum_{\lambda=0}^{\lambda^*} \binom{L(j-i)}{\lambda} \delta^\lambda (1-\delta)^{L(j-i)-\lambda} \\ &\times \sum_{t=0}^{t^*} \binom{L(M-j-s)}{t} p^t (1-p)^{L(M-j-s)}. \end{aligned}$$

The parameter λ^* is the number of symbol errors that occurred in dwell intervals with interference, and t^* is the number of symbol errors in dwell intervals with no interference. If $\nu = \lfloor (d_H - 1 - L(i+s))/2 \rfloor$, then λ^* and t^* are given by

$$\lambda^* = \min\{L(j-i), \nu\}$$

and

$$t^* = \min\{L(M-j-s), \nu - \lambda\}.$$

Performance comparisons for Hermitian and Reed-Solomon codes are given in Figure 7 for a partial-band noise channel with $\rho = 0.20$. Errors-and-erasures is employed and the parameters for the test symbols are $N = 12$ and $\gamma = 2$. As the strength of the interference is reduced (i.e., as

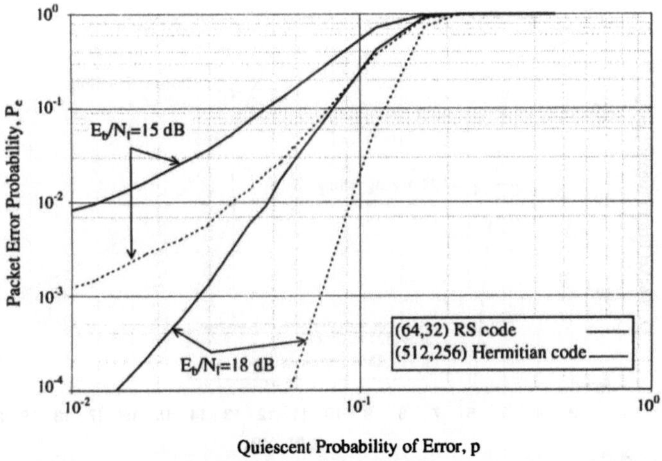

Figure 7: Packet error probabilities for a partial-band noise channel and errors-and-erasures decoding

E_b/N_I is increased), the performance improves for each type of coding. For $E_b/N_I = 18$ dB there is a larger separation between the two curves than for $E_b/N_I = 15$ dB.

For M-ary orthogonal signaling with standard noncoherent demodulation, the quiescent probability of error is given by

$$p_M = M^{-1} \sum_{j=2}^{M} \binom{M}{j} (-1)^j e^{-(1-j^{-1}) m E_{bs}/N_0}.$$

If N_0 is replaced by η_I, the same expression provides the probability of error for symbols that are hit by partial-band noise. A comparison of the performance of a system that uses binary orthogonal symbols to form M-ary code symbols and a system that uses M-ary orthogonal symbols as code symbols is shown in Figure 8 for $M = 64$ and a partial-band noise channel with $E_b/N_I = 15$ dB and $\rho = 0.25$. Each system employs errors-and-erasures decoding with side information derived from test symbols. For the system with 64-ary orthogonal modulation, there are two 64-ary test symbols per dwell interval. If both test symbols in a

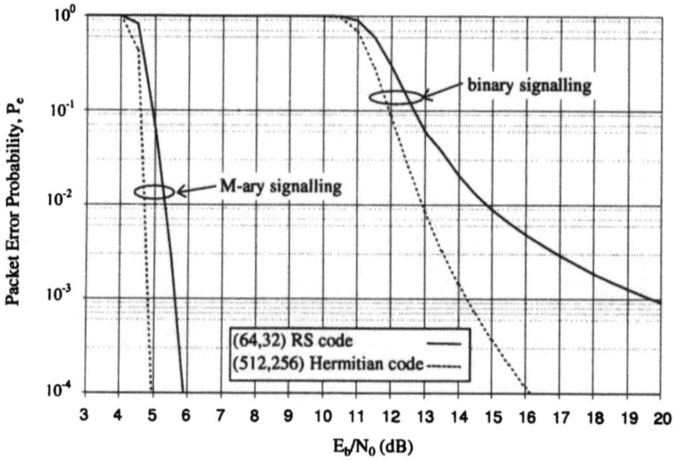

Figure 8: Packet error probabilities for errors-and-erasures decoding and a partial-band noise channel with $E_b/N_I = 15$ dB, $\rho = 0.25$, and $M = 64$

dwell interval are received in error, all symbols in that dwell interval are erased. The system with binary orthogonal modulation has twelve binary test symbols per dwell interval, so the overhead for the test symbols is the same for the 64-ary modulation as for the binary modulation. The erasure threshold is $\gamma = 2$ for the system with binary modulation.

Acknowledgements

The research reported here was supported by the Army Research Office under grant number DAAD19-99-1-0289 and by the DoD Multidisciplinary University Research Initiative (MURI) administered by the Office of Naval Research under grant number N00014-00-1-0565.

References and Selected Bibliography

[1] C. A. Baum and M. B. Pursley, "Bayesian methods for erasure insertion in frequency-hop communications systems with partial-band interference," *IEEE Transactions on Communications*, vol. 40, pp. 1231-1238, July 1992.

[2] L.-F. Chang and R. J. McEliece, "A study of Viterbi's ratio-threshold AJ technique," *1984 IEEE Military Communications Conference Record* (Los Angeles), vol. 1, pp. 182-186, October 1984.

[3] J. P. Coon, T. G. Macdonald, and M. B. Pursley, "A new method for obtaining side information in frequency-hop spread-spectrum systems," *Proceedings of the 2000 IEEE Military Communications Conference* (Los Angeles), pp. 5.2.1-5, October 2000.

[4] A. B. Cooper III, "Soft decision decoding of Reed-Solomon codes," Chapter 6 in *Reed-Solomon Codes and Their Applications*, S. B. Wicker and V. K. Bhargava (eds.), pp. 108-124, IEEE Press, New York, 1994.

[5] G. R. Cooper and R. W. Nettleton, "A spread-spectrum technique for high-capacity mobile communications," *IEEE Transactions on Vehicular Technology*, vol. VT-27, pp. 264-275, November 1978.

[6] G. D. Forney, Jr., "Generalized minimum distance decoding," *IEEE Transactions on Information Theory*, vol. IT-12, pp. 125-131, April 1966.

[7] J. H. Gass, Jr., M. B. Pursley, H. B. Russell, and J. S. Wysocarski, "An adaptive-transmission protocol for frequency-hop wireless communication networks," *Wireless Networks*, vol. 7, pp 487-495, September 2001.

[8] E. A. Geraniotis and M. B. Pursley, "Error probability bounds for slow-frequency-hopped spread-spectrum multiple-access communications over fading channels," *Conference Record, IEEE International Conference on Communications* (Denver), vol. 4, p. 76.3.1-7, June 1981.

[9] E. A. Geraniotis and M. B. Pursley, "Error-control coding for slow-frequency-hopped spread spectrum multiple-access communications via selective fading channels," *Abstracts of Papers for the 1982 IEEE International Symposium on Information Theory* (Les Arcs, France), p. 45, June 1982.

[10] E. A. Geraniotis and M. B. Pursley, "Error probabilities for slow-frequency-hopped spread-spectrum multiple-access communications over fading channels," *IEEE Transactions on Communications*, vol. COM-30, pp. 996-1009, May 1982.

[11] B. J. Hamilton, "SINCGARS system improvement program (SIP) specific radio improvements," *Proceedings of the 1996 Tactical Communications Conference* (Fort Wayne, IN), pp. 397-406, April 1996.

[12] S. W. Houston, "Modulation techniques for communication, Part I: Tone and noise jamming performance of spread-spectrum M-ary FSK and 2,4-ary DPSK waveforms," *Proceedings of the IEEE National Aerospace and Electronics Conference* (Dayton, OH), pp. 51-58, June 1975.

[13] D. E. Kammer, "SINCGARS - The new generation combat net radio system," *Proceedings of the Tactical Communications Conference* (Fort Wayne, IN), pp. 64-72, April 1986.

[14] A. W. Lam and D. V. Sarwate, "A comparison of two methods for generation of side information in frequency-hop spread-spectrum multiaccess communications," *Proceedings of the 21st Annual Conference of Information Sciences and Systems* (Johns Hopkins University), pp. 426-431, March 1987.

[15] T. G. Macdonald and M. B. Pursley, "Hermitian codes for frequency-hop spread spectrum packet radio networks," to appear in the *IEEE Transactions on Wireless Communications.*

[16] T. G. Macdonald and M. B. Pursley, "Staggered interleaving and iterative errors-and-erasures decoding for frequency-hop packet radio," submitted to the *IEEE Transactions on Wireless Communications.*

[17] R. J. McEliece and M. B. Pursley, "Error-control coding for slow-frequency-hopped spread spectrum," *Abstracts of Papers for the 1982 IEEE International Symposium on Information Theory* (Les Arcs, France), p. 42, June 1982.

[18] R. J. McEliece and E. R. Rodemich, "A study of optimal abstract jamming strategies vs. noncoherent MFSK," *Proceedings of the 1983 IEEE Military Communications Conference* (Washington, DC), vol. 1, pp. 1-6, October 1983.

[19] R. J. McEliece and W. E. Stark, "An information theoretic study of communication in the presence of jamming," *Conference Record, 1981 IEEE International Conference on Communications* (Denver), vol. 3, pp. 45.3.1-5, June 1981.

[20] R. J. McEliece and W. E. Stark, "The optimal code rate vs. a partial band jammer," *Conference Record, 1982 IEEE Military Communications Conference* (Boston), vol. 1, pp. 8.5.1-4, October 1982.

[21] R. J. McEliece and W. E. Stark, "Channels with block interference," *IEEE Transactions on Information Theory*, vol. IT-30, pp. 44-53, January 1984.

[22] R. J. McEliece and L. Swanson, "On the decoder error probability for Reed-Solomon codes," *IEEE Transactions on Information Theory*, vol. IT-32, pp. 701-703, September 1986.

[23] M. B. Pursley, "Coding and diversity for channels with fading and pulsed interference," *Proceedings of the 1982 Conference on Information Sciences and Systems* (Princeton Univ.), pp. 413-418, March 1982.

[24] M. B. Pursley, "Throughput of frequency-hopped spread-spectrum communications for packet radio networks," *Proceedings of the 1983 Conference on Information Sciences and Systems* (Johns Hopkins Univ.), pp. 550-556, March 1983.

[25] M. B. Pursley, "Packet error probabilities in frequency-hop radio networks – Coping with statistical dependence and noisy side information," *Proceedings of the 1986 IEEE Global Telecommunications Conference* (Houston), vol. 1, pp. 165-170, December 1986.

[26] M. B. Pursley, "Frequency-hop transmission for satellite packet switching and terrestrial packet radio networks," *IEEE Transactions on Information Theory*, vol. IT-32, pp. 652-667, September 1986.

[27] M. B. Pursley, "Tradeoffs between side information and code rate in slow-frequency-hop packet radio networks," *Conference Record, 1987 IEEE International Conference on Communications* (Seattle), vol. 2, pp. 947-952, June 1987.

[28] M. B. Pursley, "Reed-Solomon codes in frequency-hop communications," Chapter 8 in *Reed-Solomon Codes and Their Applications*, S. B. Wicker and V. K. Bhargava (eds.), pp. 150-174, IEEE Press, New York, 1994.

[29] M. B. Pursley and H. B. Russell, "Network protocols for frequency-hop packet radios with decoder side information," *IEEE Journal on Selected Areas in Communications*, vol. 12, pp. 612-621, May 1994.

[30] M. B. Pursley and W. E. Stark, "Antijam capability of frequency-hop spread-spectrum with Reed-Solomon coding," *1983 IEEE Military Communications Conference Record* (Washington, DC), vol. 1, pp. 7-11, October 1983.

[31] M. B. Pursley and W. E. Stark, "Performance of Reed-Solomon coded frequency-hop spread spectrum communications in partial-band interference," *IEEE Transactions on Communications*, vol. COM-33, pp. 767-774, August 1985

[32] S. Sakata, J. Justesen, Y. Madelung, H. E. Jensen and T. Hoholdt, "Fast decoding of algebraic-geometric codes up to the designed minimum distance," *IEEE Transactions on Information Theory*, vol. 41, no. 5, pp. 16772-1677, September 1995.

[33] A. Shohara, J. J. Garcia-Luna-Aceves, M. G. Lewis, B. M. Sifford, J. R. McChesney, L. G. Matheson, and M. B. Pursley, "Design Plan: SINCGARS Packet Switch Overlay," SRI International Report, SRI Project No. 1244, Contract No. DAAB07-85-C-K581, May 1986.

[34] W. E. Stark, "Performance of concatenated codes on channels with jamming," *Conference Record, 1982 IEEE International Conference on Communications* (Philadelphia), pp. 7E.4.1-5, June 1982.

[35] W. E. Stark, "Coding for frequency-hopped spread-spectrum communication with partial-band interference – Part I: Capacity and cutoff rate," *IEEE Transactions on Communications*, vol. COM-33, pp. 1036-1044, October 1985.

[36] W. E. Stark, "Coding for frequency-hopped spread-spectrum communication with partial-band interference – Part II: Coded performance," *IEEE Transactions on Communications*, vol. COM-33, pp. 1045-1057, October 1985.

[37] W. E. Stark and R. J. McEliece, "Capacity and coding in the presence of fading and jamming," *Proceedings of the National Telecommunications Conference* (New Orleans), pp. B7.4.1-5, November 1981.

[38] H. Stichtenoth, "A note on Hermitian codes over GF(q^2)," *IEEE Transactions on Information Theory*, vol. 34, no. 5, pp. 1345-1348, September 1988.

[39] D. J. Taipale and M. B. Pursley, "An improvement to generalized minimum distance decoding," *IEEE Transactions on Information Theory*, vol. 37, pp. 167-172, January 1991.

[40] J. H. van Lint and T. A. Springer, "Generalized Reed-Solomon codes from algebraic geometry," *IEEE Transactions on Information Theory*, vol. IT-33, no. 3, pp. 305-309, May 1987.

[41] A. J. Viterbi, "A robust ratio-threshold technique to mitigate tone and partial band jamming in coded MFSK systems," *Proceedings of the 1982 IEEE Military Communications Conference* (Boston), pp. 22.4.1-5, October 1982.

[42] A. J. Viterbi and I. M. Jacobs, "Advances in coding and modulation for noncoherent channels affected by fading, partial band, and multiple-access interference," in *Advances in Communication Systems: Theory and Applications*, vol. 4, A. J. Viterbi (ed.), pp. 279-308, Academic Press, New York, 1975.

[43] B. E. Whalen and J. Jimenez, "Performance comparison of Hermitian and Reed-Solomon codes," *Proceedings of the 1997 IEEE Military Communications Conference* (Monterey, CA), vol. 1, pp. 15-19, November 1997.

[44] K. Yang and P. V. Kumar, "On the true minimum distance of Hermitian codes", in *Lecture Notes in Mathematics*, vol. 1518, Springer-Verlag, pp. 99-107, *Proceedings of the International Workshop on Coding Theory and Algebraic Geometry* (Luminy, France), June 1991.

On Graph Constructions for LDPC Codes by Quasi-Cyclic Extension

R. Michael Tanner
Department of Computer Science
University of California, Santa Cruz
Santa Cruz, CA 95064
tanner@soe.ucsc.edu

Abstract

Quasi-cyclic (QC) extension is a recursive graph construction that preserves key design parameters of a "seed" graph, such as the distribution of the degrees of the nodes, while increasing graph size and girth. Algebraic QC extension graphs are described in which automorphisms of the seed graph are purposefully preserved in the extension. The girth of the extension graph will be as least as great as for the seed while the "design rate" of the code is the same. A two-stage QC extension is given for a (3,4) low-density parity-check (LDPC) code to illustrate.

Keywords: LDPC codes, graph construction, automorphism, quasi-cyclic, girth.

1 Introduction

The possibility of approaching Shannon's capacity limit on the additive white noise gaussian (AWGN) channel using iterative message-passing decoders on sparse graphs is now well established. The success both of turbo codes and low-density parity-check (LDPC) codes underline that there is considerable flexibility in the design of coding systems and message-passing decoding algorithms. In both types of codes, the commonly accepted design principles incorporate randomness in the interconnection between code constraints, either in the form of random interleavers for turbo codes, or in random sparse matrices for low-density parity check codes. The random connections form code constraint graphs that allow decoding information to be passed efficaciously by iterative algorithms, and the random graph setting has been fertile for theoretical results on the performance of coding systems (e.g., [4], [5], [6]).

Random constructions are far from ideal from an implementation point of view. Any given graph formed by randomly selecting connections may have short cycles that are deleterious to the performance of a message-passing decoder, and code system designers often screen the random graphs to check for short cycles and make adjustments (e.g., [2]). The description of the random graph itself must be stored. Its inherent entropy defies data compression and so the memory devoted to the graph description can be burdensome. Furthermore, we opine that the data movements of a message-passing decoder dictated by a random graph will become a bottleneck to decoder performance when very high data throughput is required and parallelism must be exploited in a custom integrated circuit design.

Algebraically structured graphs need not suffer from these same limitations, and as yet no theory rules out the possibility of finding algebraically-prescribed graphs that are superior to random graphs within any given class. In [8] and [9], an algebraic construction is given for regular (j, k) LDPC codes for which the constraint graphs have girths as large as twelve. For codes of lengths up to several thousands, these algebraic graph codes empirically give performance as good or better than regular random graphs of the same parameters. The algebraic codes consist of multiple cycles of some prime length. The quasi-cyclic structure will reap many of the implementation benefits associated with cyclic codes.

Graph Constructions for LDPC Codes

In this brief paper we explore an algebraically-guided recursive construction first given in [7], which we call *quasi-cyclic (QC) extension.*[1] Starting from a seed graph, each vertex in the seed graph is replaced by a set of n replicated vertices for some fixed n, and each edge in the graph is replaced by a bundle of n edges, yielding a graph that has n times as many vertices. In the simplest, trivial form one could imagine merely replicating the seed graph n times, laying the copies one on top of another. The innovation occurs when each bundle of n parallel edges is disconnected at one end and permuted before being reconnected to the n vertices at that end. Amit and Linial [1] study the expansion properties of the graph in the case where the permutation for each edge is a random permutation of n objects. They call this construction a *lifting* of the seed graph. In a quasi-cyclic extension, each permutation is constrained be a cyclic permutation of n objects.

The question is how to choose a permutation for every edge of the seed graph so as to create an extension graph that has desirable properties for an LDPC code. One heuristic measure of quality of the graph for message-passing algorithms is the girth of the graph for a fixed vertex degree distribution. A large girth assures that the first several iterations of a sum-product decoder will carry statistically independent information. The algebraic prescription of [9] is a QC extension that produces regular graphs with excellent girths for moderate lengths. These graphs have a powerful automorphism group that is transitive on edges of the graph. The girth of the QC extension must be at least as large as that of the seed graph. By linking a group of automorphisms of the seed graph with a homomorphic permutation group acting on n vertices, an extension graph can be given the same automorphisms. This is illustrated by a two-step recursive construction of two design rate $R = 1/4$ regular (3,4) LDPC codes. Starting from a trivial (3,4) complete bipartite graph as a seed, the first step is a 13-cyclic extension graph with girth eight that is a code with parameters $[52, 15, 14]$. Then this graph is used as a seed for a 53-cyclic extension yielding a graph with girth ten and a code with parameters $[2756, 691, D]$.

[1] The concept of a quasi-cyclic extension was crystallized while the author was on sabbatical at the California Institute of Technology.

2 Graphs and Quasi-Cyclic Extension

Following [7], we recall the definition of a QC extension and establish terminology. Let G be an undirected bipartite graph with a set of vertices $V = V_p \cup V_b$ consisting of parity vertices and bit vertices with a set of directed edges $E = \{e | e = (v_i, v_j), v_i \in V_p, v_j \in V_b\}$ and incidence matrix H with rows indexed by vertices from V_p and columns indexed by vertices from V_b and a "1" for every edge, "0"s elsewhere. Let g and d be the girth and diameter of G, respectively. For convenience this initial graph will be called the *seed* graph. To every vertex $v_j \in V_b$ we associate a binary variable $b_j \in \{0, 1\}$ which is the jth bit in the binary code with parity check matrix H. The vectors $(b_0, b_1, \ldots, b_{N-1})$ for which the binary sum of all the bits adjacent to each parity $v_i \in V_p$ is zero form the code C defined by G.

Definition 2.1 An *n-cyclic extension* G' of a seed G consists of a set of vertices $V' = V_p' \cup V_b'$ and edges E', with $V_p' = \{[v, k] | v \in V_p, 0 \leq k \leq n-1\}$, $V_b' = \{[v, k] | v \in V_p, 0 \leq k \leq n-1\}$, the set of ordered 2-tuples consisting of a vertex of G followed by an integer *mod n*, and edges $E' = \{([v_i, k], [v_j, (k - f(e)) mod\ n]) | (v_i, v_j) = e \in E, 0 \leq k \leq n-1\}$ for some function $f(\cdot) : E \to Z_n$, called the *offset function*, defined on the set of edges of G and mapping each edge of G to the integers *mod n*. The bits of the *n-cyclic extension* code C' are $b_{[v,k]}, v \in V_b, 0 \leq k \leq n-1$.

In brief, each vertex of the seed G is expanded and replaced by n copies and each edge of G by a bundle of n edges. The offset function $f(\cdot)$ determines a cyclic shift or offset of the bundle that occurs as it passes from an expanded bit vertex to an expanded parity vertex of G. If G has L parity vertices and N bit vertices, G' has Ln and Nn vertices, respectively. The incidence matrix H' of G' is obtained by replacing each "1" in H corresponding to an edge e by a matrix I_s, where $s = f(e)$ and I_x is an identity matrix in which each row has been cyclically left shifted by x positions. Specifying G, n, and the offset function $f(\cdot)$ determines the extension graph G'.

Any vertex v in G is replicated n times in G' with vertices that have the same degree as v. Obviously the vertex degree distribution (as the term is used in [6]) in the extension graph is the same as in the seed. The girth of the extension satisfies $g' \geq g$. For any closed cycle in G'

consisting of vertices $[(v_i)_t, k_t]$, $0 \leq t \leq l$ without a repeated edge, the series of vertices $(v_i)_t$, $0 \leq t \leq l$ form a closed cycle in G without a repeated edge.

To make $g' > g$, the offset function must be chosen so that the sum of the offsets *mod n*, adding the offset on an edge from a bit to a parity vertex and subtracting the offset on an edge from a parity to a bit vertex, will never be zero around any cycle of G of length g. If the offset values were chosen at random with a uniform distribution on the integers *mod n*, the probability of the sum being zero around any one given shortest cycle would be $1/n$. Unless n is very large, for many graphs there will exist at least one shortest cycle for which the sum is zero, and the girth is not increased. A well-chosen algebraically-specified offset function can systematically preclude such configurations.

3 Code Properties

Let L be the number of parity nodes and let k_{avg} be their average degree. Similarly let N be the number of bit nodes and j_{avg} their average degree. Counting edges both ways, $Lk_{avg} = Nj_{avg}$. If all the equations are linearly independent, the code rate is $R = 1 - L/N = 1 - j_{avg}/k_{avg}$, or $1 - j/k$ for a regular (j, k) LDPC code. This will be called the *design rate* of the code. When the code graph is generated at random, it is highly likely that the equations will be linearly independent, but this is not necessarily the case for algebraically structured codes. (For example, a length 755 code given in [9] has design rate 0.4 and actual rate 0.44.)

The design rate for the code of a QC extension graph is the same as for the seed graph, since the number of vertices of both types are increased by a factor of n. Without detailed knowledge of the offset function, little can be said about the minimum distance or weight distribution of the QC extension code. There is no increase in minimum distance when the trivial offset function $f(\cdot) = 0$ is used, in which case the QC extension is simply n separate disconnected copies of the seed graph. The ostensible extension in this instance is tantamount to the common practice of using a fixed block code C n times to send Nn bits. The effect of the offset function is to interweave the multiple copies so that bits in one block of N are then related by equations to those in other blocks. If the offset function achieves its purpose, the minimum distance will be greatly increased. One interpretation is that this is a special form of "interleaving" analogous in effect to the interleavers used for turbo codes.

4 Imposing a graph automorphism

Finding a deterministic offset function that will create a QC extension graph and code with desirable properties is *a priori* a complex combinatorial problem. The search can be focused if there is an algebraic theory guiding the specification of the offset function. For simplifying both the search and the implementation of the code, we have concentrated on graphs that have an large group of symmetries, or automorphisms. (An automorphism of a graph is a permutation of vertices that preserves edge connections; the automorphism group of a graph is the group of such permutations.)

Our goal here is to generalize the strategy of [7] and to chose an offset function that will purposefully create an automorphism for the QC extension graph. First, we note that the very definition of the QC extension imposes a cyclic shift automorphism isomorphic to the group Z_n. Since the graphs for codes are bipartite, we will restrict attention to automorphisms that respect the partition, mapping bits to bits and parities to parities.

Theorem 4.1 The automorphism group of a QC extension as in Definition 2.1 contains a subgroup isomorphic to Z_n, the group of addition *mod n*.

Proof: The permutation σ that maps an edge $e' = ([v_i, k], [v_j, (k - f(e)) mod\ n])$ of G' to the edge $\sigma(e') = ([v_i, (k+1)\ mod\ n], [v_j, (k - f(e) + 1)\ mod\ n])$ clearly has order n, $\sigma^n = 1$, and is isomorphic to Z_n.

The next question is how to impose other automorphisms on G' by an artful choice of the offset function. This can be done by linking an automorphism of the seed graph to a homomorphic permutation group acting on the integers *mod n*. [2]

[2] The QC extension can be viewed as a form product graph, the product of the seed graph with Z_n. This can be generalized by considering a product of the seed graph with any group and allowing richer homomorphisms.

Theorem 4.2 Let Π be a group of automorphisms of G and let Γ be a homomorphic group of permutations of the integers *mod* n, with homomorphism $\phi : \Pi \to \Gamma$. If for all $\pi \in \Pi$, $e \in E$, and $0 \leq k \leq (n-1)$ the offset function satisfies $\phi(\pi)((k-f(e))\, mod\, n = (\phi(\pi)(k)-f(\pi(e))\, mod\, n$, then the permutation $\pi' \in \Pi'$ defined by

$$\pi'(e') = \pi'(([v_i, k], [v_j, (k - f(e)) mod\, n])) = \quad (1)$$

$$([\pi(v_i), \phi(\pi)(k)], [\pi(v_j), \phi(\pi)((k - f(e)) mod\, n)]) = \quad (2)$$

$$([\pi(v_i), \phi(\pi)(k)], [\pi(v_j), (\phi(\pi)(k) - f(\pi(e)) mod\, n)]) \quad (3)$$

is an automorphism of G'.

Proof: The definition is sufficiently stringent that π' is guaranteed to be an automorphism if the offset function is well-defined. We need to show that the condition allows a consistent definition for the offset function. For any $\pi_1, \pi_2 \in \Pi'$ with homomorphic images $\phi(\pi_1) = \gamma_1, \phi(\pi_2) = \gamma_2$, the homomorphism ensures that $\pi_3 = \pi_1\pi_2$ has $\phi(\pi_3) = \gamma_3 = \gamma_1\gamma_2$. Therefore

$$\pi_1'\pi_2'([v_i, k], [v_j, (k - f(e)) mod\, n]) = \quad (4)$$

$$([\pi_1\pi_2(v_i), \gamma_1\gamma_2(k)], [\pi_1\pi_2(v_j), (\gamma_1\gamma_2(k - f(e))) mod\, n]) = \quad (5)$$

$$([\pi_3(v_i), \gamma_3(k)], [\pi_3(v_j), \gamma_1(\gamma_2(k) - f(\pi_2(e))) mod\, n]) = \quad (6)$$

$$([\pi_3(v_i), \gamma_3(k)], [\pi_3(v_j), (\gamma_3(k) - f(\pi_3(e))) mod\, n]) \quad (7)$$

as required. Therefore the offset function has a consistent definition over all edges in the orbit of any e under the action of Π'.

As an example, suppose the automorphism group of graph G is Π. Let the homomorphism be the trivial homomorphism $\phi(\pi) = 1$, the identity, for all $\pi \in \Pi$. Then $\phi(\pi)((k - f(e))\ mod\ n) = (k - f(e))\ mod\ n = (\phi(\pi)(k) - f(\pi(e)))\ mod\ n = (k - f(\pi(e)))\ mod\ n$, requires that $f(e) = f(\pi(e))\ mod\ n$. Effectively, $f(e)$ must be constant in any orbit, e.g., $f(e) = 0$. This is the case of a QC extension that is simply n copies of G.

As a second, more useful, example, suppose the automorphism group of graph G contains a cyclic subgroup generated by π of order t, $\pi^t = 1$. Suppose that a is relatively prime to n, and a has order t under multiplication *mod* n, $a^t = 1\ mod\ n$. Let the homomorphism be defined by $\phi(\pi) = a$. Then

$$\phi(\pi)((k - f(e)) \; mod \; n) = a(k - f(e)) \; mod \; n = ((ak - af(e)) mod \; n, \quad (8)$$

$$(\phi(\pi)(k) - f(\pi(e))) \; mod \; n = (ak - f(\pi(e))) \; mod \; n. \quad (9)$$

The condition of Theorem 4.2 will be met by letting $f(\pi(e)) = af(e) \; mod \, n$ for all edges e, and the automorphism group of G' includes the subgroup Π' generated by $\pi'(e') = ([\pi(v_i), ak], [\pi(v_j), ((ak - af(e)) mod \; n)])$. Setting a value for one $f(e_0)$ will determine the offset values for all edges in the orbit of e_0 under the action of π. This automorphism will be used for the examples in the next section.

5 A two-step extension

Forming a QC extension with an offset function chosen to impose an automorphism will be illustrated with a two-step construction. An initial 3-4 complete bipartite graph is given a 13-cyclic extension, and the resulting graph is then used as the seed graph in a 53-cyclic extension.

Ignoring for the moment the edge labels, Fig.1 shows a 3-4 complete bipartite graph in which the four vertices on the right are bit vertices and the three on the left are parity checks. The code defined by this seed graph is just a [4,3,2] parity check code; two of the equations are redundant. The graph has girth four and diameter two, and its automorphism group can be considered to be $S_3 \times S_4$, the symmetric group on 3 objects acting on the parity check nodes and the symmetric group on 4 objects acting on the bit nodes.

A 13-cyclic extension can be formed by replacing every node by a cycle of 13 nodes of the same type, parity or bit. The entire automorphism group $S_3 \times S_4$ could be imposed using the trivial $f(\cdot) = 0$. Instead, we will impose an automorphism group that retains the subgroup $\Pi_3 = Z_3 \times I$, cyclically shifting the three parities while holding the four bits fixed, and the subgroup $\Pi_4 = I \times Z_4$, which analogously cyclically shifts the four bits. The ordering of the nodes for the cyclic group is shown by the subscripts of the nodes in Fig. 1. To capture these in the QC extension, observe that 2 is a primitive element of $GF(13)$ and the multiplicative group of the non-zero integers mod 13 is generated by multiplication by 2. Therefore $a_3 = 2^4 = 3$ has order 3 and $a_4 = 2^3 = 8$

has order 4 in $GF(13)$, and these can be used as the homomorphisms for Π_3 and Π_4, respectively (isomorphisms, in this instance). Let an edge in the QC extension be indexed by $([i,k],[j,k-f((i,j))])$, $0 \leq i \leq 2$, $0 \leq j \leq 3$, $0 \leq k \leq 12$. The offset function on the edges of seed graph G will be defined by setting $f((0,0)) = 1$ and then using the homomorphisms and the conditions of Theorem 4.2 to dictate the remaining values: $f(((i+1) mod\ 3, j)) = a_3 f((i,j))\ mod\ 13 = 3f((i,j))\ mod\ 13$, i.e., $f(\pi_3(e)) = \phi(\pi_3)(f(e))\ mod\ 13$, and similarly $f((i,(j+1) mod\ 4)) = 8f((i,j))\ mod\ 13$. These two are sufficient to define $f(\cdot)$ on all edges of the 3-4 bipartite seed graph. The values are shown as the labels on the edges in Fig. 1.

The resulting code is one instance of the family of codes given in [9]. Using MAGMA [3], this QC extension is found to be a [52,15,14] (3,4) regular LDPC code. Two of the equations are again redundant, as in the seed graph. With a minimum distance of 14, this is a remarkably good code, suffering little penalty for sparseness; the best code known has parameters [52,15,17]. The QC extension has increased the graph's girth to 8 and diameter to 6.

This graph will now be used as a seed for another extension. The 13-cyclic extension has the cyclic shift automorphism generated by $\sigma, \sigma^{13} = 1$. The goal is to impose this automorphism on a further extension. In $GF(53)$, 2 is again a primitive element, and multiplication by $a'_{13} = 2^4 = 16$ modulo 53 has order 13. In defining an offset function for a 53-cyclic extension, we will let this cyclic group be the homorphism for the order 13 group generated by σ. Multiplication by $a'_4 = 2^{13} = 30$ gives a group of order 4 which can be used as the homorphic image of Π_4.

With $e = ([i,k],[j,k-f((i,j))])$ and $f((i,j))$ as defined above, an edge in the 53-cyclic extension is $([[i,k],k'],[[j,k-f((i,j))],k'-f'(e)]$, for $0 \leq i \leq 2, 0 \leq j \leq 3, 0 \leq k \leq 12, 0 \leq k' \leq 52$, and the new offset function $f'(\cdot)$ to be specified. To simplify the notation, the nested brackets can be suppressed by indexing the vertices by 3-tuples, so that an edge is written $([i,k,k'],[j,k-f((i,j)),k'-f'(e)])$. The automorphism of Π_4 will be preserved if

$$f'(([i,k],[j+1,k-f((i,j+1))]) = \\ 30f'(([i,8k\ mod 13],[j,8(k-f((i,j)))\ mod\ 13])\ mod\ 53. \quad (10)$$

The automorphism of σ will be preserved if

$$f'(([i,k+1],[j,(k-f((i,j))+1)\ mod\ 13]) = \\ 16f'(([i,k],[j,k-f((i,j))])\ mod\ 53. \quad (11)$$

These two describe the behavior of the offset function $f'(\cdot)$ on an orbit of 4(13) edges emanating from the 13 vertices expanded from a single parity vertex in Fig. 1 No ready homomorphism for Π_3 is apparent, so the definition of the offset function for three different parity vertices of Fig. 1 is not dictated by an automorphism. The offset values are instead three separate orbits with the offsets in each dictated by the Π_4 and σ automorphisms. One offset function for a sampling of edges is illustrated in Fig. 2.

This second expansion yielded a graph with a diameter of 12 and girth of 10, and code parameters [2756, 691, D]. While the girth has increased, the small growth in girth and large growth in diameter make the QC extension graph unimpressive in those terms. The code is too long for MAGMA to give meaningful bounds on the minimum distance.

6 Conclusion

The pursuit of this paper is to describe in greater detail a recursive construction for longer graphs and codes with the hope of producing high quality graphs and LDPC codes for use with iterative message-passing decoders. The main contribution is to show how algebraically-defined symmetries can be preserved, simplifying the structure and the graph description. We also show that the girth of extension graph/code will never be worse than for the seed graph/code. Our quasi-cyclic extension technique has produced a number of excellent sparse regular codes at short to moderate lengths. Better understanding of algebraic prescriptions for this versatile code construction method is needed.

References

[1] A. Amit and N. Linial, "Random Lifts of Graphs II: Edge Expansion", *Combinatorics, Probability, and Computing*, to appear.

[2] D. J. C. MacKay, "Good error correcing codes based on very sparse matrices", *IEEE Trans. on Information Theory*, vol. IT-45, No. 2, Mar. 1999, pp. 399-431.

[3] *The Magma Computational Algebra System*, University of Sydney, Australia (www.maths.usyd.edu.au:8000/u/magma/).

[4] R. J. McEliece, D. Divsalar, and H. Jin, "Coding Theorems for "Turbo-Like" Codes", *Proc. 37th Allerton Conference on Communication, Control, and Computing*, Sept. 1998 1998, pp. 210-210.

[5] T. J. Richardson and R. L. Urbanke, "The capacity of low-density parity-check codes under message- passing decoding", *IEEE Trans. on Information Theory*, vol. IT-47, No. 2, Febr. 2001, pp. 599-618.

[6] T. J. Richardson, M. A. Shokrollahi, and R. L. Urbanke, "Design of capacity-approaching irregular low-density parity-check codes", *IEEE Trans. on Information Theory*, vol. IT-47, No. 2, Febr. 2001, pp. 619-637.

[7] R. M. Tanner, "On Quasi-Cyclic Repeat-Accumulate Codes," *Proc. 37th Allerton Conference on Communication, Control, and Computing,* University of Illinois, Sept. 1999, pp. 249-259.

[8] R. M. Tanner, "A [155,64,20] Sparse Graph (LDPC) Code," *IEEE International Symposium on Information Theory* (Recent Results Session), Sorrento, June 2000.

[9] R. M. Tanner, T. E. Fuja, and D. Sridhara, "A Class of Group-Structured LDPC Codes," *Proceedings 6th ISCTA,* Ambleside, England, July 2001, pp.365-370.

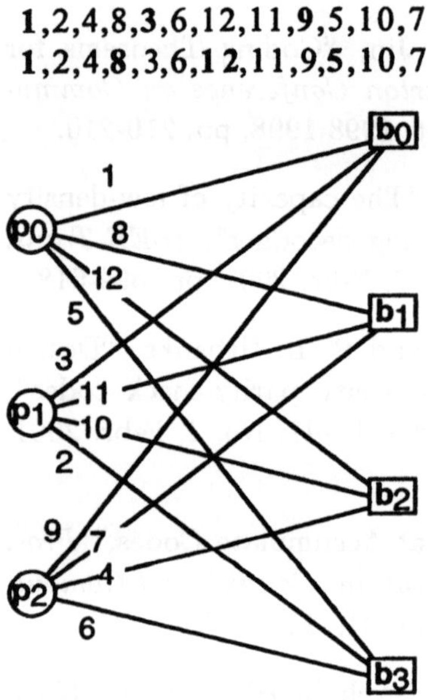

Fig. 1 Initial 3-4 Bipartite Graph with p=13 extension

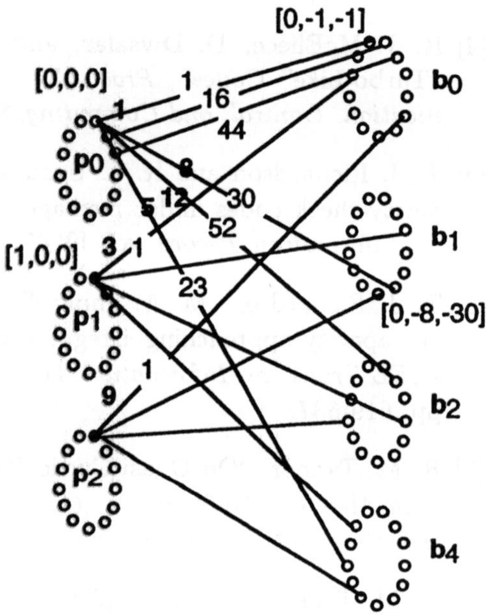

Fig. 2 Seed graph of Fig. 1 with q=53 extension

On the Channel Memory-Diversity Tradeoff in Communication Systems

Andrew P. Worthen[1] and Wayne E. Stark[2]
University of Michigan
1301 Beal Ave.
Ann Arbor, MI 48109
worthena@ieee.org, stark@eecs.umich.edu

Abstract

In this chapter we investigate the tradeoff between channel estimation and channel diversity for channels with memory. Block memory channels are analytically tractable yet provide practical models for many communication systems. We show that for finite block length transmission techniques there is a fundamental tradeoff between the channel estimator and the channel diversity. When the channel is operating at rates much below capacity, smaller channel memory is better since channel estimation is not crucial. This is because the code contains more than adequate redundancy to compensate for mismatch due to the potentially poor channel estimate as well as channel errors. However, at rates close to capacity, channel state information is crucial for successful decoding and therefore longer channel memory, which allows better estimates, is better. We draw these conclusions based on analysis of some simple discrete-input, discrete-output channels using the channel reliability function and on simulation results for more realistic channels (soft output) using low-density parity check codes with a joint iterative decoder and channel estimator.

Keywords: block-memory channels, joint channel estimation and decoding, error exponent bounds, iterative receivers, low-density parity check (LDPC) codes.

---

[1]A. Worthen is now with MIT Lincoln Laboratory, Lexington, MA.

[2]This work was supported by the the Army Research Office under grant DAAH04-96-1-0377, a National Science Foundation Graduate Research Fellowship and an AFCEA fellowship.

1 Introduction

Error control coding has become a common element in modern communications systems, which often must operate on time-varying channels that exhibit memory. Conventional wisdom has been to use coding with interleaving in order to deal with the channel memory. As communication systems have started to employ coding techniques such as turbo codes and low-density parity check codes that achieve performance close to the capacity of the channel, accurate channel estimation has become important. Optimal receivers perform joint channel estimation and decoding and their performance depends on a tradeoff between the channel estimation quality and the diversity available to the code.

Channel estimation generally works best when the memory is large because the estimator can do a significant amount of averaging over the channel noise. The mutual information between input and output for blocks of symbols is increased since the channel state is correlated across the block. Thus, from a capacity point of view, larger memory is better than smaller memory due to the improved estimation of the channel state.

However, channel capacity makes the assumption of infinite block length. The randomness of the channel characteristics has been completely averaged out. When the block length is finite, this averaging is only partially successful and "non-average" channel realizations do occur. In order to measure the effectiveness of a coding scheme against the randomness of the channel for finite block length, the reliability function of the channel is needed. The reliability function with perfect channel state information (CSI) removes the effect of channel estimation by assuming that the channel state is known. This allows it to capture the effect of the diversity, which is roughly the number of independent channel realizations within a codeword. As the memory becomes longer, the number of independent or nearly independent channel realizations in a fixed-length block is reduced leading to a larger chance of getting an abnormally bad realization of the channel states. When the channel state realization is known, shorter channel memory is better. So the channel capacity captures the effects of channel estimation, and the reliability function with CSI measures the effect of diversity.

In order to study the combined tradeoff between channel estimation and channel diversity for finite block lengths and memory, the appropriate measure is the reliability function without assuming perfect channel state information. This accounts for both the channel estimation errors and the randomness of the channel. From the reliability function, we can determine which effect dominates for different channel conditions. Qualitatively, as the rate approaches the capacity it becomes more important to have good channel estimates so larger memory is better. For lower code rates the extra redundancy in the channel code can compensate for poor channel estimates and thus less memory is required. Another interpretation is obtained for fixed rate codes by changing the signal-to-noise ratio (and thus changing the capacity). As the signal-to-noise ratio decreases, the capacity decreases and becomes closer to the code rate. In this case, larger channel memory is better. The conclusion is that there is an optimal memory length that gives the best combination of channel estimation and diversity.

In this chapter we examine the channel capacity and reliability of channels with memory and show the performance as a function of memory for some illustrative models. We also show simulation results for various memory lengths for low-density parity check codes with a decoder that jointly and iteratively estimates the data and channel. In section 2 we describe the channel models we consider. In section 3 we develop the reliability exponents for these channels. In section 4 we describe the properties of the optimal memory and verify analytically that our intuitive claims are reflected in the behavior of the bounds. We present numerical results for the reliability exponents and simulation results for low-density parity check codes with joint estimation and decoding in section 5.

2 System Model

The channel models we consider in this paper incorporate memory in such a way that the models are tractable for analysis and yet have practical implications. The source produces a set of k information bits which select a codeword $\underline{x} \in \mathcal{C} \subset \mathcal{X}^n$. The codeword is transmitted over a channel with input alphabet $\mathcal{X}$ and output alphabet $\mathcal{Y}$ characterized by its transition probability density $\mathrm{p}(\underline{y}|\underline{x})$, and the output $\underline{y} \in \mathcal{Y}^n$ is processed by the decoder to produce an estimate of $\underline{x}$ (and consequently the

k information bits).

We focus our attention on channels with block memory. A channel with transition probability $\mathrm{p}(\underline{y}|\underline{x})$ has block memory with memory length m if

$$\mathrm{p}(\underline{y}|\underline{x}) = \prod_{i=0}^{\lceil \frac{n}{m} \rceil \quad 1} \left(\int_{\mathcal{U}} \prod_{j=0}^{m \quad 1} \mathrm{p}(y_{i \cdot m+j}|x_{i \cdot m+j}, u)\, \mathrm{p}(u)\, du \right) \tag{1}$$

where u is a channel state drawn from the channel state space $\mathcal{U}$ with probability $\mathrm{p}(u)$. In this channel, a block of m consecutive transmitted code symbols are received at the channel output with the same transition probabilities, which are determined by the channel state. Then the next m are transmitted with an independently chosen channel state which determines the transition probabilities for that block.

This channel model is a good model for a number of different communication systems. In particular, it is an excellent model for frequency-hopping spread spectrum systems where the channel state reflects the fact that each hop sees an independent realization of channel conditions such as jamming or fading that is fixed for the duration of the hop. It is also a good model for any slotted system such as time division multiple access where the slot represents the block memory. Thus, we will refer to a block of symbols corresponding to the same channel state u as a hop, and will use hop length interchangeably with memory length. Note that on a block-by-block basis the channel is a memoryless channel. This observation makes analysis of the system tractable.

3 Bounds for Joint Estimation and Decoding

The error exponent technique [1, Ch. 5] gives an exponential bound on the probability of codeword error P_e for codewords of length n. The bound for any memoryless channel is

$$\begin{aligned} P_e &\leq e^{\quad nE_r(R)} \\ E_r(R) &= \max_{0\leq\rho\leq 1} \max_{\mathbf{Q}} E_0(\rho, \mathbf{Q}) \quad \rho R \ln 2 \\ E_0(\rho, \mathbf{Q}) &= \quad \ln \sum_{y\in\mathcal{Y}} \left(\sum_{x\in\mathcal{X}} Q(x)\, \mathrm{p}(y|x)^{\frac{1}{1+\rho}} \right)^{1+\rho} \end{aligned} \tag{2}$$

where $\mathbf{Q}$ is the channel input probability distribution. We can derive the error exponent for channels with block memory by considering a super-channel where each block of m symbols from the original channel forms a single input. This super-channel is a memoryless channel with input alphabet $\mathcal{X}^m$, output alphabet $\mathcal{Y}^m$, and transition probabilities $\mathrm{p}(\underline{y}|\underline{x})$. Applying the bound to the super-channel, but letting the parameters continue to refer to the original channel, gives

$$\begin{aligned} P_e &\leq e^{\ \frac{n}{m} E_r'(R)} \\ E_r'(R) &= \max_{0 \leq \rho \leq 1} \max_{\mathbf{Q}} E_0'(\rho, \mathbf{Q}) \quad \rho R m \ln 2 \\ E_0'(\rho, \mathbf{Q}) &= \quad \ln \sum_{\underline{y} \in \mathcal{Y}^m} \left(\sum_{\underline{x} \in \mathcal{X}^m} Q(\underline{x})\, \mathrm{p}(\underline{y}|\underline{x})^{\frac{1}{1+\rho}} \right)^{1+\rho} \end{aligned} \tag{3}$$

which is equivalent to the usual formulation (2) if we slightly modify the Gallager function

$$E_0(\rho, \mathbf{Q}, m) = \quad \frac{1}{m} \ln \sum_{\underline{y} \in \mathcal{Y}^m} \left(\sum_{\underline{x} \in \mathcal{X}^m} Q(\underline{x})\, \mathrm{p}(\underline{y}|\underline{x})^{\frac{1}{1+\rho}} \right)^{1+\rho} \tag{4}$$

to get

$$E_r(R, m) = \max_{0 \leq \rho \leq 1} \max_{\mathbf{Q}} E_0(\rho, \mathbf{Q}, m) \quad \rho R \ln 2. \tag{5}$$

Substituting (1) in (4) gives

$$E_0(\rho, \mathbf{Q}, m) = \quad \frac{1}{m} \ln \sum_{\underline{y} \in \mathcal{Y}^m} \left(\sum_{\underline{x} \in \mathcal{X}^m} Q(\underline{x}) \left[\int_{\mathcal{U}} \mathrm{p}(u) \prod_{i=0}^{m\ 1} \mathrm{p}(y_i|x_i, u)\, du \right]^{\frac{1}{1+\rho}} \right)^{1+\rho}. \tag{6}$$

It will often be helpful, both analytically and experimentally, to compare the performance of joint channel estimation and decoding receivers to decoders using perfect channel state information (CSI) obtained from a genie. By eliminating the effects of channel estimation, the perfect CSI receiver provides a way to measure the importance of diversity. The random coding error exponent $\bar{E}_r(R,m)$ for receivers on channels with block memory can be easily derived by adding the channel state as an additional uncorrupted channel output. The Gallager function with perfect channel state information is

$$\bar{E}_0(\rho,\mathbf{Q},m) = \frac{1}{m}\ln \int_{\mathcal{U}} \sum_{\underline{y}\in\mathcal{Y}^m} \left(\sum_{\underline{x}\in\mathcal{X}^m} Q(\underline{x})\, \mathrm{p}(\underline{y},u|\underline{x})^{\frac{1}{1+\rho}} \right)^{1+\rho} du. \quad (7)$$

Using the fact that the channel state u and the channel input $\underline{x}$ are independent,

$$\mathrm{p}(\underline{y},u|\underline{x}) = \mathrm{p}(\underline{y}|\underline{x},u)\,\mathrm{p}(u), \quad (8)$$

in (7)

$$\bar{E}_0(\rho,\mathbf{Q},m) = \frac{1}{m}\ln \int_{\mathcal{U}} \mathrm{p}(u) \sum_{\underline{y}\in\mathcal{Y}^m} \left(\sum_{\underline{x}\in\mathcal{X}^m} Q(\underline{x}) \left[\prod_{i=0}^{m-1} \mathrm{p}(y_i|x_i,u) \right]^{\frac{1}{1+\rho}} \right)^{1+\rho} du. \quad (9)$$

It is easy to obtain $\bar{E}_r(R,m)$ by replacing $E_0(\rho,\mathbf{Q},m)$ with $\bar{E}_0(\rho,\mathbf{Q},m)$ in (5).

The capacity for channels with block memory with and without CSI can also be derived using the super-channel approach. The capacity (without CSI) for a channel with hop length m is [2]

$$\begin{aligned} C_m &= \max_{\mathbf{Q}} \frac{1}{m} I(\underline{X}_m;\underline{Y}_m) \\ &= \max_{\mathbf{Q}} \frac{1}{m} \sum_{\underline{y}\in\mathcal{Y}^m} \sum_{\underline{x}\in\mathcal{X}^m} \mathrm{p}(\underline{y}|\underline{x}) Q(\underline{x}) \log_2 \frac{\mathrm{p}(\underline{y}|\underline{x})}{\mathrm{p}(\underline{y})} \end{aligned} \quad (10)$$

where C_m has units of bits per channel symbol. The capacity with CSI

$\bar{C}_m$ is

$$\begin{aligned}\bar{C}_m &= \max_{\mathbf{Q}} \frac{1}{m} I(\underline{X}_m; \underline{Y}_m, U) \\ &= \max_{\mathbf{Q}} \frac{1}{m} \int_{\mathcal{U}} \sum_{\underline{y} \in \mathcal{Y}^m} \sum_{\underline{x} \in \mathcal{X}^m} \mathrm{p}(\underline{y}|\underline{x}, u)\, \mathrm{p}(u) Q(\underline{x}) \log_2 \frac{\mathrm{p}(\underline{y}|\underline{x}, u)\, \mathrm{p}(u)}{\mathrm{p}(\underline{y}, u)}\, du,\end{aligned} \tag{11}$$

which turns out to be independent of m under some mild assumptions about the optimal input distribution [2].

4 Properties of Optimal Memory Lengths

In this section we derive properties of the channel capacity and error exponent for channels with memory. The mathematical properties of the bounds support our intuition about the channel memory-diversity tradeoff. Details of the proofs in this section are available in [3].

We will say that a hop length m is optimal for a particular rate R if it maximizes the random coding error exponent $E_r(R, m)$. In the sequel, we will often find it helpful to restrict our attention to symmetric channels.

Definition 4.1 *A discrete memoryless channel with state* u *and transition probabilities* $\mathrm{p}(y|x, u)$ *is* symmetric *if the columns of its transition probability matrix can be partitioned so that the rows of each partition are permutations of each other and the columns of each partition are permutations of each other.*

Note that since the elements of the transition matrix are functions of u, this implies that the same permutations must apply for all possible values of u. It is straightforward to show that the memoryless super-channel obtained from a symmetric block memory channel is also symmetric. It is well-known that the optimal input distribution for a symmetric discrete memoryless channel (DMC) is uniform over the input alphabet, and this turns out to be true of symmetric block memory channels as well.

Conventional wisdom suggests that memory increases the channel capacity. This is also predicted by our interpretation of the channel capacity (without side information) as primarily capturing the effects of channel estimation without the loss due to reduced diversity.

Theorem 4.1 *For a symmetric channel with block memory, the capacity is either increasing in the hop length m or independent of m.*

We outline the general idea of the proof below. The details can be found in [3]. The key is that our uncertainty about an additional symbol in a hop decreases as the hop length increases. Since the optimal input distribution is uniform, it is sufficient to show that

$$H(\underline{X}_{m+1}) \quad H(\underline{X}_{m+1}|\underline{Y}_{m+1}) \geq (1+\frac{1}{m})H(\underline{X}_m) \quad (1+\frac{1}{m})H(\underline{X}_m|\underline{Y}_m). \tag{12}$$

Because $H(\underline{X}_{m+1}) = (1+\frac{1}{m})H(\underline{X}_m)$, we need only show that

$$H(\underline{X}_{m+1}|\underline{Y}_{m+1}) \leq (1+\frac{1}{m})H(\underline{X}_m|\underline{Y}_m). \tag{13}$$

Repeated application of the chain rule for entropy shows that this is equivalent to

$$H(X_{m+1}|\underline{X}_m, Y_{m+1}, \underline{Y}_m) \leq \frac{1}{m}\sum_{i=0}^{m\ \ 1} H(X_i|\underline{X}_m^{(i)}, \underline{Y}_m), \tag{14}$$

where $\underline{X}_m^{(i)} = \{X_0, X_1, \ldots, X_{i\ \ 1}, X_{i+1}, \ldots, X_{m\ \ 1}\}$, which we can prove using the symmetry of the channel model.

The capacity is not affected by the hop length if channel estimation over long blocks does not produce additional information about the channel. Consider the binary-input, quaternary-output channel in Fig. 1. It is easy to show that the capacity for this channel is independent of the hop length, and furthermore its error exponent $E_r(R,m)$ is strictly decreasing in m for all $R > 0$.

The error exponent with perfect CSI $\bar{E}_r(R,m)$ captures the contribution of diversity. The following theorem expresses our claim that longer hop lengths reduce the diversity and therefore performance when channel estimation is not needed.

Theorem 4.2 *For symmetric channels, the reliability function with side information $\bar{E}_r(R,m)$ is either decreasing in m or independent of m.*

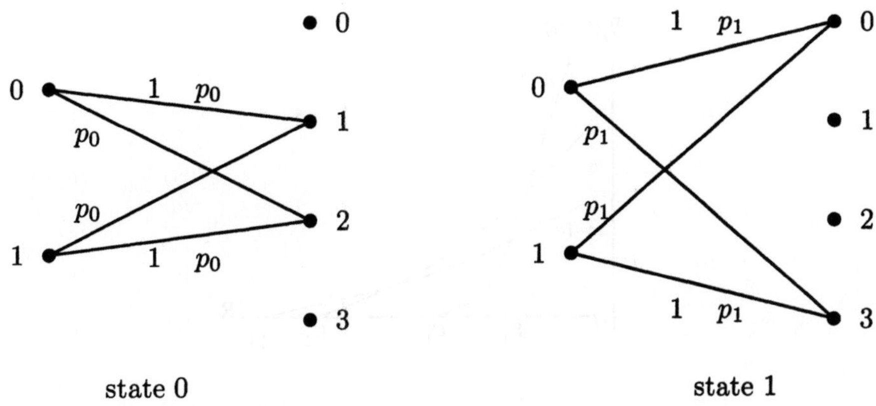

Figure 1: Example diversity dominated channel. In state 0, outputs 1 and 2 are possible and the cross-over probability is p_0. In state 1, outputs 0 and 3 are possible and the cross-over probability is $p_1 \neq p_0$.

The proof shows that, for all ρ,

$$\bar{E}_0(\rho, \mathbf{Q}, m) \geq \bar{E}_0(\rho, \mathbf{Q}, m+1) \tag{15}$$

by a function space norm argument. Then, if ρ^* maximizes $\bar{E}_0(\rho, \mathbf{Q}, m+1) \quad \rho R \ln 2$,

$$\begin{aligned} \bar{E}_r(R, m+1) = \bar{E}_0(\rho^*, \mathbf{Q}, m+1) \quad \rho^* R \ln 2 \leq \\ \bar{E}_0(\rho^*, \mathbf{Q}, m) \quad \rho^* R \ln 2 \leq \max_{0 \leq \rho \leq 1} \bar{E}_0(\rho, \mathbf{Q}, m) \quad \rho R \ln 2 = \bar{E}_r(R, m). \end{aligned} \tag{16}$$

The flip channel is a binary-input, binary-output channel with transition probability

$$\begin{aligned} p(y_i|x_i, 0) &= \begin{cases} 1 \quad p & \text{if } y_i = x_i \\ p & \text{if } y_i \neq x_i \end{cases} \\ p(y_i|x_i, 1) &= \begin{cases} p & \text{if } y_i = x_i \\ 1 \quad p & \text{if } y_i \neq x_i \end{cases}. \end{aligned} \tag{17}$$

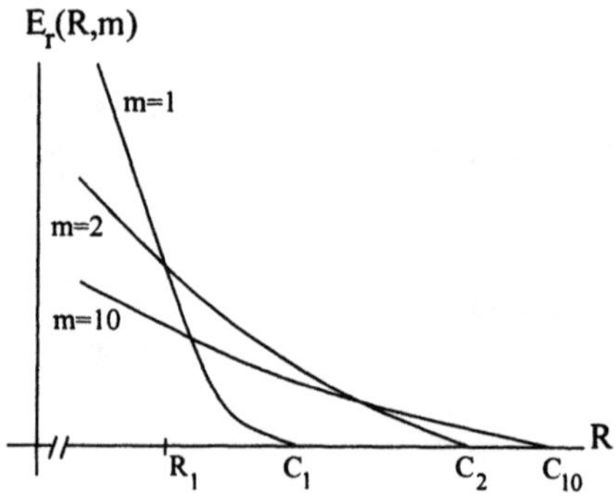

Figure 2: Sketch illustrating Theorem 4.3.

For this channel $\bar{E}_r(R,m)$ is independent of m for all R. Because the channel essentially does not have a good and bad state, diversity is not important and, therefore, we find numerically that $E_r(R,m)$ is increasing in m. Long hop lengths are favored because they allow better channel estimation.

As we would expect, $E_r(R,m) \leq \bar{E}_r(R,m)$. We can prove this for symmetric channels using Jensen's inequality and a variant of Minkowski's inequality.

The combined effects of channel estimation and diversity are captured by the error exponent without side information $E_r(R,m)$. We can use the properties of the error exponent [1, Ch. 5] to show that the optimal hop length becomes large for rates near capacity.

Theorem 4.3 *For a channel with C_m increasing in m, there is a sequence of rates R_l with $R_l \leq C_l$ such that the optimal hop length $m^* > l$ when $R > R_l$.*

Using the properties of the reliability function, $E_r(R,l) = 0$ for $R > C_l$ and $E_r(R,m) > 0$ for $R < C_m$. This implies that for any $m > l$ and $C_l \leq R < C_m$, $E_r(R,m) > E_r(R,l)$. Therefore, there must be some rate $R_l < C_l$ such that for $R > R_l$ the optimal hop length $m^* > l$. A graphical representation of this theorem is shown in Fig. 2.

Variations on theorems 4.1 and 4.2 were presented in [4] though without our interpretation.

5 Examples

Here we present numerical examples of the bounds for some channel models related to practical wireless communication systems. We also present simulation results for low-density parity check (LDPC) codes with iterative receivers that approximate joint channel estimation and decoding [5]. Due to space limitations, we omit the receiver designs, which can be found in [3].

For m larger than 1 or 2, finding the error exponent bound for continuous output channels is usually difficult because we must compute high dimensional integrals numerically. Therefore, for the error exponent bounds, we consider binary-input, binary-output variants of the channel models. The simulation results, however, are for the continuous output models unless otherwise noted.

5.1 Partial Band Jamming Channel

We begin with a model for frequency-hopping spread spectrum in partial band jamming with jamming fraction ρ_J. Assume that a hop is either entirely jammed or entirely unjammed. Then, the channel has two states J and U corresponding to jammed and unjammed hops and occurring with probabilities ρ_J and $1 \quad \rho_J$ respectively. In the jammed state J, the channel adds white Gaussian noise with one-sided power spectral density $N_0 + N_J/\rho_J$ while in the unjammed state U the noise power spectral density is just N_0. The basic modulation is taken to be coherently-detected BPSK.

For the sake of computing the bound, we consider a binary-output channel with cross-over probabilities,

$$p_J = Q\left(\sqrt{\frac{2E_s}{N_0 + N_J/\rho_J}}\right) \qquad p_U = Q\left(\sqrt{\frac{2E_s}{N_0}}\right) \tag{18}$$

The Gallager function is

$$E_0(\rho, \mathbf{Q}, m) = \frac{1}{m} \ln\left(\sum_{d=0}^{m} \binom{m}{d} \left[(1 \quad \rho_J) \cdot (1 \quad p_U)^{m \quad d} p_U^d \right.\right.$$
$$\left.\left. + \rho_J \cdot (1 \quad p_J)^{m \quad d} p_J^d \right]^{\frac{1}{1+\rho}} \right)^{1+\rho} + \rho \ln 2. \quad (19)$$

Figure 3 shows the error exponent at rates near the capacity for several values of the hop length m.

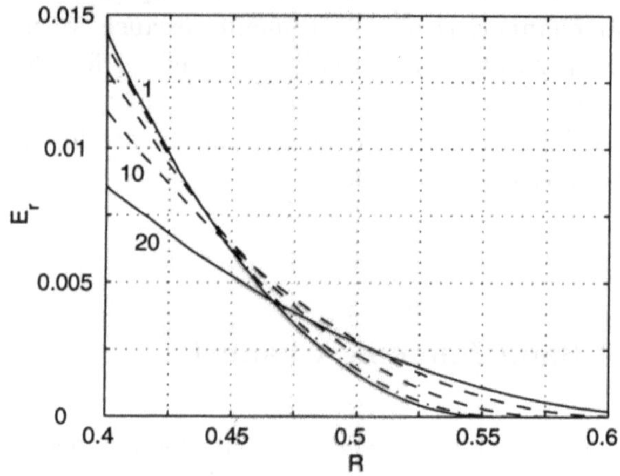

Figure 3: Reliability function $E_r(R, m)$ vs. rate R for the partial band jamming channel. The hop lengths for the curves, in order down the ordinates, are 1, 2, 5, 10, and 20 as labeled. The channel parameters are $E_s/N_0 = 7$ dB, $E_s/N_J =$ 1.5 dB, $\rho_J = 0.4$.

We see that for rates above about $R_1 = 0.43$, the optimal hop length is larger than 1. Because the error-exponents in this region are very small, this effect is only evident with reasonable error probabilities for long codewords. Figure 4 shows that the optimal hop length is about 15 symbols for the parameters given. The required SJNR with perfect channel side information increases with the hop length as expected.

In Fig. 5, and Fig. 6, we compare the qualitative behavior of the bounds to the performance of a simulation on the hard-decision channel. Al-

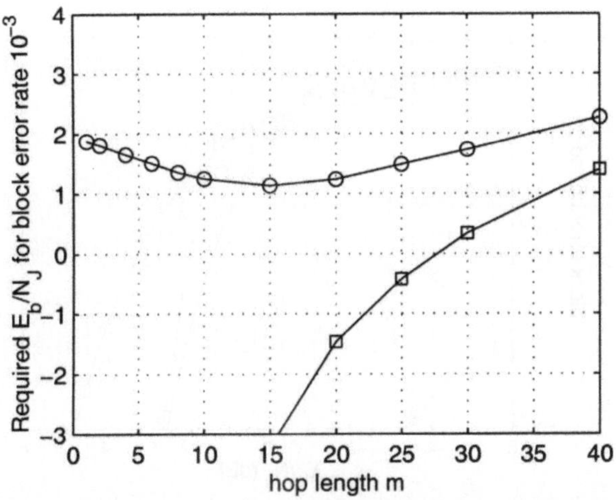

Figure 4: Required E_b/N_J for the partial band jamming channel for block error rate 10^{-3} $R = 1/2$, $n = 2520$, $E_b/N_0 = 12$ dB, and $\rho_J = 0.4$ ('o' joint estimation and decoding, '□' perfect CSI).

though the LDPC codes with iterative decoding do not have the performance predicted by the random coding bound for these short block lengths, the relative qualitative behavior of the curves is strikingly similar.

5.2 Rayleigh Fading Channel

The Rayleigh fading channel with non-coherent reception is extremely important for wireless communication systems. When frequency-hopping or time-hopping (as in a TDMA system) is used, the Rayleigh channel acts as a block memory channel with an independent complex Gaussian fading factor in each hop

$$y_i = u_{\alpha,\lfloor \frac{i}{m} \rfloor} \exp(j u_{\theta,\lfloor \frac{i}{m} \rfloor}) \cdot x_i + n_i \tag{20}$$

where $u_{\alpha,\lfloor \frac{i}{m} \rfloor}$ is a Rayleigh distributed magnitude, $u_{\theta,\lfloor \frac{i}{m} \rfloor}$ is a uniformly distributed phase, and n_i is a sample of white Gaussian noise.

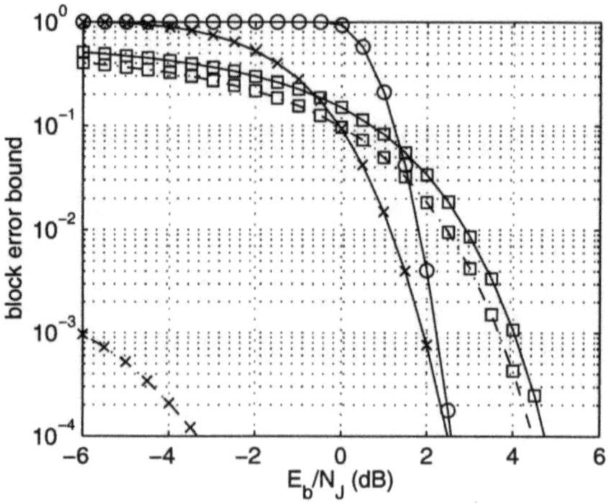

Figure 5: Performance from bounds from the error exponent for the two-state jamming channel. 'o' $m = 1$, '×' $m = 8$, '□' $m = 64$. solid–joint channel estimation and decoding, dash-dotted–perfect CSI. The parameters are $R = 1/2$, $n = 2048$, $E_b/N_0 = 10$ dB and $\rho_J = 0.4$.

Computing the random coding bound for this channel is, in general, quite demanding, although recent work has solved some cases though without special emphasis on the memory length [6][7]. We instead consider a close relative with binary inputs and outputs. The Rayleigh fading flip channel either passes the input bits through a binary symmetric channel with cross-over probability p_u which depends on the fading state u, or flips the bits and then passes them through a binary symmetric channel with cross-over probability p_u. The channel state for each hop consists of a Bernoulli random variable with probability 1/2 of taking each of its two possible values to determine whether or not to flip the bits, and an independent Rayleigh random variable u for the fading level. The cross over probability p_u depends on the fade level u

$$p_u = Q\left(\sqrt{\frac{2u^2 E_s}{N_0}}\right). \tag{21}$$

Without channel state information, we can compute the Gallager func-

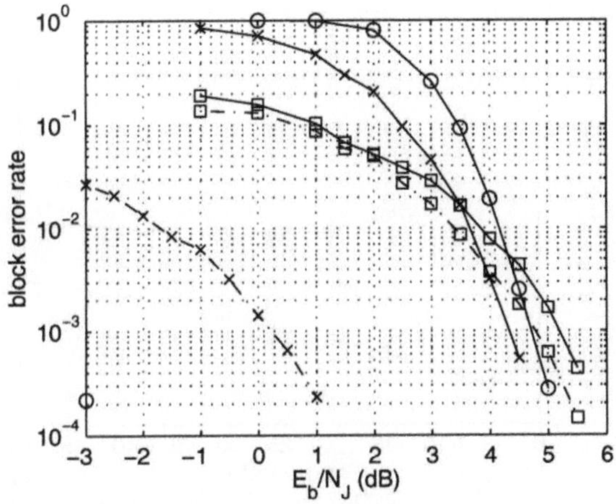

Figure 6: Simulation of performance of an LDPC code for the two-state jamming channel. 'o' $m = 1$, '×' $m = 8$, '□' $m = 64$. solid–joint channel estimation and decoding, dash-dotted–perfect CSI. The parameters are $R = 1/2$, $n = 2048$, $E_b/N_0 = 10$ dB and $\rho_J = 0.4$.

tion

$$E_0(\rho, \mathbf{Q}, m) = \frac{1}{m} \ln\left(\sum_{d=0}^{m} \binom{m}{d} \left[\frac{1}{2} \int_{\mathcal{U}} \mathrm{p}(u)((1 \quad p_u)^{m \quad d} p_u^d + (1 \quad p_u)^d p_u^{m \quad d})\, du \right]^{\frac{1}{1+\rho}} \right)^{1+\rho} + \rho \ln 2. \quad (22)$$

Figure 7 shows the required average SNR E_b/N_0 for this channel model. The optimal hop length, $m = 25$ symbols, gains about 3 dB over hop length $m = 5$.

An iterative receiver for binary low-density parity check codes on the Rayleigh fading channel with unknown fade states is described in [8][3]. A single pilot symbol is added to each hop to aid the receiver. The different phase distributions are different approaches to handling the soft phase estimation. The performance (shown in Fig. 8) is qualitatively similar to that predicted by the random coding bound for the Rayleigh

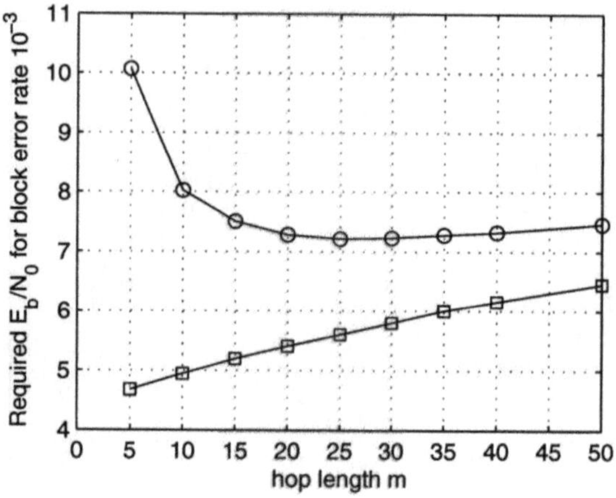

Figure 7: Required average SNR E_b/N_0 vs. hop length for Rayleigh fading flip channel models for the Rayleigh fading flip channel from the bound. '○' joint channel estimation and decoding, '□' perfect CSI. $n = 2520$, $R = 1/2$.

fading flip channel. Although in the simulation, the receiver must estimate the uniformly distributed phase state rather than just whether the bits are being flipped, the availability of soft-decisions seems to make the overall estimation task easier. Note also that the code rate for the simulation decreases slightly for small hop lengths because of the pilot symbols.

6 Conclusions

In this chapter we have investigated the tradeoff between channel estimation and channel diversity for channels with memory. We determined the reliability function for these channels as a function of the memory length. We showed that for finite block length transmission techniques there is a fundamental tradeoff between the channel estimator and the channel diversity. When the channel is operating at rates much below capacity, smaller channel memory is better since channel estimation is not crucial. However, at rates close to capacity, larger channel memory

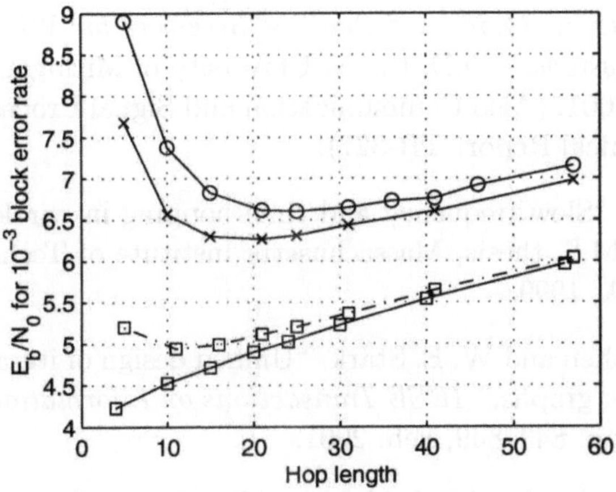

Figure 8: Required average SNR E_b/N_0 vs. hop length for Rayleigh fading channel from simulation. '□' CSI, '□' (dash-dotted) CSI with pilot symbol energy penalty, '×' quantized phase distribution receiver, 'o' bimodal phase distribution receiver. Block length 2520, rate 1/2 LDPC code, with one pilot symbol added per hop.

is better because it is important that the decoder knows the channel realization and larger memory leads to a better channel estimate. We draw these conclusions based on analysis of some simple discrete input-output channels using the channel reliability function and on simulation of more realistic soft-output channels with low-density parity check codes and a joint iterative decoder and channel estimator.

References

[1] R. G. Gallager, *Information Theory and Reliable Communication*, John Wiley & Sons, New York, 1968.

[2] R. J. McEliece and W. E. Stark, "Channels with block interference," *IEEE Transactions on Information Theory*, vol. IT-30, no. 1, pp. 44–53, Jan. 1984.

[3] A. P. Worthen, *Codes and Iterative Receivers for Wireless Communication Systems*, Ph.D. thesis, University of Michigan, Ann Arbor, MI, May 2001. (Also Communication and Signal Processing Laboratory Technical Report TR-327).

[4] E. Uysal, "Slow frequency and time hopping in wireless communications," M.S. thesis, Massachusetts Institute of Technology, Cambridge, MA, 1999.

[5] A. P. Worthen and W. E. Stark, "Unified design of iterative receivers using factor graphs," *IEEE Transactions on Information Theory*, vol. 47, no. 2, pp. 843–849, Feb. 2001.

[6] W. K. M. Ahmed and P. J. McLane, "Random coding error exponents for flat fading channels with realistic channel estimation," *IEEE Journal on Selected Areas in Communications*, vol. 18, no. 3, pp. 369–379, Mar. 2000.

[7] T. L. Marzetta and B. M. Hochwald, "Multiple-antenna communications when nobody knows the Rayleigh fading coefficients," in *Proceedings of the 35th Allerton Conference on Communication, Control, and Computing*, Sept. 1997, pp. 1033–1042.

[8] A. P. Worthen and W. E. Stark, "On iterative receivers for noncoherent channels," in *Proceedings of the 2000 International Symposium on Information Theory and Its Applications ISITA'00*, Honolulu, HI, Nov. 2000.

Duality, Dirty Paper Coding, and Capacity for Multiuser Wireless Channels

Nihar Jindal, Sriram Vishwanath, and Andrea Goldsmith
Department of Electrical Engineering
Stanford University
Stanford, CA 94303-9515
{njindal,sriram,andrea}@ee.stanford.edu

Abstract

We determine a duality between broadcast and multiple access channels that can be used to obtain the capacity region and optimal transmission strategy for one channel based on the capacity-achieving transmission strategy and region for the dual channel. This duality result in applicable to additive Gaussian noise and fading channels for several different notions of fading channel capacity, including ergodic capacity, outage capacity, and minimum rate capacity. We show that duality can be used to obtain any of these capacities for the fading broadcast channel from the same capacity on the dual MAC channel, and vice versa. We then apply this general result to obtain the minimum rate capacity of the fading multiple access channel, which also yields the ergodic and outage capacity as special cases. Next we turn our attention to broadcast channels with multiple antennas at the transmitter and receiver (the MIMO channel). Since this channel is in general non-degraded, its capacity region remains an unsolved problem. We establish a duality between the achievable region of the MIMO broadcast channel using Costa's "dirty-paper" coding and the capacity region of the MIMO multiple-access channel, which is easy to compute. We also show that the dirty paper achievable region yields the sum-rate capacity of the MIMO broadcast channel.

Keywords: Capacity, duality, fading channels, broadcast and multiple access channels, MIMO, dirty paper coding.

1 Introduction

It is my great pleasure to contribute this paper in honor of Bob McEliece on the occasion of his 60th birthday. McEliece's broad and deep impact on the fields of information and coding theory has inspired generations of theorists and practitioners. While the influence of his work on deep space communications is well known, McEliece's work has also had a significant impact on the understanding and design of wireless terrestrial systems. This paper will first review some of McEliece's contributions to wireless communications in the areas of coding, capacity, and dynamic channel allocation. The remainder of the paper focuses on new results inspired by the work of McEliece and others on capacity of fading channels and multiuser channel capacity regions.

We first consider a duality relationship between broadcast channels (BCs) and multiple access channels (MACs). Despite extensive study of both channels independently, no relationship between them has previously been discovered. We show that the Gaussian MAC and BC are essentially duals of each other. As a result, the capacity regions of the BC and the MAC with the same channel gains (i.e. the channel gain of receiver k in the BC equals the channel gain of transmitter k in the MAC) and the same noise power at every receiver (i.e. each receiver in the BC and the single receiver in the MAC has the same noise power) can be derived from one another. In particular, we show that the capacity region of a fading or constant Gaussian BC is equivalent to the union over all MAC power constraints which sum up to the BC power constraint of the capacity regions of the dual Gaussian MAC. This result allows us to directly compute the capacity of a Gaussian BC from the capacity of its dual Gaussian MAC. There also exists a direct relationship between the optimum power allocation scheme and the decoding order used to achieve points along the boundary of the MAC and BC capacity regions. Moreover, this relationship between the capacity regions and optimal transmission strategies for dual MAC and BCs holds for several different notions of fading channel capacity, including ergodic capacity, outage capacity, and minimum rate capacity.

An inverse relationship also exists such that the capacity of a fading or constant Gaussian MAC is equivalent to the intersection of Gaussian BCs with appropriately scaled channel gains and power constraint equal to the sum of the equivalently scaled MAC power constraints, over all possible scalings. Using this relationship we can compute the capacity region of a Gaussian MAC with knowledge only of the capacity region of the dual Gaussian BC. The same relationship between the optimum power allocation scheme and decoding order for the MAC and BC can be applied to this method as well, so the optimum power allocation scheme and decoding order for the MAC can be derived from knowledge of the optimum power allocation scheme for the BC.

These complimentary relationships between the ergodic capacity of the Gaussian MAC and BC are quite powerful and interesting. First they indicate a direct connection between known capacity results for dual MAC and broadcast channels. More importantly, there are several BC and MAC channels where direct calculation of the capacity region and optimal transmission strategy is unknown and may be intractable, whereas these results are known for the dual channel. In these cases we can use duality to obtain new capacity results. In particular, we will show how duality can be used to obtain the minimum rate capacity region and optimal transmission strategy for a fading MAC channel and a BC with multiple transmit and receive antennas.

The capacity of a broadcast channel with multiple antennas at the transmitter and receiver (the MIMO BC) is an open problem due to the lack of a general theory on the capacity of non-degraded broadcast channels. Pioneering work in this area by Caire and Shamai [1] developed an achievable set of rates for a broadcast channel with two transmit antennas and one receiver antenna at each node based on the "dirty paper coding" result of Costas [2]. This coding strategy allows a channel with interference known at the transmitter to achieve the same data rate as if the interference did not exist. The original coding strategy was described as writing on dirty paper, where the structure introduced by the "dirt" is exploited in the code design. Computing this corresponding set of achievable rates is extremely complex, especially for larger number of antennas at either the transmitter or the receivers.

We use duality to greatly simplify the derivation of the achievable rate set via dirty paper coding. Specifically, we show that the dirty paper region is exactly equal to the capacity region of the dual MIMO MAC, with the transmitters having the same *sum* power constraint as the MIMO BC. We also show that this set of achievable rates achieves the sum rate point of the MIMO BC channel capacity region, and lower bounds the other points in the region.

The remainder of this paper is organized as follows. Section 2 describes McEliece's contributions to wireless communications, including capacity results for both broadcast and multiple access channels. Section 3 introduces the duality between broadcast and MAC channels and shows how this can be used to compute their dual capacity regions. Section 4 introduces the different notions of capacity associated with multiuser fading channels, including the ergodic, outage, and minimum rate capacities. Sections 5 and 6 derive the ergodic and minimum rate capacity regions, respectively, for fading broadcast and MAC channels along with their duality relationship. Section 7 outlines the achievable rate region for a broadcast MIMO fading channel with dirty paper coding. Our conclusions are discussed in Section 8.

2 McEliece and the Wireless Revolution

Wireless communication systems have enjoyed exponential growth over the last decade. This success has generated intense research on fundamental performance limits and system designs for cellular and other wireless systems. McEliece has made several important contributions in this area which are not as well-known as some of his other work. In this section we describe these contributions.

Wireless channels differ from the AWGN channel model used for deep space communications in many ways. One significant difference is that wireless channels exhibit random fluctuations due to time-varying multipath and interference. An early contribution on the capacity of channels with random fluctuations was made by McEliece and Stark [3]. This work indicates that for fading channels with randomly-varying SNR γ, if γ can be measured at the receiver then the channel capacity is given by

$$C = \int B \log(1+\gamma)p(\gamma)d\gamma, \tag{1}$$

where $p(\gamma)$ is the distribution of γ. This result provoked much subsequent work on the capacity of fading channels with receiver side information, including the fact that this capacity can be achieved with a simple standard Gaussian codebook ([4] and the references therein).

McEliece also derived some important results on both multiple access and broadcast channels. In [5] McEliece and Swanson found that for wideband broadcast channels (channels operating at low symbol SNRs), a time-division strategy achieves almost the same capacity as the optimal strategy of superposition coding. The fundamental communication limits in the wideband regime for point-to-point, broadcast, and multiaccess channels has recently experienced a resurgence of interest [7, 8, 9]. However, [5] remains one of the pioneering papers in this area.

For multiple access channels, Cover, McEliece, and Posner [6] proved that asynchronicity among users does not decrease the capacity region of the Gaussian MAC. This is an important result since in practice it is difficult to synchronize the different users in a MAC channel. The lack of synchronism indicates that time-sharing cannot be used to achieve points on the capacity region boundary, as is the case for synchronous MAC channels. Thus, [6] developed an alternate method to achieve these points based on maximum likelihood decoding over time shifts of the transmitted codeword. Interestingly, this alternate method is quite similar to despreading techniques used for MAC channels with spread spectrum. The work in [6] on the asynchronous MAC capacity was later extended by Verdú to MAC channels with memory [10] and with symbol asynchronism [11].

Another important research area in wireless system design is that of dynamic channel assignment in cellular systems. Channel assignment is not typically required in CDMA systems since spread users occupy the entire system bandwidth at all times. However, most current systems use some form of time-division, especially for data (e.g. GPRS in GSM and HDR in IS-95). Under fixed frequency reuse constraints and traffic intensity, the dynamic channel allocation problem in a time-division cellular system can be viewed as a generalization of the vertex coloring problem and is thus NP-hard [12]. However, McEliece and Sivarajan determined that when the traffic load and number of channels approach infinity, the asymptotic behavior of the optimal allocation policy can be determined [13]. Moreover, the performance of this optimal dynamic allocation policy approaches that of the optimal fixed allocation policy. McEliece and Deora went on to determine a simple approximation to the optimal dynamic policy called the timid algorithm that has similar performance [14]. This work provided fundamental performance bounds on the large body of work in dynamic channel allocation [15], as well as a simple heuristic scheme that performed close to optimal in heavily loaded systems.

The remainder of this paper will focus on capacity of multiuser fading channels, building the work of McEliece described above and subsequent work by other authors in these areas.

3 Duality between Broadcast and Multiple Access Channels

In this section we establish the duality between broadcast and multiple access channels and show how this duality can be used to compute the capacity of one channel from the capacity of its dual. A discrete-time broadcast channel, where one transmitter sends information to M receivers, is described mathematically by

$$Y_j[i] = \sqrt{h_j[i]}X[i] + n_j[i], \tag{2}$$

where $X[i]$ is the transmitted signal, $Y_j[i]$ is the signal received by the jth user, $n_j[i]$ is the receiver noise sample at time i of the jth user, and $h_j[i]$ is the time-varying channel power gain at time i of the jth user. A discrete-time multiple access channel, where many transmitters send information to one receiver, is described mathematically by

$$Y[i] = \sum_{j=1}^{M} \sqrt{h_j[i]}X_j[i] + n[i], \tag{3}$$

where $X_j[i]$ is the signal transmitted at time i by the jth transmitter, $h_j[i]$ is the channel power gain between the jth transmitter and the receiver, $Y[i]$ is the received signal at time i, and $n[i]$ is the receiver noise sample at time i. We assume in our system models that the noise power of all the receivers in the BC and the single receiver in the MAC are equal to σ^2. Also, the term $h_j[i]$ is the channel power gain of receiver j in the BC (downlink) and $h_j[i]$ is also the channel power gain of transmitter j in the MAC (uplink). We call this BC the **dual** of the MAC, and vice versa. In our broadcast and MAC models the channel gains can be constant or changing with i (fading). We assume in both cases that all channel gains $h_j[i]$ are known to the transmitter(s) and receiver(s) at time i.

We now state the result, proved in [16], that the capacity region of a Gaussian BC can be characterized in terms of the capacity region of the dual MAC.

Theorem 3.1 *The capacity region of a constant Gaussian BC with power P is equal to the union of capacity regions of the dual MAC with power P_1 and P_2 such that $P_1 + P_2 = P$:*

$$\mathcal{C}_{BC}(P; h_1, h_2) = \bigcup_{0 \leq P_1 \leq P} \mathcal{C}_{MAC}(P_1, P - P_1; h_1, h_2). \tag{4}$$

In [16] it is also shown that each MAC boundary touches the BC boundary at a *different* point. More generally, each point along the BC boundary is intersected by a different MAC boundary. The capacity-achieving BC and MAC power schemes are related by an explicit duality power transformation. Moreover, the optimal decoding order to achieve any point on the boundary of the BC capacity region can be obtained by a decoding order transformation of the decoding order in the dual MAC channel that intersects the capacity region at that point. More details on the power and decoding order transformations can be found in [16].

We now show that the capacity region of the MAC can be characterized in terms of the capacity region of the dual BC. This theorem is proved in [16].

Theorem 3.2 *The capacity region of a constant Gaussian MAC is equal to the intersection of the capacity regions of the scaled dual BC over all scalings:*

$$\mathcal{C}_{MAC}(P_1, P_2; h_1, h_2) = \bigcap_{\alpha > 0} \mathcal{C}_{BC}(\frac{P_1}{\alpha} + P_2; \alpha h_1, h_2). \tag{5}$$

Moreover, the optimal power allocation and decoding order for the MAC can be obtained by the power and decoding order transformation described above. Thus, the optimal transmission strategy to achieve any point on the MAC capacity region can be obtained from the dual BC channel. Note that since MAC region is a pentagon, the BC channels characterized by $\alpha = 0$, $\alpha = h_2/h_1$ and $\alpha = \infty$ are sufficient to form the pentagon. If $\alpha = h_2/h_1$, the channel gains of both users are the same and the BC capacity region is bounded by a straight line segment because the capacity region can be achieved by time-sharing. It can be shown [16] that as $\alpha \to 0$, $R_1 \to \log(1+\frac{h_1 P_1}{\sigma^2})$ and $R_2 \to \infty$. Similarly, as $\alpha \to \infty$, $R_1 \to \infty$ and $R_2 \to \log(1+\frac{h_2 P_2}{\sigma^2})$. These two regions bound the vertical and horizontal line segments, respectively, of the MAC capacity region. All scaled BC capacity regions except the $\alpha = h_2/h_1$ channel intersect the MAC at exactly one of the two corner points of the MAC region. The $\alpha = h_2/h_1$ channel intersects the MAC region along its time-sharing line.

4 Capacity of Fading Multiuser Channels

Fading channels exhibit random amplitude variations over time. Capacity for such channels depends on what is known about the channel fading at both transmitter and receiver, and whether or not capacity can be averaged over many channel states or must be achieved in each state. We focus on the case where both transmitter and receiver track the channel state perfectly and instantaneously. We also assume a slowly fading channel relative to codeword length, so the channel is constant for the transmission of a single codeword. For this channel there are four different definitions of capacity:

- **Ergodic capacity:** The maximum rate or rate set averaged over all channel states.

- **Outage capacity:** The maximum rate that can be achieved in all non-outage fading states subject to some outage probability. Under an outage probability of zero this is called the zero-outage capacity.

- **Minimum rate capacity:** The maximum rate averaged over all fading states subject to some minimum rate required in every fading state.

- **Minimum rate capacity with outage:** The maximum rate averaged over all fading states subject to some minimum rate required in all non-outage fading states, with some nonzero outage probability.

The ergodic capacity region for fading broadcast and multiple access channels has been derived in [17] and [18] respectively. The optimal power allocation to achieve this capacity corresponds to a multi-level water-filling over both time (i.e. fading states) and users. As expected, users are allocated the most power when their channels are strong, and little, if any, power when their channels are weak. This results in a channel-dependent delay, which may not work for delay-constrained applications like voice and video.

The outage capacity region for fading broadcast and multiple access channels has been derived in [19] and [20, 21], respectively. For outage capacity each user maintains a constant rate some percentage of time with no data transmitted the rest of the time. The optimal power allocation over nonoutage must effectively invert the fading to eliminate channel variations such that a constant rate can be maintained. Clearly the ergodic capacity region exceeds the outage capacity region, since outage capacity has an additional constraint of constant rate transmission in all fading states. Since the weak channel states will effectively drive the maximum constant rate that can be maintained, outage capacity fails to fully exploit good channel states. However, outage capacity has the advantage that with fixed rate transmission, there is no channel-dependent delay.

Minimum rate capacity combines the concepts of ergodic and outage capacity to maximize the rate averaged over all channel states while maintaining some minimum rate in every fading state. Minimum rate capacity for fading broadcast channels has been derived in [22]. The optimal power allocation for the broadcast channel is a two-step process, where first the minimum power required to achieve the minimum rates in all fading states is allocated, and then the excess power to maximize average rate in excess of the minimum rate is allocated. The optimal allocation of the excess power is a multi-level water-filling based on effective noise that incorporates the minimum rate constraints. We will use these results in Section 6 to obtain the minimum rate capacity, optimal power allocation, and optimal decoding order of the MAC using duality.

5 Capacity and Duality

We now state two theorems, proved in [16], indicating the duality between the ergodic capacity region of the fading MAC and BC.

Theorem 5.1 *The capacity region of a fading Gaussian BC with power constraint $\overline{P}$ is equal to the union of ergodic capacity regions of the dual MAC with power constraints $\overline{P}_1$ and $\overline{P}_2$ such that $\overline{P}_1 + \overline{P}_2 = \overline{P}$:*

$$\mathcal{C}_{BC}(\overline{P}; H_1, H_2) = \bigcup_{0 \le \overline{P}_1 \le \overline{P}} \mathcal{C}_{MAC}(\overline{P}_1, \overline{P} - \overline{P}_1; H_1, H_2). \tag{6}$$

Theorem 5.2 *The ergodic capacity region of a fading MAC channel is equal to the intersection of the ergodic capacity regions of the scaled dual BC over all scalings:*

$$\mathcal{C}_{MAC}(\overline{P}_1, \overline{P}_2; H_1, H_2) = \bigcap_{\alpha > 0} \mathcal{C}_{BC}(\frac{\overline{P}_1}{\alpha} + \overline{P}_2; \alpha H_1, H_2). \tag{7}$$

The duality of the Gaussian MAC and BC is a general result that also holds for all definitions of fading channel capacity given in the previous section. The outage and minimum rate definitions of capacity differ from the ergodic capacity definition since they have restrictions on the *instantaneous* transmission rates (i.e. the state-by-state rates). However, as shown in [16], the duality power transformation described in the previous section preserves state-by-state rates (i.e. the same instantaneous rates are achieved in the dual channels). Thus, Theorem 5.1 and 5.2 hold for minimum rate and outage capacity as well as ergodic capacity. In the next two sections we use duality to obtain new capacity results from known results in the dual channels.

6 Minimum Rate Capacity

In this section we show that the minimum rate capacity region of the MAC and BC are duals of each other, so that the capacity region and capacity-achieving transmission strategy for one channel can be found from the other dual channel. We then use this result to find the minimum rate capacity region of the MAC using the known minimum rate capacity region of the BC obtained in [22]. The minimum rate capacity of the MAC and BC, denoted respectively by $\mathcal{C}_{MAC}^{min}(\overline{P}_1, \overline{P}_2, \mathbf{R}^*; H_1, H_2)$ and $\mathcal{C}_{BC}^{min}(\overline{P}_1 + \overline{P}_2, \mathbf{R}^*; H_1, H_2)$, is defined as the maximum ergodic capacity that can be obtained while ensuring that a set of minimum rates $\mathbf{R}^* = (R_1^*, \ldots, R_M^*)$ is maintained for all users in *all* fading states. Clearly the minimum rates themselves must be in the zero-outage capacity of the channel for the rates to be achievable in all fading states. Every feasible $\mathbf{R}^*$ defines a different minimum rate capacity region (i.e. the minimum rate capacity region is a function of $\mathbf{R}^*$ in addition to the channel and power constraints).

We now state two theorems, proved in [23], indicating the duality between the ergodic capacity region of the fading MAC and BC.

Theorem 6.1 *The minimum rate capacity region of a fading Gaussian BC with perfect channel information at the transmitter and receivers is given by:*

$$\mathcal{C}_{BC}^{min}(\overline{P}, \mathbf{R}^*; H_1, H_2) = \bigcup_{0 \le \overline{P}_1 \le \overline{P}} \mathcal{C}_{MAC}^{min}(\overline{P}_1, \overline{P} - \overline{P}_1, \mathbf{R}^*; H_1, H_2), \quad (8)$$

for all $\mathbf{R}^* \in \mathcal{C}_{BC}^{0}(\overline{P}; H_1, H_2)$.

Theorem 6.2 *The minimum rate capacity region of a fading Gaussian MAC with perfect channel i nformation at the transmitters and receiver is given by:*

$$\mathcal{C}_{MAC}^{min}(\overline{P}_1, \overline{P}_2, \mathbf{R}^*; H_1, H_2) = \bigcap_{\alpha \ge 0} \mathcal{C}_{BC}^{min}(\frac{\overline{P}_1}{\alpha} + \overline{P}_2, \mathbf{R}^*; \alpha H_1, H_2) \quad (9)$$

for all $\mathbf{R}^* \in \mathcal{C}_{MAC}^{0}(\overline{P}_1, \overline{P}_2; H_1, H_2)$.

We now use the fact that every point along the boundary of the MAC minimum rate region intersects the boundary of the BC minimum rate region for some value of α. We also know that the power allocation schemes to achieve the intersection point in the MAC and BC are related through the BC-to-MAC power transformation, so from knowledge of the BC power allocation scheme *we can find the MAC power allocation scheme f or all points along the MAC minimum rate capacity region boundary.* Because the boundary of the BC minimum rate capacity region is achieved by super position coding and because the BC-to-MAC power transformation translates a BC power allocation sche me to a MAC power allocation scheme that uses successive decoding, we know that *successive decoding in the MAC will achieve points along the boundary of its minimum rate capacity region.* This powerful result allows us to explicitly characterize the MAC minimum rate capacity region and its corresponding capacity-achieving strategy without having to directly solve for it. Specifically, these results are obtained by applying the power and decoding transformations to the known optimal transmission strategies of the dual BC obtained in [22].

7 MIMO Multiuser Channels, Duality, and Dirty Paper Coding

Multiple input multiple output (MIMO) systems have received a great deal of attention as a method to achieve very high data rates over wireless links. While the capacity region of the MIMO MAC has been fully characterized [10, 24, 25], the capacity of the MIMO BC is difficult to obtain due to a lack of general theory on the capacity of non-degraded broadcast channels. In this section we establish the duality between an achievable rate region for the MIMO BC and the MIMO MAC capacity region, which is easy to compute. This yields a lower bound on the MIMO BC capacity region. Moreover, we show that the lower bound achieves sum rate capacity, so it is fairly tight.

We consider a K-user MIMO Gaussian BC in which receiver k has r_k receive antennas and the transmitter has t transmit antennas. The $r_k \times t$ matrix $\mathbf{H_k}$ defines the channel gains from the transmitter to the r_k antennas of receiver k. Each receiver has additive white Gaussian noise with unit variance. The dual MIMO MAC channel is arrived at by converting the receivers in the BC into transmitters in the MAC and converting the t-antenna transmitter into a t-antenna receiver with additive noise of unit variance. The channel gains of the dual MAC are the same as that of the broadcast channel. Specifically, if $\mathbf{H_k}$ is the $t \times r_k$ matrix defining the channel gains from t-antenna transmitter to the kth receiver with r_k antennas in the BC, then $\mathbf{H_k}^\dagger$ is the $r_k \times t$ matrix defining the channel gains from transmitter k with r_k antennas to the t-antenna receiver in the dual MAC.

The capacity region of a general MIMO MAC was obtained in [10, 24, 25]. We now describe this capacity region for the dual MIMO MAC of our BC model. For any set of powers $(P_1, \ldots, P_K)$, the capacity of the MIMO MAC is

$$\mathcal{C}_{\text{MAC}}(P_1, \ldots, P_K; \mathbf{H}^\dagger) \triangleq \bigcup_{\{\text{Tr}(\mathbf{P}_1) \leq P_i \ \forall i\}} \Big\{ (R_1, \ldots, R_K) : \tag{10}$$

$$\sum_{i \in S} R_i \leq \frac{1}{2} \log |\mathbf{I} + \sum_{i \in S} \mathbf{H}_i^\dagger P_i \mathbf{H}_i| \ \ \forall S \subseteq \{1, \ldots, M\} \Big\} \tag{11}$$

For $P > 0$, we denote by $\mathcal{C}_{\text{union}}(P, \mathbf{H}^\dagger)$

$$\mathcal{C}_{\text{union}}(P, \mathbf{H}^\dagger) = \bigcup_{\sum_{i=0}^{K} P_i \leq P} \mathcal{C}_{\text{MAC}}(P_1, \ldots, P_K; \mathbf{H}^\dagger). \tag{12}$$

It can be easily shown that this region is the capacity region of a MAC when the transmitters have a sum power constraint instead of individual power constraints but are not allowed to cooperate.

An achievable region for the MIMO BC was first obtained in [1]. In [26], the region was extended to the more general multiple-user, multiple-antenna case using the following extension of the "dirty paper result" [2] to the vector case:

Lemma 7.1 *Consider a channel with $\mathbf{y}_k = \mathbf{H}_k \mathbf{x}_k + \mathbf{s}_k + \mathbf{n}_k$, where $\mathbf{y}_k$ is the received vector, $\mathbf{x}_k$ the transmitted vector, $\mathbf{s}_k$ the vector Gaussian interference, and $\mathbf{n}_k$ the vector white Gaussian noise. If $\mathbf{s}_k$ and $\mathbf{n}_k$ are i ndependent and non-causal knowledge of $\mathbf{s}_k$ is available at the transmitt er but not at the receiver, then the capacity of the channel is the same as if $\mathbf{s}_k$ is not present.*

This result is termed the "dirty paper result" for the following reason. If one wished to write some information on a paper covered with dirt, one could convey all the information one desires to convey by cleverly distributing the writing on the paper and by using different kinds of inks. For example, if the dirt blackens the paper, then using white ink would convey as much information as black ink would on white paper.

In other words, there is no difference between writing on a dirty paper and writing on a clean paper as long as one knows the kind of dirt on the paper. This is similar to the situation in Lemma 7.1 with the dirty paper being the channel with interference.

In the MIMO BC, this result can be applied at the transmitter when choosing codewords for different receivers. The transmitter first picks a codeword for receiver 1. The transmitter then chooses a codeword for receiver 2 with full (non-causal) knowledge of the codeword intended for receiver 1. Therefore receiver 2 does not see the codeword intended for receiver 1 as interference. Similarly, the codeword for receiver 3 is chosen such that receiver 3 does not see the signals intended for receivers 1 and 2 as interference. This process continues for all K receivers. Since the ordering of the users clearly matters in such a procedure, the following is an achievable set of rates

$$R_{\pi(i)} = \frac{1}{2}\log\frac{|\mathbf{I}+\mathbf{H}_{\pi(i)}(\sum_{j\geq i}\mathbf{\Sigma}_{\pi(j)})\mathbf{H}^{\dagger}_{\pi(i)}|}{|\mathbf{I}+\mathbf{H}_{\pi(i)}(\sum_{j>i}\mathbf{\Sigma}_{\pi(j)})\mathbf{H}^{\dagger}_{\pi(i)}|} \quad i=1,\ldots,K. \tag{13}$$

The dirty-paper region $\mathcal{C}_{\text{dirtypaper}}(P,\mathbf{H})$ is defined as the union of all such rate vectors over all covariance matrices $\mathbf{\Sigma}_1,\ldots,\mathbf{\Sigma}_K$ such that $\text{Tr}(\mathbf{\Sigma}_1+\ldots\mathbf{\Sigma}_K) = \text{Tr}(\mathbf{\Sigma}_x) \leq P$ and over all decoding order permutations $(\pi(1),\ldots,\pi(K))$. The transmitted signal is $\mathbf{x}=\mathbf{x}_1+\ldots+\mathbf{x}_K$ and the input covariance matrices are of the form $\mathbf{\Sigma}_i = \mathbf{E}[\mathbf{x}_i\mathbf{x}_i{}^{\dagger}]$.

Our next theorem, proved in [27], indicates that the capacity region of the MIMO MAC with a total power constraint of P for the K transmitters is the same as the dirty paper region of the dual MIMO BC with power constraint P. In other words, any rate vector that is achievable in the dual MAC with power constraints $(P_1,\ldots,P_K)$ is in the dirty paper region of the BC with power constraint $\sum_{i=1}^{K}P_i$. Conversely, any rate vector that is in the dirty paper region of the BC is also in the dual MIMO MAC region with the same total power constraint.

Theorem 7.1 *The dirty paper region of a MIMO BC channel with power constraint P is equal to the the capacity region of the dual MIMO MAC with sum power constraint P.*

$$\mathcal{C}_{\text{dirtypaper}}(P,\mathbf{H}) = \mathcal{C}_{\text{union}}(P,\mathbf{H}^{\dagger}).$$

Finally, the duality of the dirty paper region and MAC capacity region along with Sato's upper bound can be used to show that the dirty paper coding achieves the sum rate capacity of the MIMO broadcast channel [27]. This shows that the achievable region is a tight lower bound, at least near the sum rate point.

8 Conclusions

Duality is a power technique to connect capacity regions for broadcast and MAC channels. Duality holds for many notions of capacity, including ergodic, outage, and minimum rate capacity. Duality can be used to show the connection between capacity regions and optimal transmission strategies for BC and MAC channels. More importantly, it can be used to obtain new results previously thought intractable for the capacity region of the MAC or BC channel when results are known for the dual channel. This concept is used to obtain the minimum rate capacity region of a fading MAC channel and achievable rates on the MIMO BC channel. We expect that the generality and power of the duality results are applicable to a very wide range of multiuser capacity problems.

References

[1] G. Caire and S. Shamai, "On Achievable Rates in a Multi-antenna Broadcast Downl ink", 38th Annual Allerton Conference on Commun., Control and Computing, Monticello, I L, Oct. 4 - 6, 2000.

[2] M. Costa, "Writing on Dirty Paper", IEEE. Trans. Inf. Theory, v. 29, pp. 439-4 41, May 1983.

[3] R. J. McEliece and W. Stark, "Channels with block interference," *IEEE Trans. Info. Theory,* pp. 44-53, Nov. 1984.

[4] E. Biglieri, J. Proakis, and S. Shamai, "Fading channels: information-theoretic and communications aspects," *IEEE Trans. Info. Theory,* pp. 2619-2692, Oct. 1998.

[5] R. J. McEliece and L. Swanson, "A note on the wideband broadcast channel," *IEEE Trans. Commun.,* pp. 452-453, April 1987.

[6] T. Cover, R. J. McEliece, and E. C. Posner, "Multiple access channel capacity is attainable without synchronization," *IEEE Trans. Info. Theory,* pp. 409-413, Jan. 1981.

[7] S. Verdú, "Spectral efficiency in the wideband regime," To appear: *IEEE Trans. Info. Theory,* April 2002.

[8] S. Verdú, G. Caire, and D. Tuninetti, "Is TDMA wideband optimal," To appear: *Proc. IEEE Intl. Symp. Info. Theory,* June 2002.

[9] A. Lapidoth, I. E. Teletar, and R. Urbanke, "On wideband broadcast channels," *Proc. IEEE Intl. Symp. Info. Theory,* pp. 188, Aug. 1998.

[10] S. Verdú, "Multiple-access channels with memory with and without frame synchronism," *IEEE Trans. Info. Theory,* pp. 605-619, May 1989.

[11] S. Verdú, "The capacity region of the symbol-asynchronous Gaussian multiple-access channel," *IEEE Trans. Info. Theory,* pp. 733-751, July 1989.

[12] M.R. Garey and D.S. Johnson. *Computers and Intractability: A Guide to the Theory of NP-Completeness.* W.H. Freeman and Co., New York, 1979.

[13] R. J. McEliece and K. N. Sivarajan, "Performance limits for channelized cellular telephone systems," *IEEE Trans. Info. Theory,* pp. 21-34, Jan. 1994.

[14] R. J. McEliece and S. Deora, "Asymptotic performance analysis of a timid DCA algorithm," *Proc. Conf. Inform. Sciences and Systems,* March 1995.

[15] I. Katzela and M. Naghshineh, "Channel assignment schemes for cellular mobile telecommunication systems - a comprehensive survey," *IEEE Pers. Commun. Mag.*, pp. 10–31, June 1996.

[16] N. Jindal, S. Vishwanath, and A. J. Goldsmith, "On the duality of multiple-access and broadcast channels," *Proc. IEEE Allerton Conf. Commun. Cont. Comput.*, Oct. 2001.

[17] L. Li and A. Goldsmith, "Capacity and optimal resource allocation for fading broadcast channels: Part I: Ergodic capacity", IEEE Trans. Inform. Theory, vol. 47, pp 1083-1102, March 2001.

[18] D.N. Tse and S. Hanly, "Multi-Access Fading Channels: Part I: Polymatroid Struct ure, Optimal Resource Allocation and Throughput Capacities", IEEE Trans. Inform. Theory, v. 44, pp 2796-2815, Nov. 1998.

[19] L. Li and A. Goldsmith, "Capacity and optimal resource allocation for fading broadcast channels: Part II: Outage capacity", IEEE Trans. Inform. Theory, vol. 47, pp 1103-1127, March 2001.

[20] S.V. Hanly and D. N. Tse, "Multiaccess fading channels- Part II:Delay-limited capacities", IEEE Trans. Inform. Theory, vol. 44, pp. 2816-2831, Nov. 1998.

[21] L. Li and A. Goldsmith, "Outage capacities and optimal power allocation for fading multiple-access channels", in revision for IEEE Trans. Inform. Theory. See also *Proc. IEEE Intl. Symp. Inform. Theory,* pg. 305, June 2000.

[22] N. Jindal and A. Goldsmith, "Capacity and optimal power allocation for fading broadcast channels with minimum rates", *Proc. IEEE Globecom,* pp. 1292-1296, Nov. 2001.

[23] N. Jindal, S. Vishwanath, and A. J. Goldsmith, "Duality of multiple-access and broadcast channels," In preparation for submission to *IEEE Trans. Inform. Theory.*

[24] E. Telatar, "Capacity of Multi-antenna Gaussian Channels," European Trans. on Telecomm. ETT, 10(6):585-596, November 1999.

[25] W. Yu, W. Rhee, S. Boyd, J. Cioffi, "Iterative Water-filling for Vector Multipl e Access Channels", pp. 322, Proc. IEEE Int. Symp. Inf. Theory, (ISIT), Washington DC, June 24-29, 2001.

[26] W. Yu and J. Cioffi, "Trellis Precoding for the Broadcast Channel", Proc. IEEE Globecom, pp. 1344-1338, Nov., 2001.

[27] S. Vishwanath, N. Jindal, and A. J. Goldsmith, "On the capacity of multiple input multiple output broadcast channels," To appear: *Proc. IEEE Intl. Conf. Commun.*, April 2002.

Stability Analysis of the Turbo Decoding Algorithm Using Max-Log-MAP

Wei-Sheng Wu[1], Chung-Hsuan Wang[2], Mao-Ching Chiu[3] and Chi-chao Chao[1]

Abstract

In this study, we investigate analytically the stability and convergence properties of the iterative turbo decoding with the Max-Log-MAP algorithm for soft-output generation. With respect to the turbo decoding algorithm (TDA) using Max-Log-MAP, several equivalent forms of the mathematical formulation are derived to facilitate the stability analysis. Based on the derived formulation, we show that the TDA using Max-Log-MAP has at least one fixed point regardless of the choice of constituent codes. By some specially constructed examples, we also show that the TDA using Max-Log-MAP is in general suboptimal for symbol decision and may not even converge. Finally, from the viewpoint of the derived formulation, some conditions are given to guarantee the convergence of the iterative turbo decoding.

Keywords: Turbo codes, iterative decoding, Max-Log-MAP, stability analysis.

[1]Department of Electrical Engineering, National Tsing Hua University, Hsinchu 30013, Taiwan, R.O.C., wswu@ee.nthu.edu.tw, ccc@ee.nthu.edu.tw

[2]Department of Electronic Engineering, Chung Yuan Christian University, Chungli 32023, Taiwan, R.O.C., chwang@cycu.edu.tw

[3]Department of Electrical Engineering, National Chi Nan University, Puli, Nantou 545, Taiwan, R.O.C., mcchiu@ncnu.edu.tw

1 Introduction

Turbo codes [1] have been considered as one of the most exciting developments in the coding community, owing to their near-capacity performance achieved by the iterative decoding. However, due to the iterative decoding nature, there is no guarantee that the turbo decoding algorithm (TDA) will always converge. Actually, it has been shown that the TDA with the symbol-by-symbol maximum a posteriori probability (MAP) algorithm [2] for soft-output generation may diverge in some special cases [3]. Several related studies [3]–[10] on the convergence issue have also been done. In [4], the TDA using MAP has been proved to always have fixed points regardless of the specific constituent codes of the turbo encoder, and some conditions on the uniqueness and stability of the fixed points have been given in [5]. The TDA has also been formulated as a discrete dynamical system from a geometric point of view [5], and its phase trajectory as a function of the signal-to-noise ratio has been studied in [6]. Moreover, the TDA has been shown to be equivalent to Pearl's belief propagation algorithm as it is applied to the "belief network" of a turbo code [10]. Research to better understand why the TDA performs so well is still quite active.

For iterative decoding, many soft-output generation algorithms, which vary a lot in bit-error-rate (BER) performance and computational complexity, have been proposed such as the BCJR algorithm [2], the Max-Log-MAP algorithm [11][12], the SOVA [13][14], and the modified SOVA [15][16], etc. Though the MAP algorithm can provide the best BER performance for turbo decoding, its decoding complexity for certain applications may be considered not affordable. The Max-Log-MAP algorithm, on the other hand, has a better trade-off between the BER performance and decoding complexity among the available soft-output algorithms and can be a good alternative for high-speed VLSI or DSP implementation. Therefore, in this study, we will focus on the convergence analysis of the TDA using Max-Log-MAP.

With respect to the TDA using Max-Log-MAP, we propose three equivalent mathematical formulations to perform the stability analysis of the iterative decoding scheme. In the first formulation, the decoding procedure is viewed as the iteration of a nonlinear function, which consists of the channel outputs and extrinsic information. This nonlinear function can be proved to have at least one fixed point, hence giving the possibility of convergence for the iterative decoding because it can be shown that the TDA using Max-Log-MAP must converge to a fixed point of this nonlinear function if it does converge. Although the TDA using Max-Log-MAP may converge, it seems difficult to find the fixed points directly from the nonlinear function. We thus propose the second formulation that transforms the nonlinear function into a set of affine functions which might be helpful in finding the fixed points. Finally, to characterize the dynamics of convergence behavior, the third formulation of matrix form is derived, based on which several sufficient conditions (one of which might also be necessary) are given to guarantee the convergence of this iterative algorithm.

The rest of this study is organized as follows. Section 2 presents the system description and the iterative decoding process of the TDA using Max-Log-MAP. Various forms of the mathematical formulations together with the convergence analysis are covered in Section 3. Section 4 concludes this study and indicates several possible directions for further research. Finally, some examples illustrating the convergence issues in Section 3 are given in Appendix.

2 System Description and Iterative Decoding

Consider the turbo coding system in Figure 1, where $\boldsymbol{U} = (U_1, \cdots, U_K)^T$ stands for the input information sequence of length K and $U_j \in \{0, 1\}$ for $1 \leq j \leq K$. Let $\boldsymbol{Y}_1$ and $\boldsymbol{Y}_2$ be the channel outputs of the coded bits $\boldsymbol{C}_1$ and $\boldsymbol{C}_2$ corresponding to $\boldsymbol{U}$, and assume the channels to be independent, i.e., $\Pr\{\boldsymbol{Y}_1 = \boldsymbol{y}_1, \boldsymbol{Y}_2 = \boldsymbol{y}_2 | \boldsymbol{U} = \boldsymbol{u}\} = p_1(\boldsymbol{y}_1|\boldsymbol{u})p_2(\boldsymbol{y}_2|\boldsymbol{u})$ with $p_l(\boldsymbol{y}_l|\boldsymbol{u})$ denoting the transition probability of the lth channel and $p_l(\boldsymbol{y}_l|\boldsymbol{u}) > 0$ for $l = 1, 2$. As shown in Figure 2, the extrinsic information $\tilde{\boldsymbol{L}}_1^{(n)}$ and $\tilde{\boldsymbol{L}}_2^{(n)}$ corresponding to $\boldsymbol{U}$ are computed based on $\boldsymbol{Y}_1$ and $\boldsymbol{Y}_2$ and then exchanged between the two constituent decoders so as to allow each decoder to incorporate information extracted exclusively from the other decoder.

Let $\tilde{\boldsymbol{L}}_l^{(n)} = (\tilde{L}_{l,1}^{(n)}, \cdots, \tilde{L}_{l,K}^{(n)})^T$ for $l = 1, 2$. With the MAP algorithm for soft output generation, the lth decoder, for $l = 1, 2$, computes the reliability estimate (i.e., the soft output) $\hat{L}_{l,j}^{(n)}$ of U_j and extrinsic information

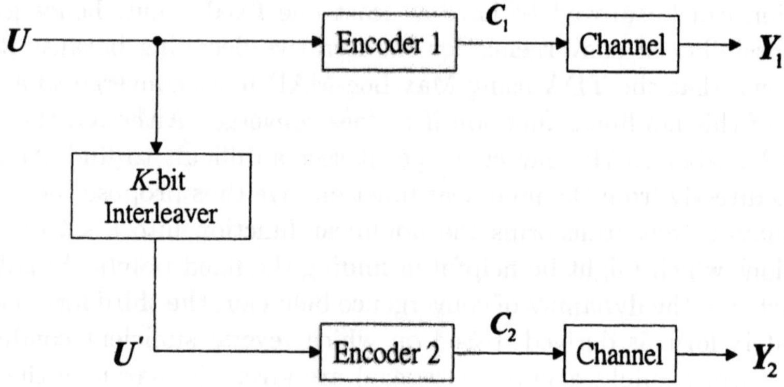

Figure 1: The turbo coding system.

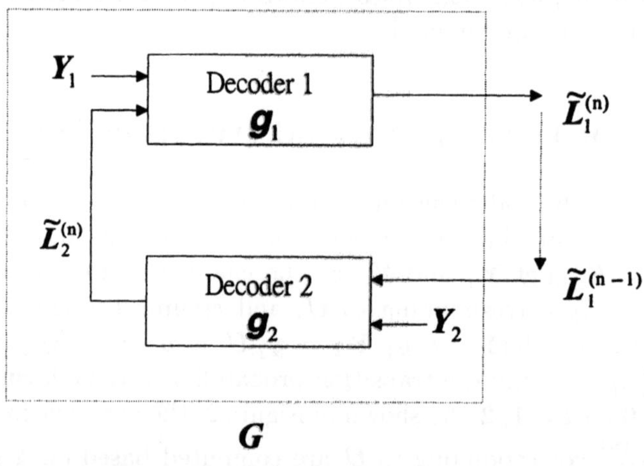

Figure 2: The iterative turbo decoding scheme.

$\tilde{L}_{l,j}^{(n)}$, for $j = 1, 2, \ldots, K$, at the nth iteration by

$$\hat{L}_{l,j}^{(n)} = \log \frac{\Pr\{\hat{U}_{l,j}^{(n)} = 1 | \boldsymbol{Y}_l = \boldsymbol{y}_l\}}{\Pr\{\hat{U}_{l,j}^{(n)} = 0 | \boldsymbol{Y}_l = \boldsymbol{y}_l\}} = \log \frac{\sum_{\boldsymbol{u}:u_j=1} p_l(\boldsymbol{y}_l|\boldsymbol{u}) \prod_{i=1}^{K} e^{u_i \Lambda_{l,i}^{(n)}}}{\sum_{\boldsymbol{u}:u_j=0} p_l(\boldsymbol{y}_l|\boldsymbol{u}) \prod_{i=1}^{K} e^{u_i \Lambda_{l,i}^{(n)}}} \quad (1)$$

$$\tilde{L}_{l,j}^{(n)} = \hat{L}_{l,j}^{(n)} - \Lambda_{l,j}^{(n)}$$

where $\hat{U}_{l,j}^{(n)}$ denotes the decision about U_j and $\Lambda_{l,j}^{(n)}$ stands for the a priori information of $\hat{U}_{l,j}^{(n)}$, i.e., $\Lambda_{l,j}^{(n)} = \log(\Pr\{\hat{U}_{l,j}^{(n)} = 1\} / \Pr\{\hat{U}_{l,j}^{(n)} = 0\})$. For iterative turbo decoding, $\tilde{L}_{1,j}^{(n-1)}$ is used for $\Lambda_{2,j}^{(n)}$, and $\tilde{L}_{2,j}^{(n)}$ for $\Lambda_{1,j}^{(n)}$, for $j = 1, 2, \cdots, K$. By applying the following approximation to (1):

$$\log(\sum_i e^{A_i}) \approx \max_i A_i, \text{ where } A_i \text{ is a real number for all } i$$

the formula for computing $\tilde{L}_{l,j}^{(n)}$ and $\hat{L}_{l,j}^{(n)}$ using the Max-Log-MAP algorithm can be derived as follows:

$$\begin{aligned} \tilde{L}_{1,j}^{(n)} &= \max_{\boldsymbol{u}:u_j=1} \left[\log(p_1(\boldsymbol{y}_1|\boldsymbol{u})) + \sum_{i=1, i\neq j}^{K} u_i \tilde{L}_{2,i}^{(n)} \right] \\ &\quad - \max_{\boldsymbol{u}:u_j=0} \left[\log(p_1(\boldsymbol{y}_1|\boldsymbol{u})) + \sum_{i=1, i\neq j}^{K} u_i \tilde{L}_{2,i}^{(n)} \right] \end{aligned} \quad (2)$$

$$\begin{aligned} \tilde{L}_{2,j}^{(n)} &= \max_{\boldsymbol{u}:u_j=1} \left[\log(p_2(\boldsymbol{y}_2|\boldsymbol{u})) + \sum_{i=1, i\neq j}^{K} u_i \tilde{L}_{1,i}^{(n-1)} \right] \\ &\quad - \max_{\boldsymbol{u}:u_j=0} \left[\log(p_2(\boldsymbol{y}_2|\boldsymbol{u})) + \sum_{i=1, i\neq j}^{K} u_i \tilde{L}_{1,i}^{(n-1)} \right]. \end{aligned} \quad (3)$$

$$\hat{L}_{1,j}^{(n)} = \tilde{L}_{1,j}^{(n)} + \tilde{L}_{2,j}^{(n)} \quad \text{and} \quad \hat{L}_{2,j}^{(n)} = \tilde{L}_{2,j}^{(n)} + \tilde{L}_{1,j}^{(n-1)}.$$

Let $\hat{\boldsymbol{L}}_l^{(n)} = (\hat{L}_{l,1}^{(n)}, \cdots, \hat{L}_{l,K}^{(n)})^T$ and $\hat{\boldsymbol{U}}_l^{(n)} = (\hat{U}_{l,1}^{(n)}, \cdots, \hat{U}_{l,K}^{(n)})^T$ for $l = 1, 2$. With $\tilde{L}_{1,j}^{(0)}$ initialized to 0 for all j, the sequence of extrinsic information is generated in the fashion of

$$\tilde{\boldsymbol{L}}_1^{(0)} \to \tilde{\boldsymbol{L}}_2^{(1)} \to \tilde{\boldsymbol{L}}_1^{(1)} \to \tilde{\boldsymbol{L}}_2^{(2)} \to \tilde{\boldsymbol{L}}_1^{(2)} \to \cdots$$

and the turbo decision about U_j at the nth iteration is determined by

$$\hat{U}_{1,j}^{(n)} = \begin{cases} 1 & \text{if } \hat{L}_{1,j}^{(n)} \geq 0 \\ 0 & \text{if } \hat{L}_{1,j}^{(n)} < 0, \end{cases} \qquad \text{for } 1 \leq j \leq K.$$

The iterative process is repeated until a consensus is reached about the value of each information bit U_i.

3 Dynamics of the TDA Using Max-Log-MAP

Rewrite (2) and (3) into the following vector form:

$$\tilde{\boldsymbol{L}}_1^{(n)} = \boldsymbol{g}_1(\tilde{\boldsymbol{L}}_2^{(n)}) \text{ and } \tilde{\boldsymbol{L}}_2^{(n)} = \boldsymbol{g}_2(\tilde{\boldsymbol{L}}_1^{(n-1)})$$

where $\boldsymbol{g}_l = (g_{l,1}, \cdots, g_{l,K})^T$ for $l = 1, 2$, and the component functions $g_{1,j}$'s and $g_{2,j}$'s are given in (2) and (3), respectively. We can then express the TDA using Max-Log-MAP as the following iterative formulation:

$$\begin{aligned} \tilde{\boldsymbol{L}}_1^{(n)} &= \boldsymbol{g}_1 \circ \boldsymbol{g}_2(\tilde{\boldsymbol{L}}_1^{(n-1)}) \triangleq \boldsymbol{G}(\tilde{\boldsymbol{L}}_1^{(n-1)}) \\ \hat{\boldsymbol{L}}_1^{(n)} &= \tilde{\boldsymbol{L}}_1^{(n)} + \tilde{\boldsymbol{L}}_2^{(n)}. \end{aligned} \tag{4}$$

In this study, the convergence of iterative turbo decoding is defined in the strict sense that $\lim_{n\to\infty} \hat{\boldsymbol{L}}_1^{(n)}$ exists rather than that $\lim_{n\to\infty} \hat{\boldsymbol{U}}_1^{(n)}$ exists. Since $\boldsymbol{g}_1$ and $\boldsymbol{g}_2$ can be shown to be continuous, it is easy to see that this definition of convergence is equivalent to that $\lim_{n\to\infty} \tilde{\boldsymbol{L}}_1^{(n)}$ exists. Hence, studying the dynamics of the TDA using Max-Log-MAP is equivalent to studying how the extrinsic information $\tilde{\boldsymbol{L}}_1^{(n)}$ change under the nonlinear iteration function $\boldsymbol{G}$, defined in (4). In Section 3.1, we will first show that $\boldsymbol{G}$ is a continuous mapping and has at least one fixed point, i.e., there exists $\boldsymbol{x}^* \in \boldsymbol{R}^K$ (the K-tuple space of the real numbers $\boldsymbol{R}$) such that $\boldsymbol{G}(\boldsymbol{x}^*) = \boldsymbol{x}^*$. Besides, if the TDA using Max-Log-MAP converges, i.e., $\lim_{n\to\infty} \tilde{\boldsymbol{L}}_1^{(n)} = \tilde{\boldsymbol{L}}_1$ exists, then $\tilde{\boldsymbol{L}}_1$ is a fixed point of $\boldsymbol{G}$, thereby implying the possibility of the TDA using Max-Log-MAP for convergence.

3.1 Fixed Points of the TDA Using Max-Log-MAP

Before proving the existence of the fixed points of $\boldsymbol{G}$, we need the following two lemmas.

Lemma 3.1 Let $f_1(\boldsymbol{x}), \cdots, f_K(\boldsymbol{x})$ be continuous functions from $\boldsymbol{R}^K$ to $\boldsymbol{R}$. Then $\max_{i \in S} f_i(\boldsymbol{x})$ is also a continuous function from $\boldsymbol{R}^K$ to $\boldsymbol{R}$, where $S = \{1, 2, \cdots, K\}$ is an index set.

Proof: Consider a point $\boldsymbol{x}_0$ in $\boldsymbol{R}^K$. Let $\max_{i \in S} f_i(\boldsymbol{x}_0) = p$ and define a subset S_1 of S by $S_1 = \{i : i \in S \text{ and } \max_{i \in S} f_i(\boldsymbol{x}_0) = p\}$. Since $f_1(\boldsymbol{x}), f_2(\boldsymbol{x}), \cdots, f_K(\boldsymbol{x})$ are continuous, it can be shown that there exists $\delta^* > 0$ such that $\max_{i \in S} f_i(\boldsymbol{x}) = \max_{i \in S_1} f_i(\boldsymbol{x})$ for all $\boldsymbol{x}$ with $||\boldsymbol{x} - \boldsymbol{x}_0|| < \delta^*$. Given $\epsilon > 0$, owing to the continuity of $f_i(\boldsymbol{x})$'s, we have that, there exists $\delta_i > 0$ such that $|f_i(\boldsymbol{x}) - p| < \epsilon$ for all $i \in S_1$ and $\boldsymbol{x}$ with $||\boldsymbol{x} - \boldsymbol{x}_0|| < \delta_i$. Let $\delta = \{\delta^*, \min_{i \in S_1} \delta_i\}$. Then, for all $\boldsymbol{x}$ with $||\boldsymbol{x} - \boldsymbol{x}_0|| < \delta$, we have

$$\left| \max_{i \in S}(f_i(\boldsymbol{x})) - \max_{i \in S}(f_i(\boldsymbol{x}_0)) \right| = \left| \max_{i \in S_1}(f_i(\boldsymbol{x})) - p \right| < \epsilon$$

hence implying the continuity of $\max_{i \in S} f_i(\boldsymbol{x})$. ■

Lemma 3.2 Consider the function $g : \boldsymbol{R}^K \to \boldsymbol{R}$, defined as

$$g(\boldsymbol{y}) = \max(A_0, A_1 + y_1, \cdots, A_K + y_K) - \max(B_0, B_1 + y_1, \cdots, B_K + y_K)$$

where $\boldsymbol{y} = (y_1, \cdots, y_K)^T$ is any point in $\boldsymbol{R}^K$ and the coefficients A_i's and B_i's are finite real numbers. Then the function g is continuous and bounded.

Proof: From Lemma 3.1, we know that $\max(A_0, A_1 + y_1, \cdots, A_K + y_K)$ and $\max(B_0, B_1 + y_1, \cdots, B_K + y_K)$ are continuous functions. Hence, g is continuous. Now let

$$A_{max} = \max(A_0, \cdots, A_K), \quad A_{min} = \min(A_0, \cdots, A_K)$$

$$B_{max} = \max(B_0, \cdots, B_K), \quad B_{min} = \min(B_0, \cdots, B_K).$$

Then we have

$$[A_{min} + \max(0, y_1, \cdots, y_K)] - [B_{max} + \max(0, y_1, \cdots, y_K)] \leq g(\boldsymbol{y})$$

$$[A_{max} + \max(0, y_1, \cdots, y_K)] - [B_{min} + \max(0, y_1, \cdots, y_K)] \geq g(\boldsymbol{y})$$

which implies

$$A_{min} - B_{max} \leq g(\boldsymbol{y}) \leq A_{max} - B_{min}.$$

Let $M = \max(|A_{min} - B_{max}|, |A_{max} - B_{min}|)$. Then we have $|g(\boldsymbol{y})| \leq M$, i.e., g is bounded. ∎

Together with the Brouwer's Theorem [17] in the fixed point theory, which states that if $\boldsymbol{B}^K$ is a K dimensional hypercube in $\boldsymbol{R}^K$ and $\boldsymbol{\Phi}$ is a continuous mapping from $\boldsymbol{B}^K$ to $\boldsymbol{B}^K$, then there exists at least one fixed point $\boldsymbol{x}^* \in \boldsymbol{B}^K$ such that $\boldsymbol{\Phi}(\boldsymbol{x}^*) = \boldsymbol{x}^*$, we can then have the following theorem for the existence of the fixed points of $\boldsymbol{G}$.

Theorem 3.1 The function $\boldsymbol{G}$ has at least one fixed point, regardless of the choice of constituent codes.

Proof: First, we claim that the function $\boldsymbol{G}$ is continuous. By defining $y_t = \sum_{i=1,i\neq j}^{K} u_i \tilde{L}_{2,i}^{(n)}$ and $z_t = \sum_{i=1,i\neq j}^{K} u_i \tilde{L}_{1,i}^{(n-1)}$ and substituting y_t and z_t into $g_{1,j}$ and $g_{2,j}$, defined in (2) and (3), respectively, the component functions $g_{1,j}$ and $g_{2,j}$, for $j = 1, \cdots, K$, have the form of the function g in Lemma 3.2. Hence, the functions $g_{1,j}$ and $g_{2,j}$, for $j = 1, \cdots, K$, are continuous by applying Lemma 3.2. The functions $\boldsymbol{g}_1$ and $\boldsymbol{g}_2$ are then continuous since component functions $g_{1,j}$ and $g_{2,j}$ are continuous. Hence, $\boldsymbol{G} = \boldsymbol{g}_1 \circ \boldsymbol{g}_2$ being the composition of continuous functions is continuous.

Second, we claim that the function $\boldsymbol{G}$ is bounded. Notice that log $(p(\boldsymbol{y}_i|\boldsymbol{u}))$ are finite real numbers since we assume the channel transition probabilities $p(\boldsymbol{y}_i|\boldsymbol{u}) \triangleq \Pr\{\boldsymbol{Y}_i = \boldsymbol{y}_i|\boldsymbol{U} = \boldsymbol{u}\} > 0$ for $i = 1,\ 2$. Besides, we know that the component functions $g_{1,j}$ and $g_{2,j}$ have the form of the function g in Lemma 3.2 after substituting y_t and z_t into $g_{1,j}$ and $g_{2,j}$, respectively. Hence, they are bounded by applying Lemma 3.2, i.e., there exist finite positive numbers $M_{1,j}$ and $M_{2,j}$ such that $|g_{1,j}(\boldsymbol{x})| \in [0, M_{1,j}]$ and $|g_{2,j}(\boldsymbol{x})| \in [0, M_{2,j}]$, for all $\boldsymbol{x}$ in $\boldsymbol{R}^K$ and $j = 1, \cdots, K$.

Finally, let $M = \max(M_{1,1}, \cdots, M_{1,K}, M_{2,1}, \cdots, M_{2,K})$. We have $|g_{1,j}(\boldsymbol{x})| \in [0, M]$ and $|g_{2,j}(\boldsymbol{x})| \in [0, M]$, for all $\boldsymbol{x}$ in $\boldsymbol{R}^K$ and $j = 1, \cdots, K$. Define a K-dimensional hypercube $\boldsymbol{B}^K = \{\boldsymbol{x} : \boldsymbol{x} \in \boldsymbol{R}^K,\ |x_i| \in [0, M],\ i = 1, \cdots, K\}$. We see that the functions $\boldsymbol{g}_1$ and $\boldsymbol{g}_2$ map the hypercube

$\boldsymbol{B}^K$ into itself. Moreover, the composition $\boldsymbol{g}_1 \circ \boldsymbol{g}_2$ also map the hypercube $\boldsymbol{B}^K$ into itself. This theorem now follows from Brouwer's Theorem applied to the continuous function $\boldsymbol{G} = \boldsymbol{g}_1 \circ \boldsymbol{g}_2$. ∎

Since now the function $\boldsymbol{G}$ has at least one fixed point, if the TDA using Max-Log-MAP converges, i.e., $\lim_{n\to\infty} \tilde{\boldsymbol{L}}_1^{(n)} = \tilde{\boldsymbol{L}}_1$, it is clear to see that $\tilde{\boldsymbol{L}}_1$ must be a fixed point of $\boldsymbol{G}$. Although Theorem 3.1 guarantees the existence of the fixed points of $\boldsymbol{G}$, it seems to be difficult to find the fixed points via $\boldsymbol{G}$ directly since $\boldsymbol{G}$ involves the nonlinear operation of $\max(\cdot)$. Denote by $(\boldsymbol{u}_j^a)^{(n)} = ((u_{j,1}^a)^{(n)}, \cdots, (u_{j,K}^a)^{(n)})$ and $(\boldsymbol{v}_j^a)^{(n)} = ((v_{j,1}^a)^{(n)}, \cdots, (v_{j,K}^a)^{(n)})$ the information vectors chosen in the first and second decoders, respectively, at the nth iteration which decode U_j as a for $a = 0, 1$. According to the decoding scenario of the Max-Log-MAP algorithm, we have

$$(\boldsymbol{u}_j^a)^{(n)} = \arg\max_{\boldsymbol{u}:u_j=a} \left[\log(p_1(\boldsymbol{y}_1|\boldsymbol{u})) + \sum_{i=1,i\neq j}^{K} u_i\, \tilde{L}_{2,i}^{(n)} \right] \tag{5}$$

$$(\boldsymbol{v}_j^a)^{(n)} = \arg\max_{\boldsymbol{v}:v_j=a} \left[\log(p_2(\boldsymbol{y}_2|\boldsymbol{v})) + \sum_{i=1,i\neq j}^{K} v_i\, \tilde{L}_{1,i}^{(n-1)} \right] \tag{6}$$

for all a and j. By substituting (5) and (6) into (4), the nonlinear formulation in (4) can be written into the following affine form:

$$\begin{aligned}
\tilde{L}_{1,j}^{(n)} &= \log\left[\frac{p_1(\boldsymbol{y}_1|(\boldsymbol{u}_j^1)^{(n)})}{p_1(\boldsymbol{y}_1|(\boldsymbol{u}_j^0)^{(n)})}\right] \\
&\quad + \sum_{i=1,i\neq j}^{K} ((u_{j,i}^1)^{(n)} - (u_{j,i}^0)^{(n)}) \log\left[\frac{p_2(\boldsymbol{y}_2|(\boldsymbol{v}_i^1)^{(n)})}{p_2(\boldsymbol{y}_2|(\boldsymbol{v}_i^0)^{(n)})}\right] \\
&\quad + \sum_{i=1,i\neq j}^{K} \sum_{l=1,l\neq i}^{K} ((u_{j,i}^1)^{(n)} - (u_{j,i}^0)^{(n)})((v_{i,l}^1)^{(n)} - (v_{i,l}^0)^{(n)})\tilde{L}_{1,l}^{(n-1)} \\
&\triangleq h_{j,0}^{(n)} + \sum_{l=1}^{K} h_{j,l}^{(n)} \tilde{L}_{1,l}^{(n-1)}, \quad \text{for all } n \text{ and } 1 \leq j \leq K
\end{aligned} \tag{7}$$

where $h_{j,l}^{(n)}$'s consist of only the channel transition probabilities and the information vectors $(\boldsymbol{u}_j^1)^{(n)}$, $(\boldsymbol{u}_j^0)^{(n)}$, $(\boldsymbol{v}_i^1)^{(n)}$, and $(\boldsymbol{v}_i^0)^{(n)}$ for all $i \neq j$.

We can then rewrite (7) into the vector form below:

$$\tilde{\boldsymbol{L}}_1^{(n)} = \boldsymbol{H}^{(n)}(\tilde{\boldsymbol{L}}_1^{(n-1)}), \quad \text{for all } n. \tag{8}$$

The above formulation (8) might be helpful in finding the fixed point of $\boldsymbol{G}$ and in devising a stop criterion for iterative decoding. Consider the following procedure:

Step 1. Set $n = 1$.

Step 2. Conduct the Max-Log-MAP decoding in decoders 2 and 1 to find $\boldsymbol{H}^{(n)}(\boldsymbol{x})$ by (7) and $\tilde{\boldsymbol{L}}_1^{(n)}$.

Step 3. Compute the fixed points of $\boldsymbol{H}^{(n)}(\boldsymbol{x})$, say $\boldsymbol{x}^*$ for example, by solving $\boldsymbol{H}^{(n)}(\boldsymbol{x})=\boldsymbol{x}$. (Since $\boldsymbol{H}^{(n)}$ is in the affine form, it is easy to compute the fixed points of $\boldsymbol{H}^{(n)}$.)

Step 4. Check if $\boldsymbol{G}(\boldsymbol{x}^*) = \boldsymbol{x}^*$. If it is true, then $\boldsymbol{x}^*$ is a fixed point of $\boldsymbol{G}$.

Step 5. If $\boldsymbol{x}^*$ is a fixed point of $\boldsymbol{G}$, then check whether $\boldsymbol{x}^*$ is attracting or called asymptotically stable [18].

Step 6. If $\boldsymbol{x}^*$ is an attracting fixed point and $\tilde{\boldsymbol{L}}_1^{(n)}$ is in the stable set [18] of $\boldsymbol{x}^*$, then stop decoding and let $\tilde{\boldsymbol{L}}_1 = \boldsymbol{x}^*$. Else increase n by 1 and go to Step 2.

Please note that the above procedure does not guarantee that the fixed points of $\boldsymbol{G}$ can always be found. On the other hand, one needs to determine if a fixed point is attracting and the stable set for true convergence. Though the procedure is truly impractical now, it might motivate some heuristic approaches in the future.

3.2 Convergence Issues of the TDA Using Max-Log-MAP

Besides the above discussion, in Appendix, some special examples are given to study the convergence behavior of the TDA using Max-Log-MAP. For $K = 1$, we observe that $\hat{L}_{1,1}^{(n)}$ is fixed as $n \geq 1$ and is the same as the result of the true MAP decoding for turbo codes. In this case, the TDA using Max-Log-MAP is thus optimal and converges in one iteration. However, for $K = 2$, we can construct an example to show the suboptimality of the TDA using Max-Log-MAP since the turbo decision is now different from that of the true MAP decoding for turbo codes. Moreover, counterexamples are also given to show that this iterative decoding algorithm may not converge, and even the turbo decision may not exist. To study the general case of $K \geq 3$, the TDA using Max-Log-MAP is formulated into another equivalent form described below.

Substituting $h_{j,l}^{(n)}$'s into (7) by the channel transition probabilities, $(\boldsymbol{u}_j^0)^{(n)}$, $(\boldsymbol{u}_j^1)^{(n)}$, $(\boldsymbol{v}_i^0)^{(n)}$, and $(\boldsymbol{v}_i^1)^{(n)}$, we have

$$\left[\tilde{L}_{1,1}^{(n)}, \tilde{L}_{1,2}^{(n)}, \cdots, \tilde{L}_{1,K}^{(n)}\right]^T = \left[\alpha_1^{(n)}, \alpha_2^{(n)}, \cdots, \alpha_K^{(n)}\right]^T + \begin{bmatrix} 0 & t_{1,2}^{(n)} & \cdots & t_{1,K}^{(n)} \\ t_{2,1}^{(n)} & 0 & \cdots & t_{2,K}^{(n)} \\ \vdots & \vdots & \ddots & \vdots \\ t_{K,1}^{(n)} & t_{K,2}^{(n)} & \cdots & 0 \end{bmatrix} \begin{bmatrix} 0 & s_{1,2}^{(n)} & \cdots & s_{1,K}^{(n)} \\ s_{2,1}^{(n)} & 0 & \cdots & s_{2,K}^{(n)} \\ \vdots & \vdots & \ddots & \vdots \\ s_{K,1}^{(n)} & s_{K,2}^{(n)} & \cdots & 0 \end{bmatrix} \begin{bmatrix} \tilde{L}_{1,1}^{(n-1)} \\ \tilde{L}_{1,2}^{(n-1)} \\ \vdots \\ \tilde{L}_{1,K}^{(n-1)} \end{bmatrix} \quad (9)$$

where $t_{j,i}^{(n)} = (u_{j,i}^1)^{(n)} - (u_{j,i}^0)^{(n)}$ for all $i \neq j$, $s_{i,l}^{(n)} = (v_{i,l}^1)^{(n)} - (v_{i,l}^0)^{(n)}$ for all $l \neq i$, and

$$\begin{aligned} \alpha_j^{(n)} &= \log\left[\frac{p_1(\boldsymbol{y}_1|(\boldsymbol{u}_j^1)^{(n)})}{p_1(\boldsymbol{y}_1|(\boldsymbol{u}_j^0)^{(n)})}\right] \\ &\quad + \sum_{i=1, i\neq j}^{K} ((u_{j,i}^1)^{(n)} - (u_{j,i}^0)^{(n)}) \log\left[\frac{p_2(\boldsymbol{y}_2|(\boldsymbol{v}_i^1)^{(n)})}{p_2(\boldsymbol{y}_2|(\boldsymbol{v}_i^0)^{(n)})}\right]. \end{aligned}$$

Let $\boldsymbol{\alpha}^{(n)} = (\alpha_1^{(n)}, \cdots, \alpha_K^{(n)})^T$. We can further write (9) as

$$\tilde{\boldsymbol{L}}_1^{(n)} = \boldsymbol{\alpha}^{(n)} + \boldsymbol{A}^{(n)} \tilde{\boldsymbol{L}}_1^{(n-1)} \quad (10)$$

where entries of $\boldsymbol{A}^{(n)}$ are of integer values. Without loss of generality, assume that $\boldsymbol{\alpha}^{(n)} = \boldsymbol{\alpha}$ and $\boldsymbol{A}^{(n)} = \boldsymbol{A}$ for $r+1$ consecutive iterations, say $n = m, \cdots, m+r$, where m and r are some nonnegative integers. We can reduce (10) to

$$\begin{aligned} \tilde{\boldsymbol{L}}_1^{(m+d)} &= \boldsymbol{\alpha} + \boldsymbol{A}\tilde{\boldsymbol{L}}_1^{(m+d-1)} = \left(\boldsymbol{I} + \boldsymbol{A} + \cdots + \boldsymbol{A}^d\right)\boldsymbol{\alpha} + \boldsymbol{A}^{d+1}\tilde{\boldsymbol{L}}_1^{(m-1)} \\ &= \left(\sum_{i=0}^{d} \boldsymbol{A}^i\right)\boldsymbol{\alpha} + \boldsymbol{A}^{d+1}\tilde{\boldsymbol{L}}_1^{(m-1)}, \quad \text{for } d = 0, 1, \cdots, r. \end{aligned} \quad (11)$$

Denote by $\boldsymbol{z}$ the difference of the extrinsic values of mth and $(m-1)$th iterations, i.e., $\boldsymbol{z} = \tilde{\boldsymbol{L}}_1^{(m)} - \tilde{\boldsymbol{L}}_1^{(m-1)}$. By the substitution of

$$\boldsymbol{z} = \boldsymbol{\alpha} + (\boldsymbol{A} - \boldsymbol{I})\tilde{\boldsymbol{L}}_1^{(m-1)} \quad \text{and} \quad \tilde{\boldsymbol{L}}_1^{(m+d)} - \tilde{\boldsymbol{L}}_1^{(m+d-1)} = \boldsymbol{A}^d \boldsymbol{z} \tag{12}$$

for $d = 0, 1, \cdots, r$, we can further reduce (11) to

$$\tilde{\boldsymbol{L}}_1^{(m+d)} = \left(\sum_{i=0}^{d} \boldsymbol{A}^i \right) \boldsymbol{z} + \tilde{\boldsymbol{L}}_1^{(m-1)}, \quad \text{for } d = 0, 1, \cdots, r. \tag{13}$$

Before addressing the conditions for convergence, some useful results are given below.

Lemma 3.3 *[19][20]* For a square matrix $\boldsymbol{A}$ of integer entries, $\sum_{i=0}^{\infty} \boldsymbol{A}^i$ converges to $(\boldsymbol{I} - \boldsymbol{A})^{-1}$ if and only if $\boldsymbol{A}$ is nilpotent, i.e., there exists a positive integer d (called the index) such that $\boldsymbol{A}^d = \boldsymbol{O}$, the zero matrix.

Lemma 3.4 Let a matrix $\boldsymbol{A} \in \boldsymbol{R}^{K \times K}$ have the minimal polynomial $m(x) = (x - \lambda_1)^{t_1} \cdots (x - \lambda_r)^{t_r}$ and the characteristic polynomial $\text{char}_{\boldsymbol{A}}(x) \triangleq \det(\boldsymbol{A} - x\boldsymbol{I}) = (-1)^n (x - \lambda_1)^{m_1} \cdots (x - \lambda_r)^{m_r}$, where $\lambda_1, \cdots, \lambda_r$ are distinct eigenvalues and $K = m_1 + m_2 + \cdots + m_r$. Also, let n_i be the geometric multiplicity of λ_i, i.e., $n_i \triangleq \dim(\text{Ker}(\boldsymbol{A} - \lambda_i \boldsymbol{I}))$, the dimension of the null space of $\boldsymbol{A} - \lambda_i \boldsymbol{I}$. Then for a vector $\boldsymbol{z}$ of proper dimension, $\lim_{d \to \infty} [\left(\sum_{i=0}^{d} \boldsymbol{A}^i \right) \boldsymbol{z}]$ exists if and only if $\boldsymbol{z} \in \text{span}\{\boldsymbol{b} : \boldsymbol{b}$ is a generalized eigenvector for an eigenvalue λ of $\boldsymbol{A}$ with $|\lambda| < 1\}$, i.e., the vector space spanned by all the generalized eigenvectors whose corresponding eigenvalues are of absolute values less than 1.

Proof: Let $f(x) = \sum_{i=0}^{d} x^i$. We can represent $\boldsymbol{A}$ as the Jordan canonical form [19], i.e., $\boldsymbol{A} = \boldsymbol{P}\boldsymbol{J}\boldsymbol{P}^{-1}$, where $\boldsymbol{P} = [\boldsymbol{v}_1, \cdots, \boldsymbol{v}_K]$ and $\{\boldsymbol{v}_1, \cdots, \boldsymbol{v}_K\}$ is the Jordan basis of $\boldsymbol{R}^K$. Let $\tilde{\boldsymbol{z}} = \boldsymbol{P}^{-1}\boldsymbol{z}$, i.e., $\boldsymbol{z} = \tilde{z}_1 \boldsymbol{v}_1 + \cdots + \tilde{z}_K \boldsymbol{v}_K$. Then

$$\begin{aligned}
\left(\sum_{i=0}^{d} \boldsymbol{A}^i \right) \boldsymbol{z} &= f(\boldsymbol{A})\boldsymbol{z} = \boldsymbol{P} f(\boldsymbol{J}) \boldsymbol{P}^{-1} \boldsymbol{z} = \boldsymbol{P} f(\boldsymbol{J}) \tilde{\boldsymbol{z}} \\
&= \boldsymbol{P} \begin{bmatrix} f(\boldsymbol{J}_{11}) & 0 & \cdots & 0 \\ 0 & f(\boldsymbol{J}_{12}) & \cdots & 0 \\ \vdots & \vdots & \ddots & \vdots \\ 0 & 0 & \cdots & f(\boldsymbol{J}_{dn_d}) \end{bmatrix} \tilde{\boldsymbol{z}}
\end{aligned}$$

$$= \boldsymbol{P} \begin{bmatrix} [\, f(\boldsymbol{J}_{11})\ 0 \cdots 0\,]\, \tilde{\boldsymbol{z}} \\ \vdots \\ [\, 0 \cdots 0\ f(\boldsymbol{J}_{ij})\ 0 \cdots 0\,]\, \tilde{\boldsymbol{z}} \\ \vdots \\ [\, 0 \cdots 0\ f(\boldsymbol{J}_{dn_d})\,]\, \tilde{\boldsymbol{z}} \end{bmatrix}.$$

Hence, $\lim_{d\to\infty}[\left(\sum_{i=0}^{d} \boldsymbol{A}^i\right) \boldsymbol{z}\,]$ exists if, and only if, every $\lim_{d\to\infty}\{[0 \cdots 0\ f(\boldsymbol{J}_{ij})\ 0 \cdots 0]\ \tilde{\boldsymbol{z}}\}$ exists, for all i, j. Without loss of generality, let us consider the first one

$$[f(\boldsymbol{J}_{11})\ 0 \cdots 0]\ \tilde{\boldsymbol{z}}$$

$$= \begin{bmatrix} f(\lambda_1) & f'(\lambda_1) & \frac{f''(\lambda_1)}{2!} & \cdots & \frac{f^{(t_1-1)}(\lambda_1)}{(t_1-1)!} & 0 & \cdots & 0 \\ 0 & f(\lambda_1) & f'(\lambda_1) & \cdots & \frac{f^{(t_1-2)}(\lambda_1)}{(t_1-2)!} & 0 & \cdots & 0 \\ 0 & 0 & f(\lambda_1) & \cdots & \frac{f^{(t_1-3)}(\lambda_1)}{(t_1-3)!} & 0 & \cdots & 0 \\ \vdots & \vdots & \vdots & \ddots & \vdots & & & \vdots \\ 0 & 0 & 0 & \cdots & f(\lambda_1) & 0 & \cdots & 0 \end{bmatrix} \begin{bmatrix} \tilde{z}_1 \\ \vdots \\ \tilde{z}_{t_1} \\ \vdots \\ \tilde{z}_K \end{bmatrix}$$

$$= \begin{bmatrix} \left(\sum_{i=0}^{d} \lambda_1^i\right) \tilde{z}_1 + \left(\sum_{i=1}^{d} i\lambda_1^{i-1}\right) \tilde{z}_2 + \cdots + \frac{f^{(t_1-1)}(\lambda_1)}{(t_1-1)!}\, \tilde{z}_{t_1} \\ \vdots \\ \left(\sum_{i=0}^{d} \lambda_1^i\right) \tilde{z}_{t_1-2} + \left(\sum_{i=1}^{d} i\lambda_1^{i-1}\right) \tilde{z}_{t_1-1} + \left(\frac{\sum_{i=2}^{d} i(i-1)\lambda_1^{i-2}}{2!}\right) \tilde{z}_{t_1} \\ \left(\sum_{i=0}^{d} \lambda_1^i\right) \tilde{z}_{t_1-1} + \left(\sum_{i=1}^{d} i\lambda_1^{i-1}\right) \tilde{z}_{t_1} \\ \left(\sum_{i=0}^{d} \lambda_1^i\right) \tilde{z}_{t_1} \end{bmatrix}.$$

If $|\lambda_1| \geq 1$, then $\lim_{d\to\infty}\{\,[\, f(\boldsymbol{J}_{11})\ 0 \cdots 0\,]\ \tilde{\boldsymbol{z}}\,\}$ exists if and only if $\tilde{z}_1 = \cdots = \tilde{z}_{t_1} = 0$. Otherwise, i.e., $|\lambda_1| < 1$, then $\lim_{d\to\infty}\{\,[\, f(\boldsymbol{J}_{11})\ 0 \cdots 0\,]\ \tilde{\boldsymbol{z}}\,\}$ always exists. The conditions for the existence of $\lim_{d\to\infty}\{\,[\,0 \cdots 0\, f(\boldsymbol{J}_{ij})\ 0 \cdots 0\,]\ \tilde{\boldsymbol{z}}\,\}$, for all i, j, are the same. Thus, we can conclude that $\lim_{d\to\infty}[\left(\sum_{i=0}^{d} \boldsymbol{A}^i\right) \boldsymbol{z}\,]$ exists if and only if $\boldsymbol{z} \in \mathrm{span}\{\boldsymbol{b} : \boldsymbol{b}$ is a generalized eigenvector for an eigenvalue λ of $\boldsymbol{A}$ with $|\lambda| < 1\}$. ∎

Notice that if $\tilde{\boldsymbol{L}}_1^{(l+1)} = \tilde{\boldsymbol{L}}_1^{(l)}$ for some positive integer l, then we have $\lim_{n\to\infty} \tilde{\boldsymbol{L}}_1^{(n)} = \tilde{\boldsymbol{L}}_1^{(l)}$ since $\tilde{\boldsymbol{L}}_1^{(l)}$ is now a fixed point of $\boldsymbol{G}$. Aided with the above results, four sufficient conditions for the convergence of the TDA using Max-Log-MAP are given below.

1. If $\boldsymbol{A} = \boldsymbol{O}$, then $\tilde{\boldsymbol{L}}_1^{(m+1)} - \tilde{\boldsymbol{L}}_1^{(m)} = \boldsymbol{A}\boldsymbol{z} = \boldsymbol{0}$. Hence, if $\boldsymbol{\alpha}^{(n)} = \boldsymbol{\alpha}$ and $\boldsymbol{A}^{(n)} = \boldsymbol{A}$ for $n = m,\ m+1$, then the TDA using Max-Log-MAP converges in m iterations and $\lim_{n\to\infty} \tilde{\boldsymbol{L}}_1^{(n)} = \tilde{\boldsymbol{L}}_1^{(m)} = \boldsymbol{z} + \tilde{\boldsymbol{L}}_1^{(m-1)}$.

2. If $\boldsymbol{\alpha}^{(n)} = \boldsymbol{\alpha}$ and $\boldsymbol{A}^{(n)} = \boldsymbol{A}$ for $n = m, m+1, \cdots, m+d$ and $\boldsymbol{A}$ is nilpotent with index d, then the TDA using Max-Log-MAP converges in $(m+d-1)$ iterations and $\lim_{n\to\infty} \tilde{\boldsymbol{L}}_1^{(n)} = \tilde{\boldsymbol{L}}_1^{(m+d-1)} = \left(\sum_{i=0}^{d-1} \boldsymbol{A}^i\right) \boldsymbol{z} + \tilde{\boldsymbol{L}}_1^{(m-1)}$.

3. By (12), if $\boldsymbol{z} \in \mathrm{Ker}(\boldsymbol{A}^d)$, i.e., the null space of $\boldsymbol{A}^d$, then $\tilde{\boldsymbol{L}}_1^{(m+d)} - \tilde{\boldsymbol{L}}_1^{(m+d-1)} = \boldsymbol{A}^d\boldsymbol{z} = \boldsymbol{0}$. Hence, if $\boldsymbol{\alpha}^{(n)} = \boldsymbol{\alpha}$ and $\boldsymbol{A}^{(n)} = \boldsymbol{A}$ for $n = m, m+1, \cdots, m+d$, and $\boldsymbol{z} \in \mathrm{Ker}(\boldsymbol{A}^d)$, then the TDA using Max-Log-MAP converges in $(m+d-1)$ iterations and $\lim_{n\to\infty} \tilde{\boldsymbol{L}}_1^{(n)} = \tilde{\boldsymbol{L}}_1^{(m+d-1)} = \left(\sum_{i=0}^{d-1} \boldsymbol{A}^i\right) \boldsymbol{z} + \tilde{\boldsymbol{L}}_1^{(m-1)}$.

4. By (13), if $\boldsymbol{\alpha}^{(n)} = \boldsymbol{\alpha}$ and $\boldsymbol{A}^{(n)} = \boldsymbol{A}$ for $n \geq m$, and $\boldsymbol{z} \in \mathrm{span}\{\boldsymbol{v} : \boldsymbol{v}$ is a generalized eigenvector for an eigenvalue λ of $\boldsymbol{A}$ with $|\lambda| < 1\}$, then the TDA using Max-Log-MAP converges eventually.

In the above, Condition 4 for the convergence of the TDA using Max-Log-MAP is actually a necessary and sufficient condition, if the information vectors $(\boldsymbol{u}_j^1)^{(n)}$, $(\boldsymbol{u}_j^0)^{(n)}$, $(\boldsymbol{v}_j^1)^{(n)}$, and $(\boldsymbol{v}_j^0)^{(n)}$ chosen by the Max-Log-MAP decoding are fixed after sufficiently large iterations. We restate it as the following theorem.

Theorem 3.2 If the chosen information vectors $(\boldsymbol{u}_j^1)^{(n)}$, $(\boldsymbol{u}_j^0)^{(n)}$, $(\boldsymbol{v}_j^1)^{(n)}$, and $(\boldsymbol{v}_j^0)^{(n)}$, defined in (5) and (6), for $j = 1, \cdots, K$, are fixed for $n \geq m$, then the TDA using Max-Log-MAP converges if and only if $\boldsymbol{z} \in \mathrm{span}\{\boldsymbol{b} : \boldsymbol{b}$ is a generalized eigenvector of $\boldsymbol{A}$ with $|\lambda| < 1\}$, where $\boldsymbol{z}$ is defined in (12).

4 Concluding Remarks

In this study, we investigate analytically the stability and convergence properties of iterative turbo decoding with the Max-Log-MAP algorithm for soft-output generation. First we derive three equivalent mathematical formulations of the TDA using Max-Log-MAP and analyze the dynamics of this iterative decoding algorithm. It is also proved that the TDA using Max-Log-MAP always has fixed points, regardless of the specific constituent codes of the turbo encoder. Besides, we show by examples that the TDA using Max-Log-MAP is optimal only when $K = 1$. When $K \geq 2$, it is in general a suboptimal decoding algorithm and may not even converge. Finally, some sufficient conditions are given for the convergence of the TDA using Max-Log-MAP.

This study is still preliminary and leaves much room for further research. As stated in Theorem 3.2, if the information vectors chosen in the first and second decoders, defined in (5) and (6), are fixed after sufficiently large iterations, then Condition 4 will become a necessary and sufficient condition for the convergence of the TDA using Max-Log-MAP. We have observed from computer simulations that the chosen information vectors are always fixed after sufficiently large iterations under convergence, yet we have no workable theoretical proof up to now. Further study on the asymptotical stability and stable set of the fixed points could be helpful in developing a practical stop condition for iterative decoding. We hope one day the knowledge from this study will be useful in improving the convergence properties of turbo decoding.

Appendix

In this appendix, using similar methodology in [3], we will show that the TDA using Max-Log-MAP is a suboptimal decoding algorithm, and may not even converge.

First, we briefly mention the MAP decoding for turbo codes, given in [3], which decodes the ith bit of $\boldsymbol{U}$ according to the true ratio of a posteriori probabilities

$$R_j = \frac{\sum_{\boldsymbol{u}:u_j=1} p_1(\boldsymbol{y}_1|\boldsymbol{u})p_2(\boldsymbol{y}_2|\boldsymbol{u})}{\sum_{\boldsymbol{u}:u_j=0} p_1(\boldsymbol{y}_1|\boldsymbol{u})p_2(\boldsymbol{y}_2|\boldsymbol{u})}, \qquad j = 1, \cdots, K. \tag{A.1}$$

Then

$$\bar{U}_j = \begin{cases} 1 & \text{if } R_j \geq 1 \\ 0 & \text{if } R_j < 1, \end{cases} \qquad j = 1, \cdots, K.$$

Now, consider the following cases. For $K = 1$, the MAP decoding for turbo codes as given by (A.1) is

$$L_1 = \log R_1 = \log\left(\frac{p_1(\boldsymbol{y}_1|u=1)p_2(\boldsymbol{y}_2|u=1)}{p_1(\boldsymbol{y}_1|u=0)p_2(\boldsymbol{y}_2|u=0)}\right).$$

On the other hand, using (2) and (3), we have

$$\begin{aligned}\tilde{L}_{1,1}^{(n)} &= \log(p_1(\boldsymbol{y}_1|u=1)/p_1(\boldsymbol{y}_1|u=0))\\ \tilde{L}_{2,1}^{(n)} &= \log(p_2(\boldsymbol{y}_2|u=1)/p_2(\boldsymbol{y}_2|u=0)), \quad \text{for } n \geq 1.\end{aligned}$$

Hence, $\hat{L}_{1,1}^{(n)} \triangleq \tilde{L}_{1,1}^{(n)} + \tilde{L}_{2,1}^{(n)} = L_1$ for all $n \geq 1$, i.e., for $K = 1$, the TDA using Max-Log-MAP converges to the MAP decoding for turbo codes in one iteration. We say that TDA using Max-Log-MAP is an optimal decoding algorithm for $K = 1$.

For $K = 2$, we will show by examples that the TDA using Max-Log-MAP can behave unexpectedly. We will construct two examples (with $K = 2$) by carefully choosing the values of the log likelihoods, defined as $\log(\Pr\{\boldsymbol{Y} = \boldsymbol{y}|\boldsymbol{U} = \boldsymbol{u}\})$. It was shown in [3] that no matter what values we preassign to the log likelihoods, there exists a potential received vector $\boldsymbol{y}$ which validates them. In order to simplify the notation, we define

$$A_i \triangleq \log(p_i(\boldsymbol{y}_i|\boldsymbol{u} = (00))) \tag{A.2}$$
$$B_i \triangleq \log(p_i(\boldsymbol{y}_i|\boldsymbol{u} = (01))) \tag{A.3}$$
$$C_i \triangleq \log(p_i(\boldsymbol{y}_i|\boldsymbol{u} = (10))) \tag{A.4}$$
$$D_i \triangleq \log(p_i(\boldsymbol{y}_i|\boldsymbol{u} = (11))), \quad i = 1, 2. \tag{A.5}$$

Substituting them into (2) and (3) with $K = 2$, we have

$$\tilde{L}_{1,1}^{(n)} = \max(C_1, D_1 + \tilde{L}_{2,2}^{(n)}) - \max(A_1, B_1 + \tilde{L}_{2,2}^{(n)}) \tag{A.6}$$
$$\tilde{L}_{1,2}^{(n)} = \max(B_1, D_1 + \tilde{L}_{2,1}^{(n)}) - \max(A_1, C_1 + \tilde{L}_{2,1}^{(n)}) \tag{A.7}$$
$$\tilde{L}_{2,1}^{(n)} = \max(C_2, D_2 + \tilde{L}_{1,2}^{(n-1)}) - \max(A_2, B_2 + \tilde{L}_{1,2}^{(n-1)}) \tag{A.8}$$
$$\tilde{L}_{2,2}^{(n)} = \max(B_2, D_2 + \tilde{L}_{1,1}^{(n-1)}) - \max(A_2, C_2 + \tilde{L}_{1,1}^{(n-1)}). \tag{A.9}$$

If we combine (A.6) with (A.9), and (A.7) with (A.8), we obtain, after some algebraic manipulations, for $n \geq 1$,

$$\tilde{L}_{1,1}^{(n)} = \max(C_1, [D_1 + \max(B_2, D_2 + \tilde{L}_{1,1}^{(n-1)}) - \max(A_2, C_2 + \tilde{L}_{1,1}^{(n-1)})])$$

$$-\max(A_1, [B_1 + \max(B_2, D_2 + \tilde{L}_{1,1}^{(n-1)}) - \max(A_2, C_2 + \tilde{L}_{1,1}^{(n-1)})]) \quad (A.10)$$

$$\tilde{L}_{1,2}^{(n)} = \max(B_1, [D_1 + \max(C_2, D_2 + \tilde{L}_{1,2}^{(n-1)}) - \max(A_2, B_2 + \tilde{L}_{1,2}^{(n-1)})])$$
$$-\max(A_1, [C_1 + \max(C_2, D_2 + \tilde{L}_{1,2}^{(n-1)}) - \max(A_2, B_2 + \tilde{L}_{1,2}^{(n-1)})]). \quad (A.11)$$

Similarly, we have for $n \geq 2$ that

$$\tilde{L}_{2,1}^{(n)} = \max(C_2, [D_2 + \max(B_1, D_1 + \tilde{L}_{2,1}^{(n-1)}) - \max(A_1, C_1 + \tilde{L}_{2,1}^{(n-1)})])$$
$$-\max(A_2, [B_2 + \max(B_1, D_1 + \tilde{L}_{2,1}^{(n-1)}) - \max(A_1, C_1 + \tilde{L}_{2,1}^{(n-1)})]) \quad (A.12)$$
$$\tilde{L}_{2,2}^{(n)} = \max(B_2, [D_2 + \max(C_1, D_1 + \tilde{L}_{2,2}^{(n-1)}) - \max(A_1, B_1 + \tilde{L}_{2,2}^{(n-1)})])$$
$$-\max(A_2, [C_2 + \max(C_1, D_1 + \tilde{L}_{2,2}^{(n-1)}) - \max(A_1, B_1 + \tilde{L}_{2,2}^{(n-1)})]). \quad (A.13)$$

Below, let us focus on (A.10) and (A.12).

Example 1. In our first example, we assume that $(C_1 - D_1) < (A_1 - B_1)$ and $(A_2 - C_2) < (B_2 - D_2)$. Then, after some algebraic manipulations, we have five possible functions to compute $\tilde{L}_{1,1}^{(n)}$ from $\tilde{L}_{1,1}^{(n-1)}$ according to the values of $\tilde{L}_{1,1}^{(n-1)}$ and $\tilde{L}_{2,2}^{(n)}$ as follows:

1. If $\tilde{L}_{2,2}^{(n)} < (C_1 - D_1)$, then $\tilde{L}_{1,1}^{(n)} = (C_1 - A_1)$.
2. If $\tilde{L}_{2,2}^{(n)} > (A_1 - B_1)$, then $\tilde{L}_{1,1}^{(n)} = (D_1 - B_1)$.
3. If $(C_1 - D_1) < \tilde{L}_{2,2}^{(n)} < (A_1 - B_1)$ and $\tilde{L}_{1,1}^{(n-1)} < (A_2 - C_2)$, then $\tilde{L}_{1,1}^{(n)} = (D_1 - A_1) + (B_2 - A_2)$.
4. If $(C_1 - D_1) < \tilde{L}_{2,2}^{(n)} < (A_1 - B_1)$ and $\tilde{L}_{1,1}^{(n-1)} > (B_2 - D_2)$, then $\tilde{L}_{1,1}^{(n)} = (D_1 - A_1) + (D_2 - C_2)$.
5. If $(C_1 - D_1) < \tilde{L}_{2,2}^{(n)} < (A_1 - B_1)$ and $(A_2 - C_2) < \tilde{L}_{1,1}^{(n-1)} < (B_2 - D_2)$, then $\tilde{L}_{1,1}^{(n)} = (D_1 - A_1) + (B_2 - C_2 - \tilde{L}_{1,1}^{(n-1)})$.

Define the set $S_1 \triangleq \{((C_1 - A_1), (D_1 - B_1), (D_1 - A_1) + (B_2 - A_2), (D_1 - A_1) + (D_2 - C_2), [(D_1 - A_1) + (B_2 - C_2)]/2)\}$. Then we know that if $\tilde{L}_{1,1}^{(n)}$ converges, it must converge to a point in S_1, i.e., $\tilde{L}_{1,1} \in S_1$ if it exists.

Similarly, we also have five possible functions to compute $\tilde{L}_{2,1}^{(n)}$ from $\tilde{L}_{2,1}^{(n-1)}$ according to the values of $\tilde{L}_{1,2}^{(n-1)}$ and $\tilde{L}_{2,1}^{(n-1)}$ as follows:

1. If $\tilde{L}_{1,2}^{(n-1)} < (A_2 - B_2)$, then $\tilde{L}_{2,1}^{(n)} = (C_2 - A_2)$.
2. If $\tilde{L}_{1,2}^{(n-1)} > (C_2 - D_2)$, then $\tilde{L}_{2,1}^{(n)} = (D_2 - B_2)$.
3. If $(A_2 - B_2) < \tilde{L}_{1,2}^{(n-1)} < (C_2 - D_2)$ and $\tilde{L}_{2,1}^{(n-1)} < (B_1 - D_1)$, then $\tilde{L}_{2,1}^{(n)} = (C_2 - B_2) - (B_1 - A_1)$.
4. If $(A_2 - B_2) < \tilde{L}_{1,2}^{(n-1)} < (C_2 - D_2)$ and $\tilde{L}_{2,1}^{(n-1)} > (A_1 - C_1)$, then $\tilde{L}_{2,1}^{(n)} = (C_2 - B_2) - (D_1 - C_1)$.
5. If $(A_2 - B_2) < \tilde{L}_{1,2}^{(n-1)} < (C_2 - D_2)$ and $(B_1 - D_1) < \tilde{L}_{2,1}^{(n-1)} < (A_1 - C_1)$, then $\tilde{L}_{2,1}^{(n)} = (C_2 - B_2) - (D_1 - A_1 + \tilde{L}_{2,1}^{(n-1)})$.

Define the set $S_2 \triangleq \{((C_2 - A_2), (D_2 - B_2), (C_2 - B_2) - (B_1 - A_1), (C_2 - B_2) - (D_1 - C_1), [(C_2 - B_2) - (D_1 - A_1)]/2)\}$. Then we know that if $\tilde{L}_{2,1}^{(n)}$ converges, it must converge to a point in S_2, i.e., $\tilde{L}_{2,1} \in S_2$ if it exists. Therefore, we know that if the nth turbo log likelihood estimate $\hat{L}_{1,1}^{(n)} = (\tilde{L}_{1,1}^{(n)} + \tilde{L}_{2,1}^{(n)})$ converges, it must converges to a point in $S \triangleq \{x_1 + x_2 : x_1 \in S_1,\ x_2 \in S_2\}$, i.e., the limiting turbo log likelihood estimate $\hat{L}_{1,1} \in S$ if it exists.

On the other hand, the MAP decoding for U_1 is based on the true log ratio of a posteriori probabilities

$$L_1 = \log(e^{(C_1+C_2)} + e^{(D_1+D_2)}) - \log(e^{(A_1+A_2)} + e^{(B_1+B_2)}). \quad \text{(A.14)}$$

In general, $\hat{L}_{1,1}$ is not the same as L_1. Moreover, we can choose the values of A_i ,B_i, C_i, D_i, $i = 1, 2$, properly such that $\hat{L}_{1,1}$ and L_1 have opposite signs, i.e., $\hat{U}_{1,1} \neq \bar{U}_1$. Hence, the turbo decision does not agree with the true MAP decoding for turbo codes, i.e., the TDA using Max-Log-MAP is a suboptimal decoding algorithm. ∎

In fact, the TDA using Max-Log-MAP may not even converge. Look at the following example.

Example 2. In this example, we will show that the TDA using Max-Log-MAP may not even converge. Let

$$A_1 = -6.8774, \quad B_1 = -7.0885, \quad C_1 = -8.5640, \quad D_1 = -2.4567$$
$$A_2 = -7.7023, \quad B_2 = -4.5841, \quad C_2 = -3.2658, \quad D_2 = -8.3414$$

and substitute them into (A.10)–(A.13). After some computations, we can find that

$$\tilde{L}_{1,1}^{(n)} = \begin{cases} 3.102, & \text{if } n \text{ is odd} \\ 0, & \text{if } n \text{ is even} \end{cases}$$

and

$$\tilde{L}_{2,1}^{(n)} = \begin{cases} 0.655, & \text{if } n \text{ is odd} \\ -3.757, & \text{if } n \text{ is even.} \end{cases}$$

Hence, the nth log likelihood estimate for the first bit is

$$\hat{L}_{1,1}^{(n)} = \begin{cases} 3.757, & \text{if } n \text{ is odd} \\ -3.757, & \text{if } n \text{ is even} \end{cases}$$

which implies that the nth turbo decision for the first bit is

$$\hat{U}_{1,1}^{(n)} = \begin{cases} 1, & \text{if } n \text{ is odd} \\ 0, & \text{if } n \text{ is even.} \end{cases}$$

Thus in this case, the limiting log likelihood estimate for the first bit does not exist. Even the turbo decision for the first bit does not exist, either. Therefore, the TDA using Max-Log-MAP may not always converge. ∎

References

[1] C. Berrou, A. Glavieux, and P. Thitimajshima, "Near Shannon limit error-correcting coding and decoding: Turbo-codes," in *Proc. IEEE Int. Conf. Commun.*, Geneva, Switzerland, May 1993, pp. 1064–1070.

[2] L. R. Bahl, J. Cocke, F. Jelinek, and J. Raviv, "Optimal decoding of linear codes for minimizing symbol error rate," *IEEE Trans. Inform. Theory,* vol. IT-20, pp. 284–287, Mar. 1974.

[3] R. J. McEliece, E. R. Rodemich, and J.-F. Cheng, "The turbo decision algorithm," in *Proc. 33rd Allerton Conference on Communication, Control, and Computing,* Allerton, IL, Oct. 1995, pp. 366–379.

[4] L. Duan and B. Rimoldi, "The turbo decoding algorithm has fixed points," in *Proc. IEEE Int. Symp. Inform. Theory,* Cambridge, MA, Aug. 1998, p. 277.

[5] T. Richardson, "The geometry of turbo-decoding dynamics," *IEEE Trans. Inform. Theory*, vol. 46, pp. 9–23, Jan. 2000.

[6] D. Agrawal and A. Vardy, "The turbo decoding algorithm and its phase trajectories," in *Proc. IEEE Int. Symp. Inform. Theory*, Sorrento, Italy, June 2000, p. 316.

[7] G. Caire, G. Taricco, and E. Biglieri, "On the convergence of the iterative decoding algorithm," in *Proc. IEEE Int. Symp. Inform. Theory*, Whistler, BC, Canada, Sept. 1995, p. 472.

[8] A. Sella and Y. Be'ery, "Convergence Analysis of turbo-decoding of product codes," in *Proc. IEEE Int. Symp. Inform. Theory*, Sorrento, Italy, June 2000, p. 484.

[9] A. Ambroze, G. Wade, and M. Tomlinson, "Practical aspects of iterative decoding," in *Proc. IEE Commum.,* vol. 147, pp. 69–74, Apr. 2000.

[10] D. J. C. Mackay, R. J. McEliece, and J. F. Cheng, "Turbo decoding as an instance of Pearl's "belief propagation" algorithm," *IEEE J. Select. Areas Commun.,* vol. 16, pp. 140–152, Feb. 1998.

[11] J. A. Erfanian, S. Pasupathy, and G. Gulak "Reduced complexity symbol detectors with parallel structures for ISI channels," *IEEE Trans. Commun.*, vol. 42, pp. 1661–1671, Feb./Mar./Apr. 1994.

[12] A. J. Viterbi, "An intuitive justification and a simplified implementation of the MAP decoder for convolutional codes," *IEEE J. Select. Areas Commun.*, vol. 16, pp. 260–264, Feb. 1998.

[13] J. Hagenauer, "Source-controlled channel decoding," *IEEE Trans. Commun.*, vol. 43, pp. 2449–2457, Sept. 1995.

[14] J. Hagenauer and P. Robertson, "Iterative (turbo) decoding of systematic convolutional codes with the MAP and SOVA algorithms," in *Proc. ITG Conf. on Source and Channel Coding,* Frankfurt, Germany, Oct. 1994, pp. 1–9.

[15] F. Battail, "Ponderation des symboles decodes par l'algorithme de Viterbi," *Ann. Telecommun.*, vol. 42, pp. 31-38, Jan. 1987.

[16] M. P. C. Fossorier, F. Burkert, S. Lin, and J. Hagenauer, "On the equivalence between SOVA and Max-Log-MAP decodings," *IEEE Commun. Lett.*, vol. 2, pp. 137–139, May 1998.

[17] B. Mendelson, *Introduction to Topology, 3rd ed.* New York: Dover, 1990.

[18] E. Zeidler, *Nonlinear Functional Analysis and its Applications, I: Fixed-Point Theorems.* New York: Springer-Verlag, 1985.

[19] J. Gilbert and L. Gilbert, *Linear Algebra and Matrix Theory.* New York: Academic Press, 1994.

[20] R. Horn and C. Johnson, *Matrix Analysis.* Cambridge: Cambridge University Press, 1985.

Recursive List Decoding for Reed-Muller Codes and Their Subcodes

Ilya Dumer and Kirill Shabunov[1]
College of Engineering
University of California, Riverside
Riverside, CA 92521
dumer@ee.ucr.edu
shabunov@ee.ucr.edu

Abstract

We consider recursive decoding for Reed-Muller (RM) codes and their subcodes. Two new recursive techniques are described. We analyze asymptotic properties of these algorithms and show that they substantially outperform other decoding algorithms with nonexponential complexity known for RM codes. Decoding performance is further enhanced by using intermediate code lists and permutation procedures. For moderate lengths up to 512, near-optimum decoding with feasible complexity is obtained.

Keywords: Recursive decoding, Reed-Muller codes, decoding threshold, Plotkin construction, permutations.

[1] **This research was supported by the NSF grant CCR-0097125.**

1 Introduction

Below we consider Reed-Muller (RM) codes and their subcodes. We use notation $\left\{{m \atop r}\right\}$ for RM codes of length $n = 2^m$, dimension $k = \sum_{i=0}^{r} \binom{m}{i}$ and distance $d = 2^{m-r}$. In this paper, we wish to design fast decoding algorithms that outperform other algorithms known for RM codes. To achieve this goal, we will use *recursive techniques* that make decoding decisions by combining decoding results obtained on the shorter codes.

RM codes are almost on par with the best codes on moderate lengths $n \leq 128$ and have found numerous applications thanks to fast decoding procedures. First, *majority algorithm* was developed in [1] followed later by numerous developments. Such a decoding has low complexity of the order nk and enables bounded distance decoding.

It is also important that majority decoding substantially extends bounded-distance *threshold* of $d/2$. Given an infinite sequence of codes $A_i(n_i, d_i)$, we say that a decoding algorithm Ψ achieves decoding thresholds δ_i if for any $\epsilon > 0$ only a vanishing fraction of error patterns of weight $\delta_i(1-\epsilon)$ is left uncorrected as $n_i \to \infty$. It can be proven [5] that majority algorithm achieves a threshold

$$\delta = (d \ln d)/4 \tag{1}$$

for long RM codes of fixed rate R (here and below we omit index i). For long low-rate RM codes of fixed order r, it is shown [11] that majority decoding achieves a threshold

$$\delta = n(1 - \varepsilon_r^{\text{maj}})/2,$$

where the residual term $\varepsilon_r^{\text{maj}}$ has a slowly declining order

$$\varepsilon_r^{\text{maj}} \sim (m/d)^{1/2^{r+1}} = (m/2^{m-r})^{1/2^{r+1}}, \quad m \to \infty. \tag{2}$$

Therefore δ exceeds bounded distance threshold $d/2$ approximately 2^r times and approaches the upper limit of $n/2$ for long codes.

Another efficient technique is based on recursive algorithms of [2] and [3]. The recursive structure of RM codes $\left\{{m \atop r}\right\}$ is well known [4] and is formed by the *Plotkin construction* $(\mathbf{u}, \mathbf{u}+\mathbf{v})$, where subblocks $\mathbf{u}$ and $\mathbf{v}$ are taken from the codes $\left\{{m-1 \atop r}\right\}$ and $\left\{{m-1 \atop r-1}\right\}$. It is proven in [2] and [3] that this recursive structure allows to execute bounded distance decoding with the lowest complexity order of $n\min(r, m-r)$ known for RM codes. The techniques developed below also show that recursive algorithms of [2] and [3] are on par with majority decoding and achieve the same error-correcting thresholds δ in both cases (1) and (2).

One more efficient algorithm based on permutation decoding was designed in [6] for RM codes $\left\{ {m \atop 2} \right\}$. This algorithm gives a slightly higher complexity order of $n^2 m$ while reducing the corresponding residual term $\varepsilon_2^{\mathrm{maj}}$ from (2) to the lower order of $(m/n)^{1/4}$ as $m \to \infty$.

The above algorithms can also be extended for soft decision channels. For RM codes of fixed rate R, soft decision majority decoding [11] gives a threshold of Euclidean weight $\rho = (n/m)^{1/2^{r+1}}\sqrt{n}$. Using technique of [11], it can be proven that recursive algorithms of [2] and [3] also have the same error-correcting threshold ρ. For RM codes $\left\{ {m \atop 2} \right\}$, the algorithm of [6] allows to increase the Euclidean threshold ρ to the order of $(n/m)^{1/4}\sqrt{n}$. Finally, multistage maximum-likelihood decoding is performed in [7] by designing an efficient multilevel trellis structure. ML decoding supersedes recursive algorithms. In particular, it can be proven (similarly to [6]) that ML decoding has Euclidean threshold upper bounded by the order of $n\sqrt{(r!\ln 2)/2m^r}$. However, ML decoding complexity is exponential in n.

Below we design new recursive algorithms that substantially outperform other (nonexponential) algorithms known for RM codes for both hard and soft decision channels. Our basic recursive procedure will split the RM code $\left\{ {m \atop r} \right\}$ of length n into two RM codes of length $n/2$. Decoding is then relegated further to the shorter codes until we reach basic codes and perform ML decoding with complexity $O(n\log n)$. In all intermediate steps, we only recalculate the reliabilities of the newly defined symbols.

Below in Section 2 we consider recursive structure of RM codes in more detail. Then in Sections 3 and 4 we proceed with decoding techniques and design two different versions Ψ_r^m and Φ_r^m of our recursive algorithm. In Section 5 we proceed with further improvements. In particular, decoding performance will be considerably improved by using subcodes of RM codes. Another improvement is based on using relatively short lists of codewords in the intermediate steps of the recursion. Finally, we use different permutations taken from the symmetry (automorphism) group of the code. As a result, we closely approach ML decoding performance on the blocklength of 256 and for low-rate codes of length 512.

2 Recursive structure

Recursive techniques known for RM codes are based on the *Plotkin construction*. Here any RM code $\left\{{m \atop r}\right\}$ is represented in the form $(\mathbf{u}, \mathbf{u}+\mathbf{v})$, where $\mathbf{u}$ and $\mathbf{v}$ are two subblocks of length 2^{m-1} that run through the codes $\left\{{m-1 \atop r}\right\}$ and $\left\{{m-1 \atop r-1}\right\}$, respectively. By continuing this process on codes $\left\{{m-1 \atop r}\right\}$ and $\left\{{m-1 \atop r-1}\right\}$, we obtain RM codes of length 2^{m-2} and so on. Finally, we arrive at the end nodes, which are repetition codes $\left\{{g \atop 0}\right\}$ for any $g = 1, \ldots, m-r$ and full spaces $\left\{{h \atop h}\right\}$ for any $h = 1, \ldots, r$. This is schematically shown in Fig. 1 for RM codes of length 16. In Fig. 2, we consider incomplete decomposition for codes of length 32 terminated at the biorthogonal codes and single-parity check codes. This recursive

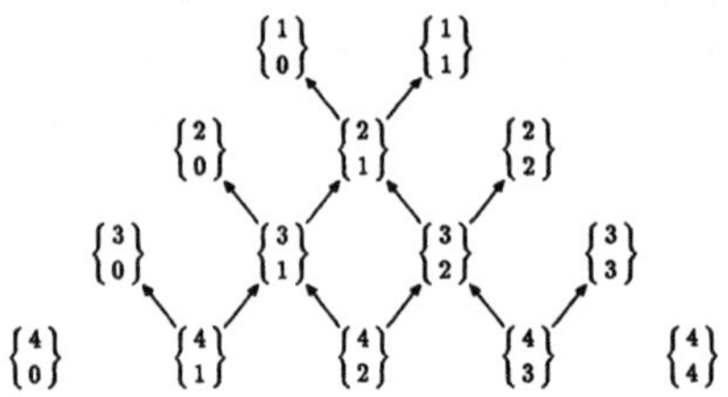

Figure 1: Full decomposition.

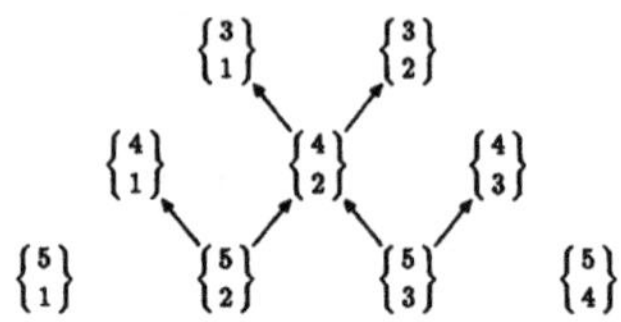

Figure 2: Partial decomposition.

structure is also exhibited in the generator matrices of RM codes. As an example, on Fig. 3 we present a generator matrix for $\left\{{7 \atop 3}\right\}$ code.

Now let $\mathbf{a}_r^m = \{a_j \mid j = 1, k\}$ be a block of information bits a_j that encodes a vector $(\mathbf{u}, \mathbf{u}+\mathbf{v})$. Then recursive procedure splits $\mathbf{a}_r^m$ into two information subblocks $\mathbf{a}_r^{m-1}$ and $\mathbf{a}_{r-1}^{m-1}$ that encode vectors $\mathbf{u}$ and $\mathbf{v}$, respectively. In this way, information subblocks are split until we arrive at the end nodes. Thus, any specific codeword can be encoded from the information strings assigned to the end nodes $\left\{{g \atop 0}\right\}$ or $\left\{{h \atop h}\right\}$.

Also, it can be proven that recursive encoding of code $\left\{{m \atop r}\right\}$ has com-

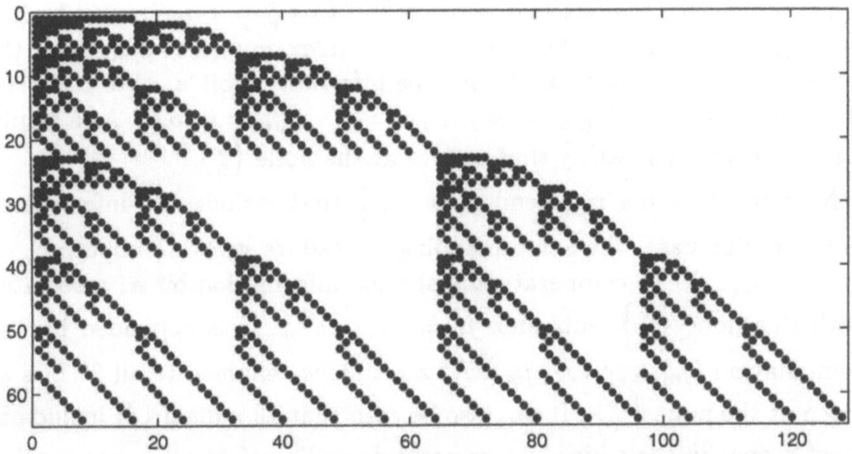

Figure 3: A generator matrix of $\left\{ {7 \atop 3} \right\}$ code. Ones are denoted by dots, zeros are not shown.

plexity

$$\psi_r^m \le n \min(r, m-r). \tag{3}$$

This observation comes from the two facts. First, the end nodes $\left\{ {g \atop 0} \right\}$ and $\left\{ {h \atop h} \right\}$ satisfy the complexity bound (3). Second, we can obtain an $(\mathbf{u}, \mathbf{u}+\mathbf{v})$-codeword using two constituent codewords taken from $\left\{ {m-1 \atop r-1} \right\}$ and $\left\{ {m-1 \atop r} \right\}$. Therefore the overall complexity satisfies inequality $\psi_r^m \le \psi_{r-1}^{m-1} + \psi_r^{m-1} + \frac{n}{2}$. Now we see that code $\left\{ {m \atop r} \right\}$ satisfies (3) if the two constituent codes do. In particular, for $r < m/2$, we obtain the bound

$$\psi_r^m \le n(r-1)/2 + nr/2 + n/2 = nr.$$

Now consider an information bit a_j associated with any left node, say $\left\{ {g \atop 0} \right\}$. Then splitting procedure also allows us to map this symbol a_j onto a specific "binary path" $\bar{j} = (j_1, \ldots, j_m)$ leading from the origin $\left\{ {m \atop r} \right\}$ to the end node $\left\{ {g \atop 0} \right\}$. To do this, we first take $j_1 = 0$ for any $a_j \in \mathbf{a}_r^{m-1}$ and $j_1 = 1$ for $\mathbf{a}_{r-1}^{m-1}$. On any step $s = 2, \ldots, m$, we take $j_{s+1} = 1$

when moving to the left (say, from $\mathbf{a}_{r-1}^{m-1}$ to $\mathbf{a}_{r-2}^{m-2}$) or $j_{s+1} = 0$ when moving to the right. Then we get a subindex $= (j_1, \ldots, j_{m-g})$ at the left-end node $\left\{ {g \atop 0} \right\}$ associated with the information bit a_j. We complete this mapping by adding g zeros $j_{m-g+1} = \cdots = j_m = 0$ to $\overline{j}_g$. As a result, we obtain the full path $\overline{j}$ that arrives at the node $\left\{ {0 \atop 0} \right\}$.

Now consider the right-end node $\left\{ {h \atop h} \right\}$ that includes 2^h information bits. In this case, the same mapping procedure gives a subpath $\overline{j}_h = (j_1, \ldots, j_{m-h})$. To enumerate any specific information bit a_j associated with this node $\left\{ {h \atop h} \right\}$, subindex $\overline{j}_h = (j_1, \ldots, j_{m-h})$ is appended by any combination $(j_{m-h+1}, \ldots, j_m)$. As a result, we enumerate all 2^h bits a_j given at the node $\left\{ {h \atop h} \right\}$. It can also be seen that all indices $\{\overline{j}\}$ include at most r ones in their binary representation. Therefore all k information bits of the whole code are mapped onto m-digital binary paths $\overline{j}$ of weight r or less.

3 New decoding techniques

Below we consider a new recursive algorithm Ψ_{rec} based on the $(\mathbf{u}, \mathbf{u}+\mathbf{v})$ construction. The received block $\mathbf{y} = (\widetilde{\mathbf{u}}, \widetilde{\mathbf{u}+\mathbf{v}})$ consists of two halves $\mathbf{y}'$ and $\mathbf{y}''$ corrupted by noise. By taking outputs y_i' and y_i'', the decoder finds the posterior probabilities of symbols u_i and u_i+v_i :

$$p_i' \stackrel{\text{def}}{=} \Pr\{u_i = 0 \mid y_i'\}, \quad p_i'' \stackrel{\text{def}}{=} \Pr\{u_i+v_i = 0 \mid y_i''\}, \quad i = 1, \ldots, n/2.$$

We first try to find the better protected codeword $\mathbf{v}$ from $\left\{ {m-1 \atop r-1} \right\}$. Then we decode the block $\mathbf{u} \in \left\{ {m-1 \atop r} \right\}$.

Step 1. To find a subblock $\mathbf{v}$ in hard-decision decoding, one would use its corrupted version $\widetilde{\mathbf{v}} = \widetilde{\mathbf{u}} + \widetilde{\mathbf{u}+\mathbf{v}}$. Using more general approach, we find $n/2$ posterior probabilities

$$p_i^{\mathbf{v}} \stackrel{\text{def}}{=} \Pr\{v_i = 0 \mid y_i', y_i''\} = p_i' p_i'' + (1 - p_i')(1 - p_i''). \tag{4}$$

Here we apply the formula of total probability to the binary sum of independent symbols u_i and u_i+v_i. Now we can use any soft-decision decoding $\Psi_{\mathbf{v}}(p_i^{\mathbf{v}})$ to find vector $\widehat{\mathbf{v}}$. This completes Step 1 of our algorithm. Vector $\widehat{\mathbf{v}}$ is then passed to Step 2.

Step 2. Now we use both symbols y_i'' and v_i to estimate symbol u_i on the right half. *Assuming that* $\widehat{\mathbf{v}} = \mathbf{v}$, we find that symbol u_i has posterior probability

$$p_i^{\wedge} \stackrel{\text{def}}{=} \Pr\{u_i = 0 \mid y_i'', \widehat{v}_i\} = \begin{cases} p_i'', & \text{if } \widehat{v}_i = 0, \\ 1 - p_i'', & \text{if } \widehat{v}_i = 1. \end{cases}$$

Now we have the two posterior probabilities p_i' and $p_i^{\wedge}$ of symbol u_i obtained on both corrupted halves. By using the Bayes' formula, we find the combined estimate

$$p_i^{\mathbf{u}} \stackrel{\text{def}}{=} \Pr\{u_i = 0 \mid p_i', p_i^{\wedge}\} = \frac{p_i' p_i^{\wedge}}{p_i' p_i^{\wedge} + (1 - p_i')(1 - p_i^{\wedge})}. \tag{5}$$

Finally, we perform soft decision decoding $\Psi_{\mathbf{u}}(p_i^{\mathbf{u}})$ and find a subblock $\widehat{\mathbf{u}} \in \left\{ {m-1 \atop r} \right\}$. So, the basic soft-decision decoding Ψ_{rec} uses procedures $\Psi_{\mathbf{v}}$, $\Psi_{\mathbf{u}}$ and outputs a decoded codeword $\widehat{\mathbf{c}}$ and the corresponding information block $\widehat{\mathbf{a}}$ as follows.

Algorithm Ψ_{rec}.
1. Calculate probabilities $p_i^{\mathbf{v}}$ according to (4).
2. Decode $\widehat{\mathbf{v}}$ using $\Psi_{\mathbf{v}}(p_i^{\mathbf{v}}), i = 1, \ldots, n/2$.
3. Calculate probabilities $p_i^{\mathbf{v}}$ according to (5).
4. Decode $\widehat{\mathbf{u}}$ using $\Psi_{\mathbf{u}}(p_i^{\mathbf{u}}), i = 1, \ldots, n/2$.
5. Output decoded components:
$\widehat{\mathbf{a}} := (\widehat{\mathbf{a}}_{\mathbf{v}} \mid \widehat{\mathbf{a}}_{\mathbf{u}}); \quad \widehat{\mathbf{c}} := (\widehat{\mathbf{u}} \mid \widehat{\mathbf{u}} + \widehat{\mathbf{v}})$

In a more general scheme Ψ_r^m, we repeat this recursion by decomposing subblocks $\widehat{\mathbf{v}}$ and $\widehat{\mathbf{u}}$ further. On all intermediate steps, we only recalculate the probabilities of the newly defined symbols. Finally, we perform ML decoding once we reach the end nodes $\left\{ {g \atop 0} \right\}$ and $\left\{ {h \atop h} \right\}$. The algorithm is described below.

Algorithm Ψ_r^m.
1. If $0 < r < m$, find $\Psi_{\text{rec}}(\mathbf{p})$ using $\Psi_{\mathbf{v}} = \Psi_{r-1}^{m-1}$ and $\Psi_{\mathbf{u}} = \Psi_r^{m-1}$.
2. If $r = 0$ decode code $\left\{ {r \atop 0} \right\}$.
3. If $r = m$ decode code $\left\{ {r \atop r} \right\}$.

In the next algorithm Φ_r^m, we terminate decoding $\Psi_{\mathbf{v}}$ at the biorthogonal codes $\left\{ {g \atop 1} \right\}$.

Algorithm Φ_r^m.
1. If $1 < r < m$, find $\Psi_{\mathrm{rec}}(\mathbf{p})$ using $\Psi_{\mathbf{v}} = \Psi_{r-1}^{m-1}$ and $\Psi_{\mathbf{u}} = \Psi_r^{m-1}$.
2. If $r = 1$ decode code $\left\{ {r \atop 1} \right\}$.
3. If $r = m$ decode code $\left\{ {r \atop r} \right\}$.

Thus, procedures Ψ_r^m and Φ_r^m have recursive structure that calls itself until ML decoding is applied on the end nodes. ML decoding of biorthogonal codes has complexity order of $n \log_2 n$. Simple analysis also shows that recalculating all posterior probabilities in (4) and (5) has complexity at most $5n$. Therefore our decoding complexity Φ_r^m satisfies recursion

$$\Phi_r^m \leq \Phi_{r-1}^{m-1} + \Phi_r^{m-1} + 5n.$$

Similarly to the derivation of (3), this recursion brings the overall complexity of Φ_r^m and Ψ_r^m to the order of $5n \log_2 n$ real operations.

4 Analysis of algorithms Ψ_r^m and Φ_r^m.

In general, procedure Ψ_r^m enters each end node by taking all paths leading to this node. It turns out that the output bit error rate (BER) significantly varies on different nodes and even on different paths leading to the same node. Therefore our first problem is to define the most error-prone paths.

Asymptotic analysis. We consider AWGN channels and assume that the all-zero codeword is transmitted as a sequence of +1s. Then n outputs y_i' and y_i'' are independent random variables (RV) with normal distribution $\mathcal{N}(1, \sigma^2)$. It can be readily seen that the posterior probabilities p_i (that is p_i' or p_i'') become independent RV with non-Gaussian distribution

$$p_i = \frac{1}{2}(1 + \varepsilon_i), \quad \text{where } \varepsilon_i = \tanh \frac{y_i}{\sigma^2}, \quad \tanh(x) \triangleq \frac{e^x - e^{-x}}{e^x + e^{-x}}. \tag{6}$$

In the next step, we obtain the RV $p_i^{\mathbf{v}}$ and $p_i^{\mathbf{u}}$. Here we rewrite equations (4) and (5) as follows.

Lemma 4.1 *The values p_i^{v} and p_i^{u} can be calculated as*

$$p_i^{\mathrm{v}} = \frac{1}{2}(1+\varepsilon_i^{\mathrm{v}}), \quad \varepsilon_i^{\mathrm{v}} = \varepsilon_i' \varepsilon_i'', \tag{7}$$

$$p_i^{\mathrm{u}} = \frac{1}{2}(1+\varepsilon_i^{\mathrm{u}}), \quad \varepsilon_i^{\mathrm{u}} = \tanh \frac{y_i' + (-1)^{\widehat{v}_i} y_i''}{\sigma^2}. \tag{8}$$

The *product* RV $\varepsilon_i^{\mathrm{v}}$ defined in (7) has a smaller expected value relative to the original estimates ε_i' and ε_i'', since $0 \leq \varepsilon_i', \varepsilon_i'' \leq 1$. Therefore the mean value of p_i^{v} converges to 0.5 in the subsequent left-hand steps. This makes decoding Ψ_{v} less reliable. On the positive side, we note that each step gives us a better protected code that has twice the relative distance of the former one. Therefore we subsequently degrade the channel while entering the new codes with higher correcting capabilities.

If $\widehat{v}_i$ is correct (i.e. $\widehat{v}_i = 0$) we have $\varepsilon_i^{\mathrm{u}} = \tanh\{(y_i' + y_i'')/\sigma^2\}$. So the second RV $\varepsilon_i^{\mathrm{u}}$ has a greater expected value. Consequently, the mean probabilities p_i^{u} increase as we move to the right. Note, however, that each new code has half the relative distance of its parent code. In other words, we subsequently improve the channel while entering the new codes with weaker correcting capabilities.

Now we proceed with an asymptotic analysis. We first consider RM codes with $m \to \infty$ and fixed order r. These codes have rates $R \to 0$. Therefore we have to consider the case $\sigma^2 \to \infty$ to obtain any *fixed* signal-to-noise ratio $1/R\sigma^2$ as $m \to \infty$. We will use the following lemma proven in [11].

Lemma 4.2 *For large noise power $\sigma^2 \to \infty$ the first two moments $E\varepsilon$ and $E\varepsilon^2$ of the random variable $\varepsilon = \tanh(y/\sigma^2)$ satisfy the relation*

$$E\varepsilon \sim E\varepsilon^2 \sim \sigma^{-2}. \tag{9}$$

In general, we wish to use the original RV y_i, ε_i', and ε_i'' and recalculate their probability density functions (pdf), using (7) and (8) to find the pdf of the new RV variables $\varepsilon_i^{\mathrm{v}}$ and $\varepsilon_i^{\mathrm{u}}$. However, the latter formulas make these recalculations very involved. Therefore we consider a simplified version of our algorithm Ψ_{rec}. Namely, given a channel symbol y with posterior probability $p = (1+\varepsilon)/2$, we define the *likelihood* of 0

$$\rho(\varepsilon) \stackrel{\text{def}}{=} \frac{2y}{\sigma^2} = \ln \frac{1+\varepsilon}{1-\varepsilon}.$$

Note that the likelihoods form independent Gaussian RV. It can be easily seen that the new RV $\varepsilon_i^{\mathbf{u}}$ obtained in (8) gives the likelihood

$$\rho(\varepsilon_i^{\mathbf{u}}) = \rho(\varepsilon_i') + \rho(\varepsilon_i'') \tag{10}$$

for any noise power σ^2. For the RV $\varepsilon_i^{\mathbf{v}}$, the corresponding recalculation results in a longer formula

$$\rho(\varepsilon_i^{\mathbf{v}}) = \ln \frac{1 + \exp \rho(\varepsilon_i') \exp \rho(\varepsilon_i'')}{\exp \rho(\varepsilon_i') + \exp \rho(\varepsilon_i'')}. \tag{11}$$

Given the asymptotic case $\sigma^2 \to \infty$, note that the RV $\rho(\varepsilon_i)$ takes small values with high probability. Therefore we replace the latter formula by its approximation valid for small $\rho(\varepsilon_i')$ and $\rho(\varepsilon_i'')$:

$$\rho(\varepsilon_i^{\mathbf{v}}) \sim \rho(\varepsilon_i')\rho(\varepsilon_i''). \tag{12}$$

It can be proven that the output bit error rate obtained on any end node $\left\{ {g \atop 0} \right\}$ with large g can be calculated using only the first two moments $E\rho$ and $E\rho^2$ of the RV $\rho = \rho(\varepsilon_i^{\mathbf{v}})$ obtained on this node. It can also be proven that the original formula (11) and its approximation (12) give the same moment $E\rho$ as $\sigma^2 \to \infty$. Also, the two formulas give the same asymptotic moments $E\rho^2$. This justifies using the above approximation in the asymptotic case.

Now we consider a simplified algorithm Ψ_{rec}, with recalculations (7) and (8) replaced by (10) and (12). Using this simplified version Ψ_{rec}, we can arrive at the following conclusions [12].

Theorem 4.1 *For $\sigma^2 \to \infty$, replacing code $\left\{ {m \atop r} \right\}$ by $\left\{ {m-1 \atop r-1} \right\}$ in the algorithm $\Psi_{\mathbf{v}}$ is equivalent to increasing the original noise power σ^2 to σ^4. Replacing code $\left\{ {m \atop r} \right\}$ by $\left\{ {m-1 \atop r} \right\}$ in the algorithm $\Psi_{\mathbf{u}}$ reduces the original noise power σ^2 to $\sigma^2/2$.*

Therefore in asymptotic setting our recursive procedure can be considered as a "propagation of the noise power". This propagation undergoes two different types of changes while we move from the parent node to the two descendant nodes. This propagation is also illustrated on Fig. 4 as an example for the $\left\{ {7 \atop 3} \right\}$ code (note, however, that procedure becomes exact only for very long codes).

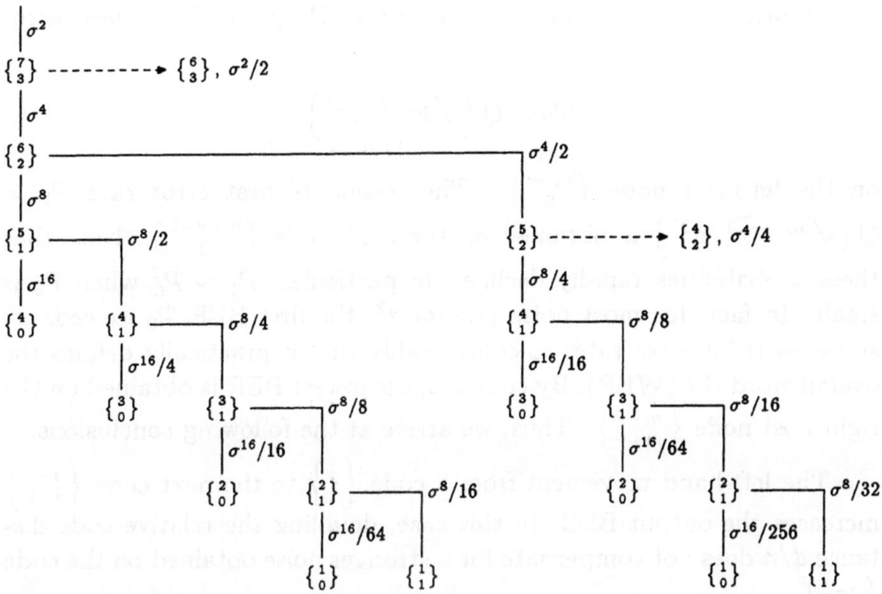

Figure 4: Propagation of the noise power during the decoding of RM $\left\{ {7 \atop 3} \right\}$ code.

Now we can find asymptotic error rate for each bit (path) a_i. Note that algorithm Ψ_r^m has the highest noise power $\sigma^{2^{r+1}}$ when it arrives at the leftmost (repetition) code $\left\{ {m-r \atop 0} \right\}$. Also, a repetition code of any length n used on the AWGN channel $\mathcal{N}(0, \sigma^2)$ has an output error probability of ML decoding

$$P = Q\left(\sqrt{n}/\sigma\right), \quad \text{where } Q(x) \triangleq \int_x^\infty e^{-u^2/2}\, du \,/ \sqrt{2\pi}. \tag{13}$$

Note that for a general binary memoryless channel with noise variance σ^2, the same estimate can also be used on any node $\left\{ {g \atop 0} \right\}$ given the following two conditions:

$$g \to \infty, \quad \sigma/2^{g/3} \to \infty. \tag{14}$$

Both conditions directly follow from the estimates obtained in [8] (p. 549) for the large deviations of the sums of independent random variables with common distribution.

Estimate (13) shows that the algorithm Ψ_r^m gives the highest error rate

$$\mathcal{P}_0 = Q\left(\sqrt{2^{m-r}}/\sigma^{2^r}\right)$$

on the leftmost node $\left\{ {m-r \atop 0} \right\}$. The second highest error rate $\mathcal{P}_1 = Q\left(\sqrt{2^{m-r+1}}/\sigma^{2^r}\right)$ is obtained on the next node $\left\{ {m-r-1 \atop 0} \right\}$. Note that these probabilities rapidly decline. In particular, $\mathcal{P}_1 \sim \mathcal{P}_0^2$ when $\mathcal{P}_0$ is small. In fact, for most noise powers σ^2, the first BER $\mathcal{P}_0$ exceeds all subsequent bit error rates so considerably that it practically defines the overall word ER (WER). By contrast, the lowest BER is obtained on the rightmost node $\left\{ {m-r \atop m-r} \right\}$. Thus, we arrive at the following conclusions:

• The left-hand movement from a code $\left\{ {j \atop i} \right\}$ to the next code $\left\{ {j-1 \atop i-1} \right\}$ increases the output BER. In this case, doubling the relative code distance d/n does not compensate for a stronger noise obtained on the code $\left\{ {j-1 \atop i-1} \right\}$.

• Moving to the right from a code $\left\{ {j \atop i} \right\}$ to the next code $\left\{ {j-1 \atop i} \right\}$ allows us to reduce the BER of the algorithm relative to the parent code $\left\{ {j \atop i} \right\}$. As a result, the lowest BER is obtained on the rightmost node $\left\{ {r \atop r} \right\}$.

In a more general setting, we can estimate asymptotic error rates for any information bit a_j. For $c = 0, 1$ and $x > 0$, we use notation $x \diamond c \stackrel{\text{def}}{=} 2^{1-c}x^{1+c}$. Given a symbol a_j arriving at the node $\left\{ {g \atop 0} \right\}$ we can consider the corresponding subpath $\overline{j}_g = (j_1, \ldots, j_{m-g})$. Then we define the product $x \diamond \overline{j}_g = (\cdots(x \diamond j_1) \diamond j_2 \cdots) \diamond j_{m-g}$ and arrive at the following statement.

Theorem 4.2 *Consider RM codes with $m \to \infty$ and fixed order r. For the information bit a_j associated with a node $\left\{ {g \atop 0} \right\}$, algorithm Ψ_r^m has bit error rate*

$$\mathcal{P}(a_j) \sim Q\left(\sigma \diamond \overline{j}_g\right), \quad g \to \infty. \tag{15}$$

Similar results hold for the algorithm Φ_r^m, which stops at the nodes $\left\{ {g \atop 1} \right\}$. This node is associated with a subblock of $g + 1$ information bits.

In this case the corresponding subindex $\overline{j}_g^1$ has weight $r-1$ or less. Therefore algorithm Φ_r^m reduces the highest noise power $\sigma^{2^{r+1}}$ to σ^{2^r} and gives substantial improvement over Ψ_r^m. More generally, we obtain the following statement.

Theorem 4.3 *Consider RM codes with $m \to \infty$ and fixed order r. For the subset of $g+1$ information bits $\{a_j\}$ associated with a node $\left\{ {g \atop 1} \right\}$, algorithm Ψ_r^m has bit error rate*

$$\mathcal{P}(\{a_j\}) \lesssim 2^g Q\left(\sigma \diamond \overline{j}_g^1\right), \quad g \to \infty.$$

In particular, the highest BER $\mathcal{P}_0$ obtained at the node $\left\{ {m-r \atop 0} \right\}$ by Ψ_r^m is now being replaced by

$$\mathcal{P}_0' \lesssim 2^{m-r+1} Q\left(\sqrt{2^{m-r}}/\sigma^{2^{r-1}}\right)$$

obtained at the node $\left\{ {m-r+1 \atop 1} \right\}$. As the block length grows, decoding Φ_r^m increasingly outperforms both the majority algorithm and recursive techniques of [2], [3]. Further, this analysis can be extended to codes of fixed rate R. In particular, the following statement holds for hard-decision decoding.

Theorem 4.4 *For RM codes with $m \to \infty$ and fixed rate R, algorithm Φ_r^m has error-correcting threshold $(d \ln d)/2$ thus:*

- *increasing* $\ln d$ *times the threshold of bounded-distance decoding;*
- *doubling the threshold* $(d \ln d)/4$ *of majority decoding.*

5 Improvements

1. Subcodes of RM codes To improve output error rate, we *set the leftmost information bits as zeros.* In this way, we arrive at the subcodes of the original code $\left\{ {m \atop r} \right\}$ that are obtained by eliminating a few least protected information bits. This expurgation starts with the node $\left\{ {m-r \atop 0} \right\}$ in procedure Ψ_r^m, and with the node $\left\{ {m-r+1 \atop 1} \right\}$ in Φ_r^m. It can be shown that after eliminating only one bit, algorithm Ψ_r^m gives the same BER on the channel whose noise power σ^2 is increased $2^{1/2^r}$ times. For the algorithm Φ_r^m, the sustainable noise power is increased $2^{1/2^{r-1}}$ times. For long codes of small order $r = 2, 3$ this amounts to a gain of 1.5 dB and 0.75 dB, respectively.

2. List decoding Decoding performance is further improved by choosing L best candidates after each decoding step. This *list decoding* $\Psi_r^m(L)$ starts at the leftmost code $\left\{{m-r \atop 0}\right\}$. Here we define posterior probabilities $p(\mathbf{v} \mid \mathbf{y})$ of both codewords $\mathbf{v}' = 0$ and $\mathbf{v}'' = 1$. These codewords are represented as two initial edges with the corresponding cost functions $\log p(\mathbf{v} \mid \mathbf{y})$. Then we decode the next code $\left\{{m-r-1 \atop 0}\right\}$. Note that codewords $\mathbf{v}'$ and $\mathbf{v}''$ give different probability distributions on this node. Therefore our new decoding is performed 2 times, separately for $\mathbf{v}'$ and $\mathbf{v}''$. The result is a full tree of depth 2 that has 4 new edges. On further steps, we keep doubling the number of paths until $2L$ paths are formed. Then we choose L paths with maximum cost functions and proceed further. In the end, the most probable path (that has maximum cost function) is chosen among L paths survived at the rightmost node. Simulation results and analytic estimates give very substantial improvements when both techniques – using the subcodes and short decoding lists – are combined. These results are presented below in Figures 5 to 8.

3. New permutation techniques Finally, the third improvement utilizes the rich *symmetry group* $GA(m)$ of RM codes that includes $2^{O(m^2)}$ permutations. First, note that even for large L algorithm $\Psi_r^m(L)$ is likely to fail if error positions substantially disagree on the two halves of the original block. By using a symmetry group, we try to find the permutations that match unknown erroneous positions in the two permuted halves. If successful, procedure $\Psi_{\mathbf{v}}$ will eliminate most errors on the permuted block $\mathbf{v}$ of length $n/2$. This process can be advanced in the next step, by finding another permutation that again gives a good match on erroneous positions left on the new halves of length $n/4$.

In particular, we use the following sets of permutations. Represent any position $i = 1, \ldots, 2^m$ in the binary form $i = (i_1, \ldots, i_m)$. We take any permutation $\pi(1), \ldots, \pi(m)$ and consider the subgroup $S \subset GA(m)$ that includes $m!$ permutations $(i_1, \ldots, i_m) \mapsto (i_{\pi(1)}, \ldots, i_{\pi(m)})$. Note that using subgroup S also changes the "folding" order used in algorithm $\Psi_{\mathbf{v}}$ (say, we fold adjacent quarters instead of halves when we permute i_1 and i_2). We can also consider $\binom{m}{r}$ permutations taking exactly one permutation with a given subset $\pi^{-1}(1), \ldots, \pi^{-1}(r)$ of the first r "folding" indices. Permutations from this subset T change the order in which we decode left-end nodes. Finally, consider a subgroup U that includes m cyclic shifts $\pi(1), \ldots, \pi(m)$. Simulation results for the moderate lengths 256 and 512 showed that using subsets T and even U allows to reduce the *combined* list of L best candidates by one decimal order. As a result, we obtained nearly ML decoding on the lengths 512 while using lists of moderate size L.

6 Simulation results

Simulation results are described in Figures 5 to 8. We start with Fig. 5 that reflects substantial improvements obtained when both techniques

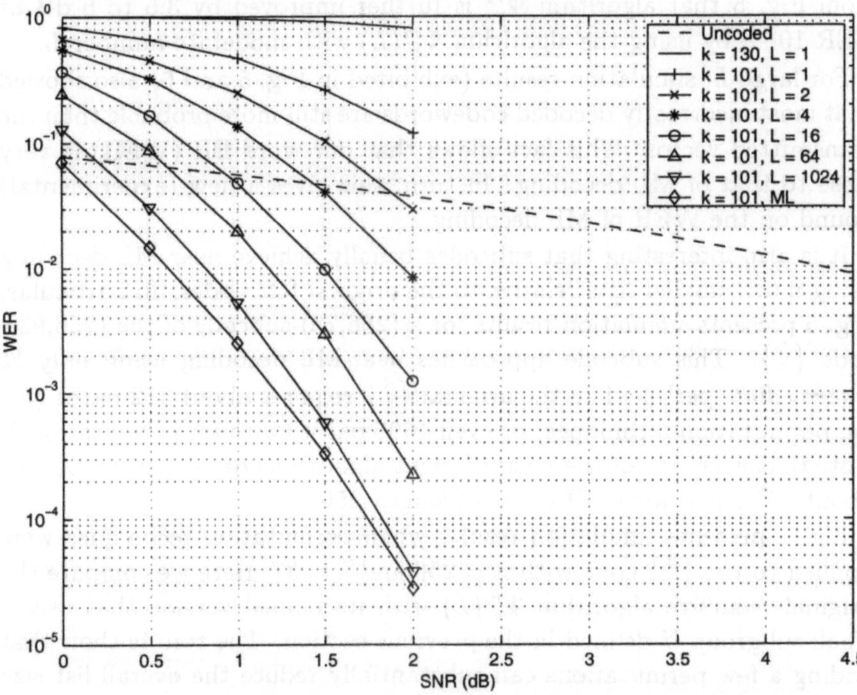

Figure 5: RM $\left\{ {9 \atop 3} \right\}$ code, $n = 512$, $k = 130$, and its subcode, $k = 101$.

– using the subcodes and short decoding lists – were combined. The first (upper) curve with $L = 1$ shows the performance of the algorithm Ψ_r^m applied to the $\left\{ {9 \atop 3} \right\}$ code with $n = 512$ and $k = 130$. This algorithm can be considered as a refined version of the former recursive techniques from [2], [3], and [10]. Namely, Ψ_r^m uses exact probability recalculations presented in formulas (7) and (8) instead of various metrics used in these papers.

The second curve with $L = 1$ shows the performance of the algorithm Ψ_r^m applied to the $(512, 101)$-subcode of the original $\left\{ {9 \atop 3} \right\}$ code. This subcode is obtained by removing 29 leftmost information bits with the highest BER. We see that the subcode gives substantial improvement in the output BER despite having a smaller code rate. All other curves on Fig. 5 correspond to the same subcode but use the bigger lists. We see from Fig. 5 that algorithm Ψ_r^m is further improved by 3.5 to 5 dB at BER 10^{-4}, by using the algorithm $\Psi_r^m(L)$ with moderate number L.

For large L, simulation results (exhibited in Fig. 5 and 6) also showed that most incorrectly decoded codewords are still more probable than the transmitted vector. This fact shows that our word ER (WER) is very close to that of ML decoding. In turn, this gives a new (experimental) bound on the WER of ML decoding.

It is also interesting that subcodes usually achieve near-ML decoding using much smaller lists relative to the original RM codes. In particular, Fig. 6 presents simulation results for a (256,78)-subcode of the (256,93)-code $\left\{ {8 \atop 3} \right\}$. This subcode approaches near-ML decoding using *only 32 intermediate paths,* while the original $\left\{ {8 \atop 3} \right\}$ requires about 512 paths (using permutation techniques) or even 4096 paths (without permutations). Note that even one of the most efficient algorithms developed in [9] uses about 10^5 paths for BCH codes of length 256.

Fig. 7 presents simulation results when permutation techniques were applied to the $\left\{ {8 \atop 2} \right\}$ code with $n = 256$ and $k = 37$. Here we compare the original recursive algorithm $\Psi_r^m(L)$ with its refined version that uses a small subgroup U defined in the previous section. The results show that adding a few permutations can substantially reduce the overall list size (taken over all permutations). For this specific code, the refined version reduces approximately 4 times the number of trials used in $\Psi_r^m(L)$ to obtain near-ML decoding. Similar results show that for codes of length 512 the complexity of near-ML decoding is reduced tenfold.

Finally, in Fig. 8, we summarize the results for all nontrivial RM codes

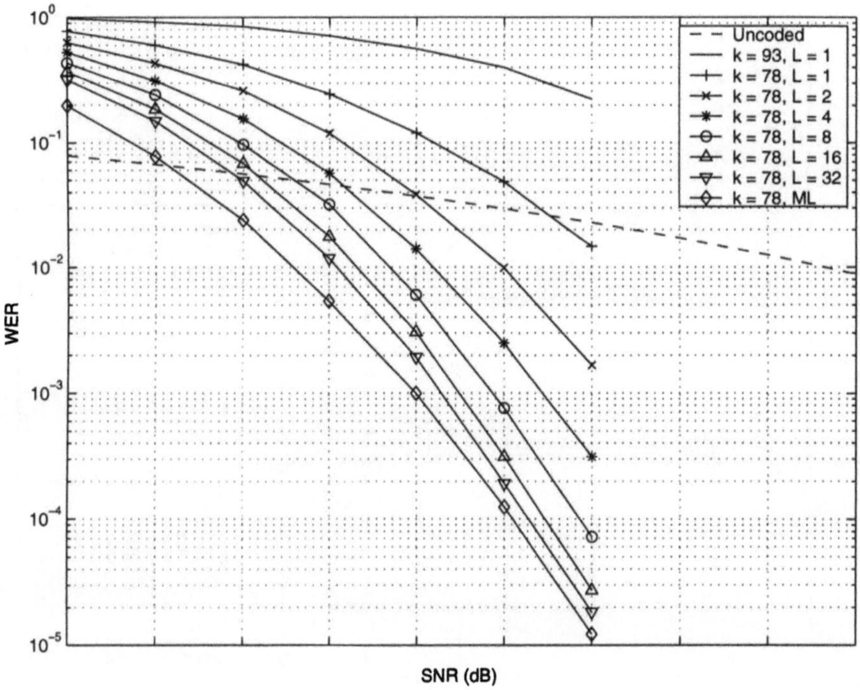

Figure 6: RM $\left\{ {8 \atop 3} \right\}$ code, $n = 256$, $k = 93$, and its subcode, $k = 78$.

of length 256 and for the code $\left\{ {9 \atop 2} \right\}$ of length 512. This figure presents almost exact experimental bounds on the error probability of ML decoding, along with the minimum lists L that were used to meet ML-decoding performance. Here we also use permutation techniques to reduce this size L. An interesting open problem is to provide a theoretical explanation as to why permutation decoding allows to substantially reduce the overall size L of the lists over the basic recursive algorithms Ψ_r^m and Φ_r^m.

7 Conclusion

Our main conclusion is that recursive decoding of RM codes combines good performance and low complexity on *moderate* blocklengths up to

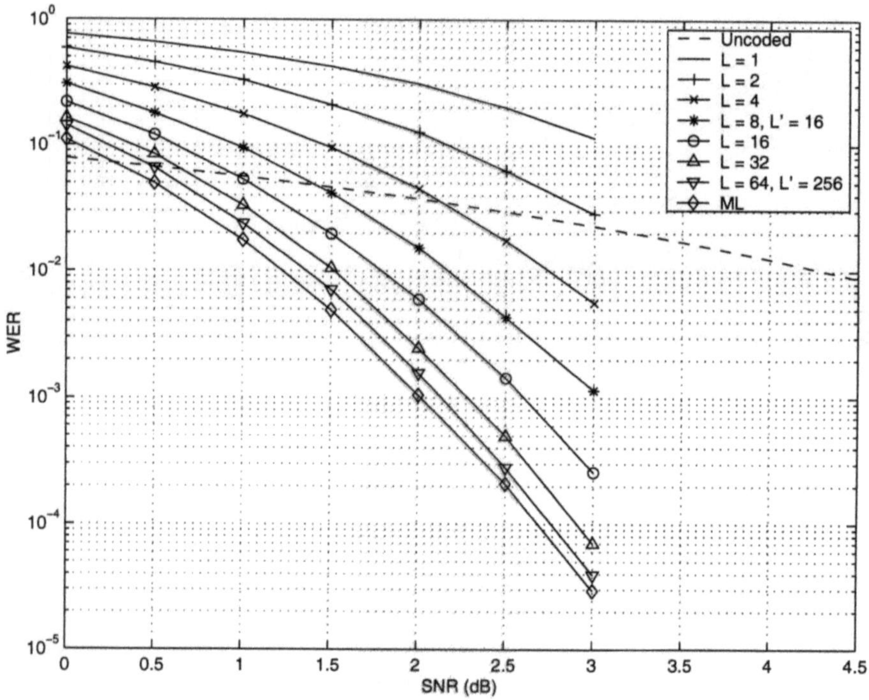

Figure 7: RM $\left\{ {8 \atop 2} \right\}$ code, $n = 256$, $k = 37$. L' is the number of paths required to achieve approximately the same WER when no permutations are used.

512. In turn, this allows us to partially fill the gap left by optimum maximum likelihood (ML) decoding and suboptimal iterative decoding. Note that the former has unfeasible complexity for nontrivial codes even on relatively short blocks of hundreds bits, while the latter becomes very efficient for turbo codes and low parity-check codes only on the blocks of thousands bits. An important open problem is whether recursive techniques can enable fast near-ML decoding for the lengths of 1024 and 2048. A positive solution to this problem would allow to completely fill the gap in the blocklengths left by the best algorithms known to date.

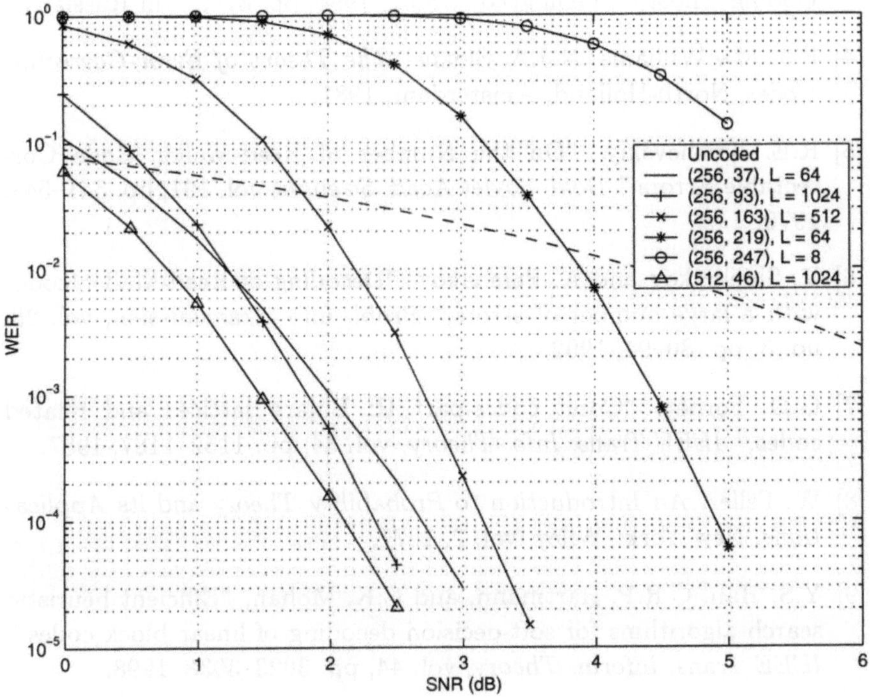

Figure 8: Experimental lower bounds for RM $\{{8 \atop 2}\}, \ldots, \{{8 \atop 6}\}$, and $\{{9 \atop 2}\}$ codes.

References

[1] I.S. Reed, "A class of multiple error correcting codes and the decoding scheme," *IEEE Trans. Info. Theory*, vol. IT-4, pp. 38–49, 1954.

[2] S.N. Litsyn, "On decoding complexity of low-rate Reed-Muller codes," *Proc. 9th All-Union Conf. on Coding Theory and Info. Transmission*, Part 1, Odessa, USSR, pp. 202–204, 1988 (in Russian).

[3] G.A. Kabatyanskii, "On decoding of Reed-Muller codes in semicontinuous channels," *Proc. 2nd Int. Workshop "Algebr. and Comb.*

Coding Theory", Leningrad, USSR, 1990, pp. 87–91 (in Russian).

[4] F.J. MacWilliams, N.J.A. Sloane, *The Theory of Error-Correcting Codes*, North-Holland, Amsterdam, 1981.

[5] R.E. Krichevskiy, "On the Number of Reed-Muller Code Correctable Errors," *Dokl. Soviet Acad. Sciences*, vol. 191, pp. 541–547, 1970.

[6] V. Sidel'nikov and A. Pershakov, "Decoding of Reed-Muller codes with a large number of errors," *Probl. Info. Transmission*, vol. 28, no. 3, pp. 80–94, 1992.

[7] G.D. Forney, "Coset codes-part II: Binary lattices and related codes," *IEEE Trans. Info. Theory*, vol. 34, pp. 1152–1187, 1987.

[8] W. Feller, *An Introduction to Probability Theory and its Applications*. New York: Wiley, vol. 2, 1971.

[9] Y.S. Han, C.R.P. Hartmann, and C.K. Mohan, "Efficient heuristic search algorithms for soft-decision decoding of linear block codes," *IEEE Trans. Inform. Theory*, vol. 44, pp. 3023–3038, 1998.

[10] G. Schnabl and M. Bossert, "Soft-decision decoding of Reed-Muller Codes as generalized multiple concatenated codes," *IEEE Trans. Info. Theory*, vol. 41, pp. 304–308, 1995.

[11] I. Dumer and R. Krichevskiy, "Soft Decision Majority Decoding of Reed-Muller Codes," *IEEE Trans. Info. Theory*, vol. 46, pp. 258–264, Jan. 2000.

[12] I. Dumer, "Recursive decoding of Reed-Muller codes," *Proc. 37th Allerton Conf. on Commun., Cont., and Comp.*, Monticello, IL, Sept. 22–24, 1999, pp. 61–69.

Adaptive Soft-Decision Decoding In Two Dimensions

Xiao-Hong Peng
Roke Manor Research Ltd
Roke Manor, Romsey
Hampshire SO51 0ZN, UK

x.peng@btinternet.com

Paddy G. Farrell
Paul Coulton
DCS
Lancaster University,
Lancaster LA1 4YR, UK

p.g.farrell@lancaster.ac.uk

Abstract

To achieve efficient soft-decision decoding for two-dimensional product and array codes, an adaptive decoding scheme is proposed. In the scheme, the special structure of these codes has been fully exploited, resulting in a method for effective use of combined hard- and soft-decision decoding of component codes. As a result, a substantial reduction in decoding complexity is obtained, especially when channel conditions are good. We will show that the decoding algorithm presented can guarantee bounded soft distance performance, while simulation results indicate even better decoding performance and an improvement over similar previous decoding methods. Turbo decoding is about 1dB better in performance, but at the cost of a very significant increase in complexity.

Keywords: Array codes, product codes, soft-decision decoding, low-complexity decoding and adaptive decoding.

1 Introduction

Maximum-likelihood soft-decision decoding can improve system performance significantly over hard-decision decoding, but at the cost of increased decoding complexity. The complexity of such soft-decision decoding, mainly involving an exhaustive serial or parallel search, is exponentially dependent on the code parameters. Although some efficient decoding algorithm such as trellis decoding [1][2] can be used to reduce the complexity, for long codes with large dimensions it is very difficult to implement any trellis-based maximum-likelihood decoding algorithm on a full code trellis, especially for IC chip hardware implementation [3]. Consequently, there have been consistent efforts made towards reducing the computational work of soft-decision decoding with as little loss in performance as possible. Various approaches to this end dedicated to either exploiting the specific structure of the code, or simplifying the decoding metric and aiming for sub-optimal decoding performance, such as information set decoding [4], generalised minimum distance decoding [5], Chase decoding [6], maximum-radius decoding [7], soft syndrome decoding [8], and statistical decoding [9].

Array and product codes are constructed using component codes in two dimensions. The component codes are relatively short and simple, while the full code constructed can be of large dimensions. This property leads quite naturally to the idea that it is possible to have a substantial reduction in decoding complexity if the code can be decoded through processing its component codes rather than the full code in the conventional way. Weldon [10] employed weighted erasure decoding (WED) of binary product codes, which corrects all error patterns of up to Hamming weight $(d-1)/2$ (where d is the minimum distance of the product code) by decoding the component codes in two steps. Reddy and Robinson [11] suggested a more general decoder for both random and burst error-correction by adapting Forney's generalized minimum distance decoding (GMD) algorithm [5].

Compared to other codes with the same error-control capacity, array codes are relatively simple both in structure and decoding methods [12]. Their hard-decision decoding performance, however, is not as good as one might expect, mainly due to the lower rates of product codes. This can be, however, compensated for by the relative ease of efficient use of channel information and code structure in soft-decision decoding for array codes. An initial proposal for this purpose is the error-erasure-decoding algorithm suggested by Wainberg [13]. Later, Yu and Costello [14] developed a general soft-decision method using GMD to decode product codes. The decoding performance and the geometric properties associated with this method have been investigated [15], resulting in an improvement over the effective error coefficient. In general, the decoding performance for product codes can be enhanced considerably by employing a soft-decision method, but, as a trade-off, the decoding complexity increases inevitably at the same time.

In this paper, we investigate low-complexity soft-decision decoding methods for binary product and array codes that take advantage of the code structure and of hard-decision techniques. In the proposed methods, soft-decision decoders of component codes, rather than the full code, are used to reduce the soft-decision computational burden. Also, the information from hard-decision decoding is utilized to decide whether or not the soft-decision decoding on certain rows or columns of the received array is necessary, according to channel conditions. This leads to an adaptive decoding scheme that provides a good compromise between decoding performance and complexity.

In Section 2, we present the new soft-decision decoding algorithm for general two-dimensional product codes. This algorithm can correct all the error patterns guaranteed by minimum soft distance of the code and many error patterns with higher error weight. We demonstrate through simulation results and examples how the adaptive features of the decoding algorithm contribute to the reduction of decoding complexity compared to existing and previous decoding algorithms. We conclude the paper in Section 3.

2 Decoding of Product Codes

An adaptive soft-decision decoding algorithm is available and works well for row-and column parity codes [16], a simple class of array codes. Here we extend the idea to general two-dimensional product codes, aiming to perform bounded soft-minimum-distance decoding with a new algorithm of significantly low complexity. Before presenting the new decoding algorithm it is convenient to review the soft decision metric and related terms, which are linked to corresponding scenarios in the decoding algorithm.

2.1 Soft Decision Metric

We assume the use of antipodal transmitted signals, so the transmitted codeword of code C is defined as $\boldsymbol{c} = (c_0, c_1, \cdots, c_{n-1})$ (n is the code length), where

$$c_i = \begin{cases} 1 & \text{if it represents a 'zero' bit} \\ -1 & \text{if it represents a 'one' bit} \end{cases} \tag{1}$$

and the received vector as $\boldsymbol{r} = (r_0, r_1, \cdots, r_{n-1})$. At the receiver, the received symbol r_i (i = 0, 1, ..., n-1) is quantized in the soft-decision demodulator over the closed interval [-1, 1] of Q possible channel outputs, i.e.

$$r_i = 1 - \frac{2q}{(Q-1)}, \quad q = 0, 1, \cdots, Q-1,$$
(2)

where q denotes a certain quantization level and Q is the total number of quantization levels. This results in only a small degradation in performance when Q, which in practice is always a power of 2, is 8 or greater [17]. In this paper we choose Q = 8, so the eight possible channel outputs in respect of q are given by:

$$\begin{cases} q & = 0 \quad 1 \quad 2 \quad 3 \quad 4 \quad 5 \quad 6 \quad 7 \\ r_i & = 1 \quad \frac{5}{7} \quad \frac{3}{7} \quad \frac{1}{7} \quad \frac{-1}{7} \quad \frac{-3}{7} \quad \frac{-5}{7} \quad -1 \end{cases}$$

This soft-decision quantization scheme is shown in Fig. 1.

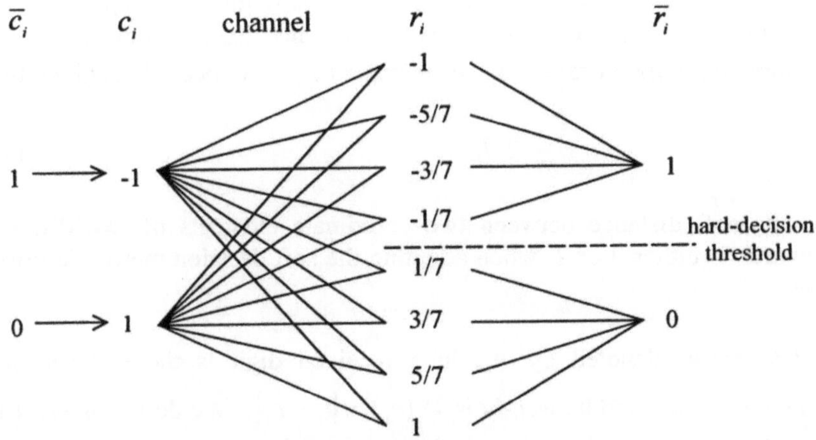

Fig. 1: A soft-decision quantization scheme ($Q = 8$)

The symbols used in Fig.1 are defined as:

$\overline{c}_i$ ---- the transmitted binary symbol,

c_i ---- "the soft-decision " value of $\overline{c}_i$

r_i ---- the soft-decision received symbol,

$\overline{r}_i$ ---- the hard-decision value of r_i.

Q ---- the number of quantization levels

Note that the soft-decision value for each element of the codeword is either 1 for a 'zero' bit (corresponding to $q = Q$-1), or -1 for a 'one' bit (corresponding to $q = Q$-1).

The *soft distance* between the received array, r, and a codeword, c, is defined as

$$d_s(r,c) = \sum_{i=0}^{n-1} |r_i - c_i| \tag{3}$$

where r_i take on one of the Q soft-decision quantization values according to (2), while c_i take on the value 1 or −1 according to whether their hard values are 0 or 1, respectively. This definition for soft distance applies to any component code of the array as well. For any linear block code with minimum Hamming distance d_h, its *minimum soft distance*, d_s is given by

$$d_s = 2\,d_h \tag{4}$$

since the soft distance between two coordinate elements of two different codewords is either 0 or 2 when adopting the soft decision metric described above.

The *error*, denoted by e_i, in a received digit is the soft-decision difference $r_i - c_i$, and its *weight* is $W(e_i) = |r_i - c_i|$. We define an error to be a *soft error* if the hard-decision value of the received symbol containing such error is the same as that of the symbol transmitted, otherwise to be a *hard error* in binary data transmission. For instance, in 8-level quantization a soft error weighs 0, 2/7, 4/7 or 6/7; and a hard error weighs 8/7, 10/7, 12/7 or 2. An error with weight 2 is called a *full hard error*. The boundary dividing the received symbols with a soft error and a hard error is marked by the hard-decision threshold shown in Fig.1.

Definition 1: For the received array $\boldsymbol{r}$, *minimum-soft-distance (MSD) decoding* is a decoding algorithm that always decodes $\boldsymbol{r}$ to a codeword $\boldsymbol{c}^*$ at the minimum soft distance from it, i.e., $d_s(\boldsymbol{r}, \boldsymbol{c}^*) \le d_s(\boldsymbol{r}, \boldsymbol{c}^i)$ for all possible *i*, where $\boldsymbol{c}^i \in C$ and *C* is the code subspace.

2.2 Decoding Algorithm

Consider an (*n*, *k*, *d*) two-dimensional product code *C* which is formed by two component codes C_1 and C_2 with parameters (n_1, k_1, d_1) and (n_2, k_2, d_2). It can also be arranged in a rectangular array, as shown in Fig.2. The parameters satisfy: $n = n_1 n_2$, $k = k_1 k_2$, and $d = d_1 d_2$, where *n*, *k*, and *d* (with and without suffix) represent block length, dimension and code minimum distance, respectively. Without loss of generality, we call C_1 and C_2 the row component code and column component code, respectively.

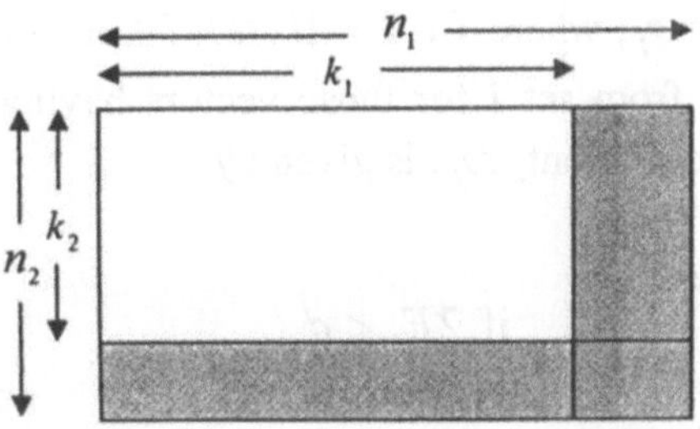

Fig.2: A product code array

For convenience, we present the decoding method for binary linear product codes. Assume that a codeword of C is transmitted over a time-discrete memoryless channel with Q-ary equally spaced quantization output, for the received vector $\boldsymbol{r}$. Using the structure of product codes, the received vector $\boldsymbol{r}$ can also be represented either by the composition of row component vectors (simply row vectors) $\boldsymbol{r}_i = (r_{i,0}, r_{i,1}, \cdots, r_{i,n_1-1})$ $(i = 0, 1, \cdots, n_2 - 1)$, or by the composition of column component vectors (simply column vectors) $\boldsymbol{r}_j = (r_{0,j}, r_{1,j}, \cdots, r_{n_2-1,j})$ $(j = 0, 1, \cdots, n_1 - 1)$.

The syndrome, $\boldsymbol{s}$, of a received vector, $\boldsymbol{r}$, is defined by $\boldsymbol{s} = \bar{\boldsymbol{r}}H^T$ and shows whether $\bar{\boldsymbol{r}}$ is a codeword ($\boldsymbol{s} = \boldsymbol{0}$) or not ($\boldsymbol{s} \neq \boldsymbol{0}$), where $\bar{\boldsymbol{r}}$ is the hard-decision value of $\boldsymbol{r}$ and H is the parity-check matrix of the code used.

The decoding procedure is divided into two parts, decoding of row vectors and decoding of column vectors. Let H_1 be the parity-check matrix of C_1 and H_2 the parity-check matrix of C_2. The decoding algorithm for the whole procedure is described as follows.

Decoding Algorithm:

Stage 1: Compute the hard-decision syndrome for each of the component vectors (rows and columns) of the received array. If all the vectors have a zero syndrome go to Stage 5; otherwise, if every symbol in the received array has value either 1 or −1, carry out hard-decision decoding for the received array; otherwise, if all the row vectors have a zero syndrome go to Stage 3 (b); otherwise,

Stage 2: Decode with an appropriate soft-decision decoder the row vectors that have a non-zero syndrome, and assign a weighting coefficient

ω_l for the decoded vector $\boldsymbol{r}_l$, where $l \in I = \{i : 0 \leq i \leq n_2 - 1\}$, taking all possible values from set I for those vectors having a non-zero syndrome. The coefficient, ω_l, is given by

$$\omega_l = \begin{cases} 1 - \dfrac{2E_l}{d_{s1}} & \text{if } 2E_l < d_{s1} \\ \dfrac{2}{d_{s1}(Q-1)} & \text{if } 2E_l \geq d_{s1} \end{cases} \tag{5}$$

where d_{s1} is the minimum soft distance of C_1 and E_l is the estimated soft error weight of the l-th row vector.

Stage 3: (a) Let column vectors after Stage 2 be $\boldsymbol{r}'_j$ $(j = 0, 1, \cdots, n_1 - 1)$. If the syndromes of $\boldsymbol{r}'_j$ for all j are zero, go to Stage 5, otherwise,

(b) decode the row vectors that have a zero syndrome as computed in Stage 1, by simply setting all the symbols of the vectors to their hard-decision values, and assign the weighting coefficient, according to (5), to them.

Stage 4: Decode with an appropriate soft-decision decoder the column vectors having a non-zero syndrome in the first k_1 column of the array.

Stage 5: Set all the symbols not valued either 1 or −1 in the first k_2 rows and first k_1 columns of the array to their hard-decision values. Decoding is now complete.

The decoding algorithm has demonstrated that only a selected group of component vectors, instead of all them in conventional methods, are soft-decision decoded. The number of such component vectors selected depends on channel conditions and consequently on the hard-decision results from computing the syndromes of both row and column vectors, at various stages of decoding. On top of this feature, as the decoding algorithm is based on decoding received vectors in each dimension with component-code decoders, it will result in substantial reduction in decoding complexity, as will be shown in sub-section (*C*). However, the decoding result from processing the component vectors during the initial stages (e.g., decoding the row component vectors) does not ensure the error control capacity guaranteed by the minimum distance of the full code. Therefore, a weighting coefficient reflecting the degrees of uncertainty of the result is assigned to each of the decoded row vectors. The criterion for setting the weighting coefficients is that they lie between 1 and $4/[d_{1s}(Q-1)]$ for $2E_i < d_{1s}$, and $2/[d_{1s}(Q-1)]$ for $2E_i \geq d_{1s}$.

To help understand how data are processed during decoding of row vectors and column vectors, the data formats in two cases are given below:

1) After Stage 2 (completion of decoding the row vectors with a non-zero syndrome) the row vectors become $\boldsymbol{v}_i$ $(i = 0, 1, \cdots, n_2 - 1)$, such that

$$\boldsymbol{v}_i = \begin{cases} \boldsymbol{r}_i \text{ (uncoded)} & \text{if } \boldsymbol{s}_i = \bar{\boldsymbol{r}}_i H_1^T = \boldsymbol{0} \\ \omega_i \hat{\boldsymbol{c}}_i & \text{otherwise} \end{cases} \tag{6}$$

   where $\hat{\boldsymbol{c}}_i \in C_1$ is the estimated component codeword in the i-th row by the row decoder, and $\bar{\boldsymbol{r}}_i$ and $\bar{\boldsymbol{r}}'_j$ are hard-decision values of $\boldsymbol{r}_i$ and $\boldsymbol{r}'_j$, respectively.

2) After Stage 3 (before decoding column vectors) the elements of the column vectors, $\boldsymbol{u}_j = (u_{0,j}, u_{1,j}, \cdots, u_{n_2-1,j})$ $(j = 0, 1, \cdots, n_1 - 1)$, can be expressed as

$$u_{i,j} = \begin{cases} r_{i,j} \text{(uncoded)} & \text{if } \boldsymbol{s}_i = \bar{\boldsymbol{r}}_i H_1^T = \boldsymbol{0} \text{ and } \begin{cases} \boldsymbol{s}'_j = \bar{\boldsymbol{r}}'_j H_2^T = \boldsymbol{0} \text{ for all } j \\ \text{or} \\ r_{ij} = 1 \text{ or -1 for all } j \end{cases} \\ \omega_i \hat{c}_{i,j} & \text{otherwise} \end{cases} \tag{7}$$

   where $\hat{c}_{i,j}$ is the element of the codeword of the row component code.

Given in (4) the relationship between the minimum soft distance and minimum Hamming distance of the code, the weighting coefficient shown in (5) can also be expressed as $\omega_l = 1 - E_l / d_1$ if $E_l < d_1$, otherwise, $\omega_l = 1/[d_1(Q-1)]$, where d_1 is the minimum Hamming distance of the row component code.

It is now necessary to prove that this decoding algorithm can correct all the error patterns whose weight is bounded by the code soft minimum distance. For comparison purposes, we adopt the GMD decoder [5] for decoding all appropriate component vectors. Before stating Theorem 1, we have three lemmas as follows.

Lemma 1 gives the GMD decoding rule [5] in an alternative version.

Lemma 1 [14]: For any (n, k) block code with minimum distance d and the given quantized received vector $\boldsymbol{r}$, there is at most one codeword $\tilde{\boldsymbol{c}}_l$ such that

$$\dot{\boldsymbol{r}} \cdot \tilde{\boldsymbol{c}}_l > n - d \tag{8}$$

where $\dot{\boldsymbol{r}} = \boldsymbol{r}/|r_m|$ and r_m $(r_m \neq 0)$ is the element of $\boldsymbol{r}$ which has the greatest absolute value.

Lemma 2: An $(n_1 n_2, k_1 k_2)$ binary linear product code with Hamming minimum distance $d_h = d = d_1 d_2$ is able to correct any error patterns, $\boldsymbol{e}$, with weight, $W(\boldsymbol{e})$, bounded by

$$W(\boldsymbol{e}) \leq \left\lfloor d_1 d_2 - \frac{1}{Q-1} \right\rfloor \tag{9}$$

where Q is the number of quantization levels, and $\lfloor x \rfloor$ denotes the largest integer equal to or less than x.

Proof: The minimum soft distance of the code is $d_s = 2d_1 d_2$, according to (4). Assume that an error pattern of weight $W(\boldsymbol{e}) \leq \lfloor d_1 d_2 - 1/(Q-1) \rfloor$ occurs. In other word, the soft distance between the received vector $\boldsymbol{r}$ and a codeword $\boldsymbol{c}_{ls}$ is $d_s(\boldsymbol{r}, \boldsymbol{c}_{ls}) = W(\boldsymbol{e}) < d_1 d_2$. Suppose that there is another codeword and its soft distance from $\boldsymbol{r}$,

$d_s(\boldsymbol{r}, \boldsymbol{c}_{ks})$, is also less than $d_1 d_2$. Then, from the triangle inequality, we have

$$\begin{aligned} d_s(\boldsymbol{c}_{ls}, \boldsymbol{c}_{ks}) &\le d_s(\boldsymbol{r}, \boldsymbol{c}_{ls}) + d_s(\boldsymbol{c}_{ks}, \boldsymbol{r}) \\ &< d_1 d_2 + d_1 d_2 \\ &< 2 d_1 d_2 \end{aligned}$$

But this contradicts the assumption that $d_s(\boldsymbol{c}_{ls}, \boldsymbol{c}_{ks}) \ge 2 d_1 d_2$. ❑

<u>Lemma 3:</u> Suppose that after decoding the i-th row vector, the estimated error weight of the row is E_i and the coefficient assigned to the decoded row is ω_i. The actual error weight in the i-th row vector, $W_i(\boldsymbol{e})$, satisfies:

$$\begin{aligned} &W_i(\boldsymbol{e}) = d_{s1}(1-\omega_i)]/2 && \text{if } 2E_i < d_{s1} \text{ (correctly decoded)} \\ &W_i(\boldsymbol{e}) \ge [d_{s1}(1+\omega_i)]/2 && \text{if } 2E_i \ge d_{s1} \text{ (wrongly decoded)} \\ &W_i(\boldsymbol{e}) = 0 \text{ or} \ge d_{s1} && \text{if } \boldsymbol{s}_i = \bar{\boldsymbol{r}}_i H_1^T = \boldsymbol{0} \text{ and } r_{i,j} = 1 \text{ or -1 for all } j \end{aligned} \quad (10)$$

for $i = 0, 1, \ldots, n_2 - 1$.

Proof: The first result can be derived directly from (5), as in the case of correctly decoding we have $W_i(\boldsymbol{e}) = E_i$. For the second result, if the i-th row vector is wrongly decoded, it means that the decoder finds that $d_s(\boldsymbol{c}_i, \boldsymbol{r}_i) > d_s(\boldsymbol{r}_i, \hat{\boldsymbol{c}}_i)$, where $\boldsymbol{c}_i$, $\boldsymbol{r}_i$ and $\hat{\boldsymbol{c}}_i$ are transmitted row component codeword, received row vector and decoded row component codeword, respectively. As $d_s(\boldsymbol{c}_i, \boldsymbol{r}_i) = W_i(\boldsymbol{e})$ and $d_s(\boldsymbol{r}_i, \hat{\boldsymbol{c}}_i) = E_i$, it implies $W_i(\boldsymbol{e}) > E_i$. Given the condition for the second result, i.e., $E_i \ge d_{s1}/2$, we have $W_i(\boldsymbol{e}) \ge d_{s1}/2 + 1/(Q-1) = d_{s1}[1 + 2/d_{s1}(Q-1)]/2 = [d_{s1}(1+\omega_i)]/2$. For the third result, when $\boldsymbol{r}_i$ satisfies both

$s_i = \bar{r}_i H_1^T = 0$ and $r_{i,j} = 1$ or -1 for all j, it must either be error-free or have become another codeword. ❑

Theorem 1: The Decoding Algorithm for product codes corrects all error patterns whose weight satisfies $W(e) \leq \lfloor d_1 d_2 - 1/(Q-1) \rfloor$.

The proof of Theorem 2 is presented in the Appendix.

Example 1: Consider the (35,16,6) product code. The parameters and parity-check matrices of the component codes are given below:

$$\text{Row component code: } \begin{cases} n_1 = 7 \\ k_1 = 4 \\ d_1 = 3 \end{cases} \Rightarrow \text{the (7, 4, 3) Hamming code,}$$

$$H_1 = \begin{bmatrix} 1\,0\,0\,1\,0\,1\,1 \\ 0\,1\,0\,1\,1\,1\,0 \\ 0\,0\,1\,0\,1\,1\,1 \end{bmatrix}$$

$$\text{Column component code: } \begin{cases} n_2 = 5 \\ k_2 = 4 \\ d_2 = 2 \end{cases} \Rightarrow \text{the (5, 4, 2) single-parity-code,}$$

$$H_2 = [1\ 1\ 1\ 1\ 1]$$

Assuming $Q = 8$, the maximum correctable error weight of the code is, according to (9), $d_1 d_2 - 1/(Q-1) = 3 \times 2 - 1/7 = 41/7$. In the same way, it can be worked out that the row and column component codes can correct all the error patterns of weight up to $d_1 - 1/(Q-1) = 3 - 1/7 = 20/7$ and $d_2 - 1/(Q-1) = 2 - 1/7 = 13/7$, respectively.

Received array

							$W_i(e)=\sum_{j=0}^{n_1-1}(1-r_{i,j})$	$s_i=\bar{r}_i H_1^T$
1	1	1	1	1	1	1	0	000
1	1	1	1	1	1	1	0	000
1	-5/7	1	1	1	-5/7	-5/7	36/7	000
1	1	1	1	1	1	1	0	000
1	3/7	1	1	1	1	1	4/7	000
	$s=1$				$s=1$	$s=1$		

$$\sum_{i=0}^{n_2-1} W_i(e)=\frac{40}{7}$$

⇓

Row decoding

	1	1	1	1	1	1	1
	1	1	1	1	1	1	1
	ω_2	$-\omega_2$	ω_2	ω_2	ω_2	$-\omega_2$	$-\omega_2$
	1	1	1	1	1	1	1
	ω_4	ω_4	ω_4	ω_4	ω_4	ω_4	ω_4
$\bar{u}_j H_2^T =$	0	1	0	0	0	1	1

$\omega_2=1/21 \quad \omega_4=17/21$

Column decoding ⇒

k_1 (columns), k_2 (rows)

1	1	1	1
1	1	1	1
1	1	1	1
1	1	1	1

In Example 1, an all-one array is transmitted and the received array has a total error weight of 40/7. Note that the transmitted codeword symbols are $c_i \in \{1,-1\}$, as defined in (1), so the hard-decision value of a received symbol used for syndrome computation, $\bar{r}_i$, is 0 if $r_i > 0$, or 1 if $r_i < 0$. After computing the syndromes and checking the values of the received symbols in Stage 1of the decoding algorithm, it is found that there are 3 column vectors having a non-zero syndrome, and that not every received symbol has value either 1 or -1. However, it is shown that all the row vectors of the received array have a zero syndrome, so move directly to Stage 3(b). In Stage 3, the decoding is simply to set all the symbols of the row vectors to their hard-decision values. As a result, the third row is decoded as the component codeword (1 –1 1 1 1 –1 –1), and the fifth row as (1 1 1 1 1 1 1). The rest of the row vectors are unchanged. The coefficients assigned to the estimated codewords in the third row and the fifth row are $\omega_2 = 2/[d_{s1}(Q-1)] = 2/(6\times 7) = 1/21$ (as $2E_2(72/7) > d_{s1}(42/7)$), and $\omega_4 = 1-2E_4/d_{s1} = 1-(2\times 4/7)/6 = 17/21$ (as $2E_4(8/7) < d_{s1}(42/7)$), respectively, according to (5). In Stage 4, the second column in the first $k_1 = 4$ columns of the array has a non-zero syndrome. It can be decoded with either a minimum soft distance (MSD) decoder or a GMD decoder. In the latter case, the column satisfies $u_1 \cdot \tilde{c}_2 > n_2 - d_2 = 3$, where $\tilde{c}_2 \in C_2$ and $\tilde{c}_2 = (1111111)$. The remaining 3 columns in the first 4 columns are decoded in Stage 5 in a simple hard-decision processing manner, i.e., replacing ω_2 or ω_4 where appropriate with a '1'. Finally, the decoding algorithm results in a correct information array as the output for this example.

In Example 2, again we consider the (35,16,6) product code, and assume that an all-one array has been transmitted. We will show that the new algorithm can correct the error patterns whose weight is over the bounded figure given in (9). The total error weight in this example is 104/7, far more than the bounded figure: 41/7. Following the computation of syndromes and checking the symbol values of the received array, it indicates that the first row vector needs to be decoded since it has a non-zero syndrome and some symbols valued neither 1 nor –1. In Stage 2, the first row vector is decoded and assigned a weighting coefficient $\omega_0 = 1-2E_0/d_{1s} = 1-(2\times 20)/7/6 = 1/21$ (as $2E_0(40/7) < d_{s1}(42/7)$). Syndrome checking for column vectors at Stage 3(a) reveals that all syndromes are zero, hence go directly to Stage 5. In Stage 5, once again, the first 4 columns are decoded by simply setting all the symbols involved to their hard-decision values, leading to a correctly decoded information array. It is easy to show that the operation carried out in Stage 5, given that all the column vectors have a zero syndrome, produces the same result as MSD decoding. Note that the same decoding result can be obtained by directly applying MSD decoding on the full array received. However, if we use a GMD method [14] (referred to as the old algorithm below) to decode the received array in Example 2, we will have a different decoding result. In this method, hard-decision decoding information (syndrome) is not considered, and all the row vectors are decoded and assigned a weighting coefficient. Column decoding with GMD decoders in the old decoding algorithm will have a tie on $\boldsymbol{u}_j \cdot \boldsymbol{c}_2$ for each column vector, ending up with a random selection of possible codewords for each column.

Example 2:

Received array

							$W_i(e)=\sum_{j=0}^{n_1-1}(1-r_{i,j})$	$s_i=\bar{r}_i H_1^T$
-1/7	5/7	1	1	5/7	-1/7	1	20/7	011
1	5/7	3/7	3/7	3/7	5/7	1/7	22/7	000
1	5/7	3/7	5/7	3/7	3/7	1	16/7	000
3/7	1	5/7	3/7	5/7	5/7	3/7	18/7	000
3/7	1	3/7	3/7	3/7	1/7	1/7	28/7	000

$$\sum_{i=0}^{n_2-1} W_i(e)=\frac{104}{7}$$

⇓

Row decoding

Old algorithm

ω_0	ω_0	ω_0	ω_0	ω_0	ω_0	ω_0
ω_1	ω_1	ω_1	ω_1	ω_1	ω_1	ω_1
ω_2	ω_2	ω_2	ω_2	ω_2	ω_2	ω_2
ω_3	ω_3	ω_3	ω_3	ω_3	ω_3	ω_3
ω_4	ω_4	ω_4	ω_4	ω_4	ω_4	ω_4

New algorithm

ω_0	ω_0	ω_0	ω_0	ω_0	ω_0	ω_0
1	5/7	3/7	3/7	3/7	5/7	1/7
1	5/7	3/7	5/7	3/7	3/7	1
3/7	1	5/7	3/7	5/7	5/7	3/7
3/7	1	3/7	3/7	3/7	1/7	1/7

$\omega_0 = 1/21;\ \omega_1 = \omega_4 = 0;$
$\omega_2 = 5/21;\ \omega_3 = 1/7$

$$\bar{r}_j' H_2^T = 0 \quad 0 \quad 0 \quad 0 \quad 0 \quad 0 \quad 0$$

⇓ Column decoding ⇓

Old algorithm (k_1 columns × k_2 rows)

1	1	1	1
-1	1	-1	1
1	1	1	1
-1	1	-1	1

New algorithm (k_1 columns × k_2 rows)

1	1	1	1
1	1	1	1
1	1	1	1
1	1	1	1

2.3 Complexity and Performance

The main difference between the old and new algorithms is that the old algorithm decodes all the row and column vectors of the receive array with soft-decision decoders, without considering the information from hard-decision decoding, whilst the new algorithm only decodes the selected row and column vectors based on their syndrome status and other information. This difference will bring improvements to soft-decision decoding of two-dimensional array codes in both decoding complexity and performance.

The computational saving with the new decoding algorithm is due to the use of hard-decision decoding results for both row and column vectors, in other words, in two dimensions. As we can see from the decoding algorithm, hard-decision syndromes have been computed at various stages throughout the decoding procedure. This provides the decoders with extra information reflecting channel conditions at very low computational costs, and allows soft-decision decoders to operate only in the situation where soft-decision decoding would outperform hard-decision decoding. For instance, in Example 1, only one column vector is decoded with the soft-decision decoder and all the row vectors and the rest of column vectors are decoded by hard-decision operations. Similarly in Example 2, only one row vector is soft decoded. Clearly, the reduction in decoding complexity made by the new decoding algorithm is significant, and the decoding strategy designed intends to adapt to different channel environments

For the purpose of computing the decoding complexity, we count the total number of real additions and comparisons and ignore other kinds of operations such as the checking of logical conditions, memory addressing, taking absolute values and modulo-2 additions.

Since we use a GMD decoder [5] to decode component vectors for both the old and new decoding algorithms, we assume, for simplicity, that decoding a row vector will involve on average half the all possible 2^{k_1} component codewords for working out the estimated codeword $\hat{c}_{ri}$ such that $r_i\hat{c}_{ri} > n_1 - d_1$. Each of such operations requires $n_1 - 1$ real additions. To find $\hat{c}_{ri}$, therefore, $(2^{k_1}/2)(n_1 - 1)$ real additions and $2^{k_1}/2$-1 real comparisons are required. Similarly, decoding a column vector requires

$(2^{k_2}/2\text{-}1)(n_2-1)$ real additions and $(2^{k_2}/2\text{-}1)$ real comparisons, respectively. With the old decoding algorithm, all the n_2 row vectors and the first k_1 column vectors (containing information symbols) need to be decoded for any received array, so its decoding complexity count is given by

$$\begin{aligned} N_{old} &= [(2^{k_1}/2)(n_1-1)+(2^{k_1}/2-1)]n_2+[(2^{k_2}/2)(n_2-1)+(2^{k_2}/2-1)]k_1 \\ &= (2^{k_1}n_1/2-1)n_2+(2^{k_1}n_2/2-1)k_1 \end{aligned} \tag{11}$$

For the new decoding algorithm, assume that on average $n_r\,(0\le n_r\le n_2)$ row vectors and $k_r\ (0\le k_r\le k_1)$ column vectors are decoded, respectively, for each received array. When the channel quality gets better, less row and column vectors, sometimes non of them, require soft-decision decoding. So its decoding complexity count is given by

$$N_{new} = (2^{k_1}n_1/2-1)n_r+(2^{k_1}n_2/2-1)k_r \tag{12}$$

In Table 1, a list of the four parameters, n_r, k_r, N_{new} and N_{old} versus the signal-to-noise ratio E_b/N_0 is given. Those results are obtained through the simulation in which 10^7 codewords of the (35,16,6) product code were transmitted for each dB value via the AWGN channel with Q = 8 level output.

Table 1: n_r, k_r, N_{old} and N_{new} versus E_b/N_0

E_b/N_0 (dB)	1	2	3	4	5	6	7	8
n_r	3.6	3.3	2.9	2.1	1.7	1.3	1.0	0.8
k_r	3.8	3.5	3.1	2.6	2.0	1.5	1.2	1.1
N_{new}	760	643	567	439	347	263	206	176

N_{old}	871	871	871	871	871	871	871	871

It is worth pointing out that although a maximum-likelihood decoding performance can also be achieved by using a trellis [2], its decoding complexity for this example would consistently be $N_{trellis} = 1968$ for the whole range of E_b / N_0, while N_{new} varies dynamically from 760 down to 176 according to channel conditions.
To show the computational gains achieved by using the new algorithm, a graph of R = N_{old} / N_{new} versus E_b / N_0 is given in Fig.3.

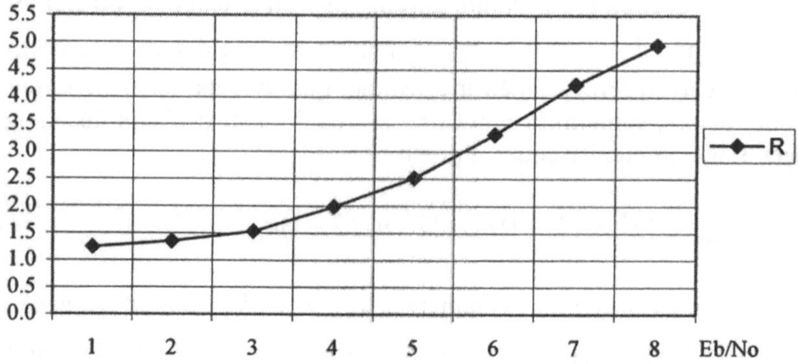

Fig.3: Decoding complexity saving in terms of R= N_{old} / N_{new}

As to the decoding performance, we have shown that the new decoding algorithm can correctly decode all the error patterns as long as their weight is no greater than that guaranteed by the minimum soft distance, given in (9). It is also capable of correcting the error patterns whose weight is beyond the bounded figure, as shown in Example 2. In the old algorithm, however, all the received vectors have to be decoded and assigned a weighting coefficient. In some situations where the vectors contain only soft errors (equivalent of zero syndrome), assigning the weighting coefficient to them will bring down the original absolute values and, in effect, the credibility of their symbols. This would reduce the chances for them to be successfully decoded under certain decoding criteria, e.g., the GMD, as shown in

Example 2. With the new algorithm, those situations can be identified and avoided under most circumstances.

The decoding performances of the old and the new decoding algorithms were simulated for the (35,16,6) product code, as shown in Fig.4. Assume that binary antipodal signaling is transmitted over the AWGN channel with eight-level output quantization.

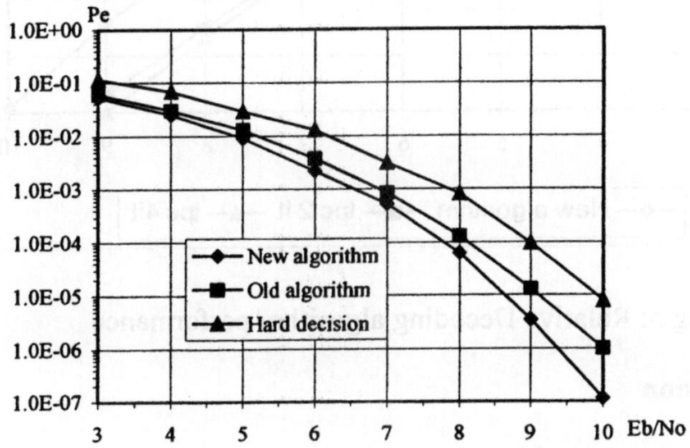

Fig.4: Decoding performance of (35, 16) product code

Two-dimensional array and product codes can also be turbo decoded, using the Log-MAP algorithm on row and column component codes alternately [18]. Figure 5 shows the performance of this algorithm for the (35, 16, 6) product code, for I = 1, 2 and 4 iterations (improvement is negligible with more) compared to that of the new algorithm described above. It can be seen that turbo decoding provides about 1dB extra gain with 2 iterations, and 1.3dB with 4 iterations, at an error rate of 10^{-5}. The complexity, however, computed on the same basis as above, stays consistently at $1176\times I$ (i.e., 2352 for I = 2and 4704 for I = 4) over the whole range of E_b / N_0, which is very significantly more than the complexity of the new algorithm.

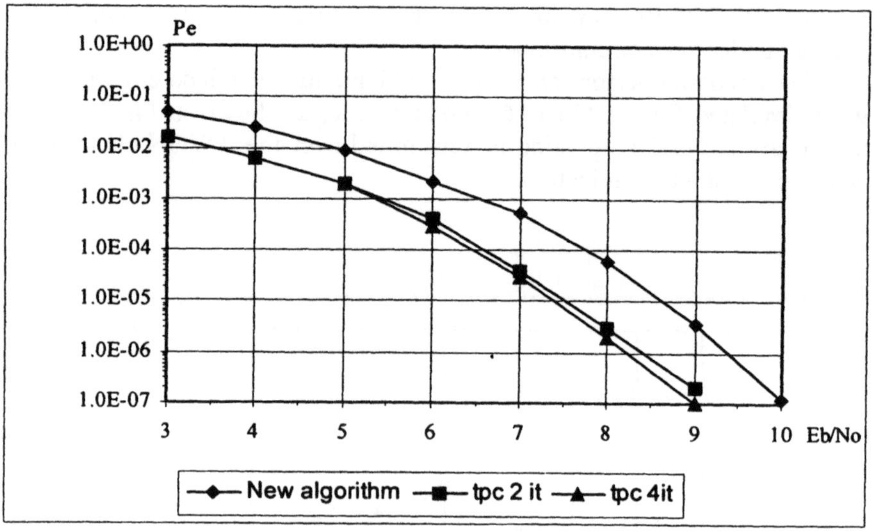

Fig 5: Relative Decoding algorithm performance

3. Conclusion

We have described a new low-complexity soft-decision decoding algorithm for product and array codes. In conventional soft-decision decoding, a decoding algorithm works without considering changes in channel characteristics and therefore has a consistently high decoding complexity. Our results have shown that the new algorithm can achieve low and dynamical decoding complexities, by taking advantage of hard-decision decoding on both row and column vectors of the received array. The information provided by hard-decision decoding and symbol value checking reflects the channel conditions and is used by the decoder to decide when soft-decision decoding is required and when it is not. This adaptive feature performs particularly well when channel quality becomes good.

We have also shown that the new algorithm can correctly decode all error patterns with weight bounded by the minimum soft distance of the code, and many error patterns of greater weight. As demonstrated in examples and simulation results, the new decoding algorithm improves on the old one in both complexity and performance. Compared to maximum-likelihood decoding, or to turbo decoding, the new algorithm's advantage in complexity saving is overwhelming, in contrast to its loss in performance. In certain scenarios this may be a valuable trade-off.

Appendix: Proof of Theorem 1.

Theorem 1: The Decoding Algorithm for product codes corrects all error patterns whose weight satisfies $W(e) \le \lfloor d_1 d_2 - 1/(Q-1) \rfloor$.

Proof: We will prove that given $\boldsymbol{u}_j$ the column vector after decoding row component vectors, $\boldsymbol{u}_j \cdot \tilde{\boldsymbol{c}}_{2j} > n_2 - d_2$ for each of the transmitted columns $\tilde{\boldsymbol{c}}_{2j} \in C_2$ when the condition specified in (9) is met. Let I_c, I_w and I_o be the index sets for row vectors that are correctly decoded, wrongly decoded and, satisfying $\boldsymbol{s}_i = \bar{\boldsymbol{r}}_i H_1^T = \boldsymbol{0}$ and $r_{i,j} = 1$ or - 1 for all j, respectively, where $\{I_c\} + \{I_w\} + \{I_o\} = \{0, 1, \ldots, n_2 - 1\}$. From Lemma 2 and Lemma 3, we have

$$\sum_{i=0}^{n_2-1} W_i(e) = \sum_{i\in I_c} d_{s1}(1-\omega_i)/2 + \sum_{i\in I_w} d_{s1}(1+\omega_i)/2 + \sum_{i\in I_o} d_{s1} f \le \frac{d_1 d_2 (Q-1) - 1}{Q-1}$$

where $f = (0,1)$. As $d_{s1} = 2d_1$, then

$$\sum_{i\in I_c} d_1(1-\omega_i) + \sum_{i\in I_w} d_1(1+\omega_i) + \sum_{i\in I_o} 2d_1\left(\frac{1}{2} + f - \frac{1}{2}\right) \le d_1 d_2 - \frac{1}{Q-1}$$

Dividing the both sides by d_1, we have $\left(\sum_{i\in I_c} 1 + \sum_{i\in I_w} 1 + \sum_{i\in I_o} 1 = n_2\right)$

$$n_2 - \sum_{i\in I_c} \omega_i + \sum_{i\in I_w} \omega_i + \sum_{i\in I_o} (2f-1) \le d_2 - \frac{1}{d_1 Q - 1}$$

$$\sum_{i\in I_c} \omega_i - \sum_{i\in I_w} \omega_i + \sum_{i\in I_o} (1-2f) \ge n_2 - d_2 + \frac{1}{d_1 Q - 1}$$

As $(1-2f)=\begin{cases}1 & \text{if } r_{i,j}=\widetilde{c}_{2i,j}(f=0)\\ -1 & \text{if } r_{i,j}\neq\widetilde{c}_{2i,j}(f=1)\end{cases}$

thus $\sum_{i\in I_c}\omega_i-\sum_{i\in I_w}\omega_i+\sum_{i\in I_o}(1-2f)=\boldsymbol{u}_j\cdot\widetilde{\boldsymbol{c}}_{2j}$

therefore $\boldsymbol{u}_j\cdot\widetilde{\boldsymbol{c}}_{2j}>n_2-d_2$. ❑

References:

[1] L.R. Bahl, J. Cocke, F. Jelinek and J. Raviv, " Optimal decoding of linear codes for minimizing symbol error rate," *IEEE Trans. Inform. Theory*, IT-20, pp.284-287, 1974.

[2] J.K. Wolf, "Efficient maximum likelihood decoding of linear block codes using a trellis," *IEEE Trans. Inform. Theory*, Vol. IT-24, pp.76-80, Jan. 1978.

[3] H.T. Moorthy, S. Lin, and G.T. Uehara, G. "Good trellises for IC implementation of Viterbi decoders for linear block codes," *IEEE Trans. on Comms.*, **COM-45**, pp. 52-63, 1997.

[4] E. Prange, "The use of information sets in decoding cyclic codes," *IRE Trans*. IT-18, pp.55-59, 1962.

[5] G.D. Forney, *Concatenated Codes, Cambridge*, MA: M.I.T. Press, 1966(a).

[6] D. Chase, "A class of algorithms for decoding block codes with channel measurement information," *IEEE Trans. Inform. Theory*, IT-18, pp.170-179, 1974.

[7] L.D. Rudolph, C.R.P. Hartmann, T.-Y. Hwang and N.Q. Duc," Algebraic analog decoding of linear binary codes," *IEEE Trans. Inform. Theory*, IT-25, pp.430-440, July 1979.

[8] J. Snyders and Y. Be'ery, "Maximum likelihood soft decoding of binary block codes and decoders for the Golay codes," *IEEE Trans. Inform. Theory*. IT-35, pp.963-975, 1989.

[9] M.P.C. Fossorier and S. Lin, "Soft-decision decoding of linear block codes based on ordered statistics," *IEEE Trans. Inform. Theory*, IT-41, pp.1379-1396. Sept. 1995.

[10] E.J. Weldon, "Decoding binary block codes on Q-ary output channels," *IEEE Trans. Inform. Theory*, IT-17, pp.713-718, 1971.

[11] S.M. Reddy and J.P. Robinson, "Random error and burst correction by iterated codes," *IEEE Trans. Inform. Theory*, IT-18, pp.182-185, Jan. 1972.

[12] P.G. Farrell, "An survey of array error control codes," *Europ. Trans. Telecomm.* and Related Tech. (ETT), vol. 3, No.5, pp.441-451 , 1992.

[13] S. Wainberg, "Error-erasure decoding of product codes," *IEEE Trans. Inform. Theory*, IT-18, pp.821-823, 1972.

[14] C.C.H. Yu and D. Costello, Jr., "Generalised minimum distance decoding algorithms for Q-ary output channels," *IEEE Trans. Inform. Theory*, IT-26, pp.238-243, March 1980.

[15] O. Amrani and Y. Be'ery, "Soft decision decoding of binary product codes," preprint.

[16] X.-H. Peng and P.G. Farrell, "Adaptive soft-decision decoding for array codes," *Europ. Trans. Telecomm.*, vol.12, No. 6, pp.487-500, Nov.-Dec. 2001.

[17] G.C. Clark and J.B. Cain, *Error Correcting Codes for Digital Communications*, Plenum Press, 1981.

[18] Hagenauer, J. Offer E., and Papke, L., ``Iterative Decoding of Binary Block and Convolutional Codes", *IEEE Transactions on Information Theory*, vol. 42, No 2, pp. 429-445, March 1996.

On the Theory of Linear Trellises

Ralf Koetter
Coordinated Science Laboratory
University of Illinois at Urbana-Champaign
Urbana, IL 61801, U.S.A.
`koetter@csl.uiuc.edu`

Alexander Vardy
Department of Electrical Engineering
Department of Computer Science
University of California, San Diego
La Jolla, CA 92093, U.S.A.
`vardy@kilimanjaro.ucsd.edu`

January 31, 2002

Abstract

Trellis linearity, first considered by McEliece in 1996, turns out to be crucial in the study of tail-biting trellises. In this chapter, basic structural properties of linear trellises are investigated. A rigorous definition of linearity is given for both conventional and tail-biting trellises. An algorithm that determines in polynomial time whether a given trellis is linear is then derived. The relationship between linear trellises and the trellis product construction is discussed. The key result of this chapter is that a trellis

1. Introduction

Trellis representations of linear block codes have received much attention in the last decade [2, 9, 5, 10, 13, 15]. Such representations illuminate code structure and often lead to efficient trellis-based decoding algorithms. Today, we have not only the conventional trellises, whose theory is by now well developed [10, 13], but also the tail-biting trellises.

Numerous examples are known [2, 7, 11] where the complexity of a tail-biting trellis is much lower than the complexity of the best possible conventional trellis for the same code. Tail-biting trellises for block codes were studied by several authors [2, 3, 11, 12], including McEliece [1]. However, such trellises are not yet well understood: we certainly know much less about tail-biting trellises than about conventional trellises.

McEliece [10] introduced in 1996 the class of *simple linear trellises.* He thus made an important distinction between trellis structures that possess certain linearity properties and those that do not. This distinction was immediately forgotten, and remained overlooked in most of the subsequent literature on trellises, both conventional and tail-biting.

Indeed, one does not need linearity in the study of conventional trellises. It is well known that, given a specific coordinate ordering, there exists a unique, up to isomorphism, minimal conventional trellis for any linear block code. The minimal trellis for a linear code $\mathbb{C}$ simultaneously minimizes all of the conceivable measures of trellis complexity, and can be easily constructed from a generator matrix or a parity-check matrix for $\mathbb{C}$. This minimal trellis is, in fact, linear [9, 10], but it is not necessary to know this in order to construct it or to establish its properties.

On the other hand, linearity turns out to be crucial in the study of tail-biting trellises. The class of tail-biting trellises, which includes the conventional trellises as a (somewhat degenerate) subclass, is so broad that not much can be said about tail-biting trellises in general. In contrast, linear tail-biting trellises have rich and beautiful theory, some of which is developed in this chapter. The study of this theory paves the way to the construction of minimal linear tail-biting trellises [7], even though such trellises are not unique.

The rest of this chapter is organized as follows. We start with a definition of both conventional and tail-biting trellises in the next section. We then define *reduced* and *one-to-one* trellises. In addition to the usual labeling of edges in a trellis, we also need to deal with labelings of *vertices*.

Thus we introduce the concepts of a *labeled trellis* and a *label code*. We define a *representation code* of a labeled trellis T as a subset of the label code of T, from which the entire trellis can be uniquely reconstructed.

In Section 3, we give a rigorous definition for the notion of linear (or group) trellises. Loosely speaking, trellis linearity means that the set of edge/vertex label sequences in the trellis is closed under componentwise operation in the appropriate algebraic domain — a field or a group. We shall see, however, that it is also crucial to require that every linear trellis is reduced. Several simple, but fundamental, properties of linear trellises are also discussed in Section 3. For example, we show that the vertex classes and the edge classes in a linear trellis are linear spaces.

In Section 4, we investigate the following problem: given a trellis T, how can one determine whether T is linear? Specifically, the question is whether the vertices of T can be labeled in a linear fashion. We provide an algorithm that constructs an appropriate labeling of the vertices if the trellis is linear, and halts if it is not. An interesting outcome of this investigation is that trellis linearity is inherently a graph-theoretic property: we give an example of two trellises over the trivial alphabet $\mathcal{A} = \{0\}$ one of which is linear while the other is not.

In Section 5, we discuss the *product construction* and *elementary trellises.* The product construction was introduced in [9]; it was later employed in [2] in the context of tail-biting trellises. As already observed by McEliece [10], any conventional trellis that results from this construction is linear. We shall see that this remains true for tail-biting trellises.

While it is easy to see that the product construction yields linear trellises, it is by no means clear if there exist linear trellises that cannot be obtained by the product construction. In Section 6, we present the key result of this chapter: we prove that a trellis — either tail-biting or conventional — is linear *if and only if* it may be obtained by the product construction. Thus every linear trellis factors into a product of elementary trellises over a field, and every abelian-group trellis factors into elementary trellises over a group.

We conclude in Section 7 by establishing a number of properties of linear trellises that follow directly from the factorization theorem of the previous section. In particular, we show that a linear tail-biting trellis that is non-mergeable must also be biproper. It follows that a generator matrix for a minimal linear tail-biting trellis must have the minimal-span form.

2. Preliminaries

We start with the definitions of conventional and tail-biting trellises. We also introduce a number of concepts related to tail-biting trellises, that will be used throughout this chapter.

An edge-labeled directed *graph* is a triple $(V, E, \mathcal{A})$, consisting of a set V of *vertices*, a set $\mathcal{A}$ called *alphabet*, and a set E of ordered triples (v, a, v'), with $v, v' \in V$ and $a \in \mathcal{A}$, called *edges*.

Definition 2.1. *A conventional trellis $T = (V, E, \mathcal{A})$ of depth n is an edge-labeled directed graph with the following property: the vertex set V can be partitioned as*

$$V = V_0 \cup V_1 \cup \cdots \cup V_n \tag{1}$$

such that every edge in T begins at a vertex of V_i and ends at a vertex of V_{i+1}, for some $i = 0, 1, \ldots, n-1$. The sets $V_0, V_1, \ldots, V_n$ are called the vertex classes of T. The ordered index set $\mathcal{I} = \{0, 1, \ldots, n\}$ induced by the partition in (1) is called the time axis for T.

If every vertex in T lies on at least one path from a vertex in V_0 to a vertex in V_n, we say that T is *reduced*. The sequence of edge labels along each path of length n in T defines an ordered n-tuple over the label alphabet $\mathcal{A}$. We say that T *represents* a block code $\mathbb{C}$ of length n over $\mathcal{A}$ if the set of all such n-tuples is precisely the set of codewords of $\mathbb{C}$.

Tail-biting trellises have been traditionally used as a means of terminating a convolutional code without incurring a rate loss. More recently, tail-biting trellises for *block codes* were considered in [1, 2, 6, 11, 15] and other works. Such trellises may be viewed (cf. [2, 15]) as a generalization of a conventional trellis to a circular time-axis.

Definition 2.2. *A tail-biting trellis $T = (V, E, \mathcal{A})$ of depth n is an edge-labeled directed graph with the following property: the vertex set V can be partitioned into n vertex classes*

$$V = V_0 \cup V_1 \cup \cdots \cup V_{n-1} \tag{2}$$

such that every edge in T either begins at a vertex of V_i and ends at a vertex of V_{i+1}, for some $i = 0, 1, \ldots, n-2$, or begins at a vertex of V_{n-1} and ends at a vertex of V_0. The time axis for T is $\mathcal{I} = \{0, 1, \ldots, n-1\}$.

Due to the tail-biting nature of the definition, it is convenient to identify $\mathcal{I}$ with $\mathbb{Z}_n$, the ring of integers modulo n. Thus, when dealing with tail-biting trellises, all index arithmetic is implicitly performed modulo n.

A *cycle* of length m in a tail-biting trellis T is a subgraph of T consisting of a *closed* path through m vertices. It is easy to see that the length of any cycle is a multiple of the depth n, and we will be mostly concerned with cycles of length *exactly* n. From now on, whenever we refer to cycles in T we mean cycles of length n, unless specified otherwise.

Clearly, any cycle in T contains exactly one vertex from each vertex class, and each vertex in a cycle has degree two. The notion of a cycle in a tail-biting trellis is analogous to the notion of a path in a conventional trellis. In particular, we say that a tail-biting trellis T is *reduced* if every vertex *and every edge* in T belongs to at least one cycle. Note that in contrast to conventional trellises, the edges of a reduced tail-biting trellis are explicitly required to belong to a cycle. For instance, the following trellis

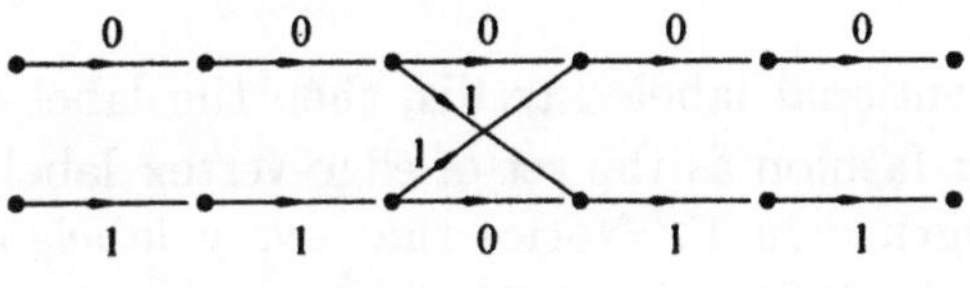

is reduced as a conventional trellis, but not as a tail-biting* trellis, although every vertex belongs to a cycle. Cutting this trellis at any position produces a reduced conventional trellis.

The set of edge labels along a cycle in T is an n-tuple $\{a_0, a_1, \ldots, a_{n-1}\}$ over the label alphabet $\mathcal{A}$. Postulating that all cycles in T start at a vertex of V_0, every cycle defines a vector $(a_0, a_1, \ldots, a_{n-1}) \in \mathcal{A}^n$. We refer to such vectors as *edge-label sequences* in T. We say that T *represents* a block code $\mathbb{C}$ over $\mathcal{A}$ if the set of all such edge-label sequences is the set of codewords of $\mathbb{C}$. Note that if we were to postulate that cycles in T start at a different time, then the *same trellis* would represent a different code, namely a cyclic shift of $\mathbb{C}$.

We let $\mathbb{C}(T)$ denote the code represented by a trellis T, either conventional or tail-biting. A trellis is said to be *one-to-one* if there is a one-to-one correspondence between cycles in T and codewords in $\mathbb{C}(T)$. For

*When depicting tail-biting trellises, we implicitly identify the leftmost vertex class in a figure with the rightmost vertex class. Thus as a tail-biting trellis, the trellis above consists of two cycles and has only 10 vertices.

such trellises, we denote by $\Gamma_T(c)$ the unique cycle of T that corresponds to the codeword $c \in \mathbb{C}(T)$. We write $\Gamma(c)$ for $\Gamma_T(c)$ if no confusion arises.

In addition to the usual labeling of *edges* in a trellis, we will be also interested in certain labelings of *vertices* in a trellis. Specifically, suppose that, for all i, each vertex in a vertex class V_i is labeled by a sequence of length l_i over $\mathcal{A}$, where $l_i \geq \lceil \log_{|\mathcal{A}|} |V_i| \rceil$. We also require that all vertex labels within the same vertex class are distinct. We will use the term *labeled trellis* to refer to a trellis endowed with a vertex labeling of this kind.

Let $l = l_0 + l_1 + \cdots l_{n-1}$. Then every cycle Γ in a labeled tail-biting trellis T defines an ordered sequence of length $n + l$ over $\mathcal{A}$, consisting of the labels of edges and vertices in Γ. We will refer to such a sequence as a *label sequence* in T.

Definition 2.3. *The set of all the label sequences in a labeled tail-biting trellis T is called the label code of T and denoted by $\mathbb{S}(T)$. The label code is a block code of length $n + l$ over $\mathcal{A}$.*

If T is a conventional labeled trellis, then the label code $\mathbb{S}(T)$ is defined in a similar fashion as the set of edge-vertex label sequences of all the *paths* of length n in T. Notice that *every* labeled trellis T represents its label code $\mathbb{S}(T)$ in a one-to-one manner: there is a one-to-one correspondence between cycles (or paths) in T and codewords in $\mathbb{S}(T)$.

Given a labeled trellis T, it is straightforward to determine its label code by simply reading-off the label sequences of all the cycles (paths) in T. On the other hand, if T is reduced, then it is also possible to uniquely determine T from its label code. Indeed, given $\mathbb{S}(T)$, we can construct T cycle-by-cycle (or path-by-path); this works since every edge in T belongs to at least one cycle. Thus we may describe a reduced labeled trellis T by specifying the label code $\mathbb{S}(T)$.

Further notice that a *subset* of $\mathbb{S}(T)$ is often sufficient to completely describe T. Each codeword in $\mathbb{S}(T)$ is a label sequence that uniquely determines a cycle in T. If we take enough codewords so that every edge of T is contained in one of the corresponding cycles, then the entire trellis may be reconstructed from these codewords.

Definition 2.4. *We say that a code $\mathbb{S} \subseteq \mathbb{S}(T)$ represents the trellis T if the cycles (or paths) corresponding to the codewords of $\mathbb{S}$ cover all the edges of T. We shall call $\mathbb{S}$ a representation code for T.*

The following example illustrates this definition. When specifying label sequences, we write vertex labels in a smaller font and underline them.

Example 2.1. Consider the labeled trellis T over the alphabet $\mathcal{A} = \mathbb{F}_2$, depicted below. As $|V_0| = 1$, this trellis may be thought of either as a conventional or as a tail-biting trellis. Clearly, the trellis T is reduced

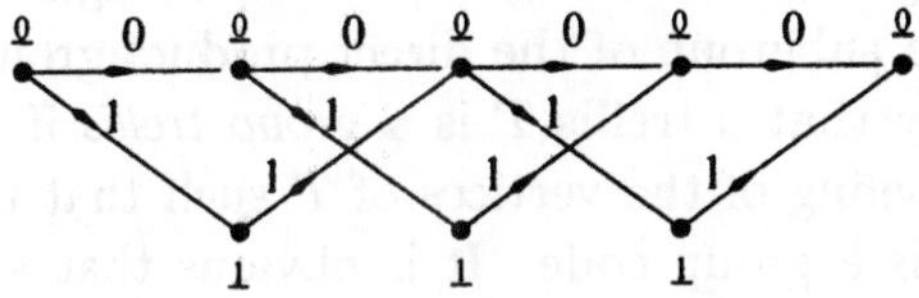

and represents the code $\mathbb{C}(T) = \{(0000), (0011), (0110), (1100), (1111)\}$, so that $|\mathbb{S}(T)| = |\mathbb{C}(T)| = 5$. However, the set of only three codewords:

$$\mathbb{S} = \{(\underline{0}0\underline{0}0\underline{0}0\underline{0}0), (\underline{0}1\underline{1}1\underline{0}1\underline{1}1), (\underline{0}0\underline{0}1\underline{1}1\underline{0}0)\} \subset \mathbb{S}(T) \qquad (3)$$

is sufficient to represent the trellis. Notice that the code $\mathbb{S}$ is not unique. For example, replacing the first two label sequences in (3) by $(\underline{0}0\underline{0}0\underline{0}1\underline{1}1)$ and $(\underline{0}1\underline{1}1\underline{0}0\underline{0}0)$, we obtain another representation code for T. □

3. Definition and properties of linear trellises

We henceforth impose an algebraic structure on the trellis alphabet: we assume that $\mathcal{A} = \mathbb{F}_q$ is the finite field with q elements. Thus all the edge and vertex labels in a labeled trellis T are symbols or sequences over $\mathbb{F}_q$. It follows that the edge-label code $\mathbb{C}(T)$ and the label code $\mathbb{S}(T)$ are codes over $\mathbb{F}_q$. The case where $\mathbb{S}(T)$ is a *linear* code is of particular interest.

Definition 3.1. *A labeled trellis $T = (V, E, \mathbb{F}_q)$ is linear over $\mathbb{F}_q$, or simply linear, if T is reduced and $\mathbb{S}(T)$ is a linear code over $\mathbb{F}_q$. An (unlabeled) trellis T is said to be linear if there exists a vertex labeling of T such that the resulting labeled trellis is linear.*

This definition applies to both conventional and tail-biting trellises. As we shall see, a linear trellis always represents a linear code, although linear codes may be also represented by nonlinear trellises [7]. We assume that edge labeling is an inherent part of a trellis, whereas vertex labeling may be chosen freely. In contrast, system-theoretic treatments assume that the labeling of vertices, or states, is also an inherent part of a state-space realization (*cf.* [3, 4]). However, in the case of tail-biting trellises the interpretation of a vertex as a state is much less clear, since the notions of past and future do not apply for a circular time-axis.

Remark. Another important class of trellises corresponds to the case where $\mathbb{S}(T)$ is a group code. A group code of length n over a group $\mathcal{G}$ is defined [4] as a subgroup of the direct product group $\mathcal{G}^n$. As in Definition 3.1, we say that a trellis T is a *group trellis* if T is reduced and there exists a labeling of the vertices of T such that the corresponding label code $\mathbb{S}(T)$ is a group code. It is obvious that every linear trellis is also a group trellis over the additive group of $\mathbb{F}_q$, but not vice versa. Group trellises naturally arise in two different ways. It is possible that the trellis alphabet $\mathcal{A}$ is itself a group which does not admit an invertible multiplicative structure. For example, group trellises for codes over $\mathbb{Z}_4$ are of this type; linear trellises for $\mathbb{Z}_4$-linear codes do not exist. It is also possible that vertices of a trellis T for a linear code over $\mathbb{F}_q$ cannot be labeled in such a way that $\mathbb{S}(T)$ is a linear code over $\mathbb{F}_q$, but can be labeled so that $\mathbb{S}(T)$ is a group code over the additive group of $\mathbb{F}_q$. Group trellises of this kind were considered in [2]. In this chapter, we will consider group trellises only briefly. We note that for binary and ternary codes, the classes of group trellises and linear trellises coincide. For larger alphabets, we shall see that the theory of group trellises is very similar to the theory of linear trellises, at least if the underlying group is abelian. In particular, all the results in this chapter hold essentially without change for group trellises over an abelian group. □

Certain properties of linear trellises are apparent directly from Definition 3.1. It is obvious that a reduced labeled trellis T is linear if and only if a linear combination of any two label sequences in T is a label sequence in T. Further, if a labeled trellis T is linear, then the vertex labels within each vertex class constitute a linear space; this space is just a projection of $\mathbb{S}(T)$ on the corresponding position. Specifically, for each $y \in \mathbb{S}(T)$, let

$$\nu_i(y) \stackrel{\text{def}}{=} \text{label of the vertex } \{\Gamma(y) \cap V_i\} \qquad \text{for } i = 0,1,\ldots,n-1 \tag{4}$$

$$\varepsilon_i(y) \stackrel{\text{def}}{=} \text{label of the edge } \{\Gamma(y) \cap E_i\} \qquad \text{for } i = 0,1,\ldots,n-1 \tag{5}$$

where $\Gamma(y)$ is the unique cycle in T that corresponds to the label sequence y. Then the set of labels for the vertices of V_i is the image of $\mathbb{S}(T)$ under the mapping $\nu_i(\cdot)$, for all $i = 0,1,\ldots,n-1$.

Since the labels within a vertex class are distinct by definition, one can identify the vertices of a labeled trellis T with their labels, and we will often adopt this convention. Thus we say that, in a linear trellis, the vertex classes $V_0, V_1, \ldots, V_{n-1}$ are linear spaces, and write $\nu_i : \mathbb{S}(T) \mapsto V_i$. With this convention, the edge classes $E_0, E_1, \ldots, E_{n-1}$ also become linear spaces. Specifically, for all $i = 0, 1, \ldots, n-1$, the edge-class E_i is as the image of $\mathbb{S}(T)$ under the linear mapping $\tau_i : \mathbb{S}(T) \mapsto E_i$ defined by:

$$y \in \mathbb{S}(T) \;\mapsto\; \tau_i(y) \stackrel{\text{def}}{=} \Big(\nu_i(y), \varepsilon_i(y), \nu_{i+1}(y)\Big) \tag{6}$$

Notice that the assumption that T is reduced is essential here. Otherwise, the mapping $\tau_i : \mathbb{S}(T) \mapsto E_i$ is not necessarily onto, and the corresponding edge-class E_i may be nonlinear.

The following useful property of linear trellises follows directly from the observations in the foregoing paragraph. Loosely speaking, a linear combination of any two *paths* in a linear tail-biting trellis T, whether closed or not, is another path in T. More precisely, consider all paths of length n in a tail-biting trellis T, starting at a vertex of V_0. Every such path defines an ordered sequence consisting of the edge labels along the path. We refer to such sequences as *edge-label paths*, and denote the set of all the edge-label paths in T by $\overline{\mathbb{C}}(T)$. If T is labeled, then every path in T also defines a vertex-edge label sequence, which we call a *label path* in T. The set of all label paths is denoted $\overline{\mathbb{S}}(T)$.

Proposition 3.1. *If T is linear, then $\overline{\mathbb{C}}(T)$ and $\overline{\mathbb{S}}(T)$ are linear codes: $\overline{\mathbb{C}}(T)$ is a union of cosets of $\mathbb{C}(T)$, and $\overline{\mathbb{S}}(T)$ is a union of cosets of $\mathbb{S}(T)$.*

Proof. Follows immediately from the fact that T is reduced and every edge class in T is a linear space. ∎

Observe that the converse statement is true for $\overline{\mathbb{S}}(T)$, but not for $\overline{\mathbb{C}}(T)$. For instance, consider a nonlinear conventional trellis T for a linear code $\mathbb{C}$. Then $\overline{\mathbb{C}}(T) = \mathbb{C}(T)$ is obviously linear, while T is not.

Recall that a subcode $\mathbb{S} \subseteq \mathbb{S}(T)$ is a representation code for T if the cycles (paths) corresponding to the label sequences in $\mathbb{S}$ cover all the edges of T. If $\mathbb{S}$ is linear, then it can be specified by a generator matrix. We say that a matrix G over $\mathbb{F}_q$ is a *representation matrix* for T if it generates $\mathbb{S}$. A representation matrix G is said to be *complete* if it generates the entire label code $\mathbb{S}(T)$, or equivalently if $\operatorname{rank} G = \dim \mathbb{S}(T) = \dim \mathbb{C}(T)$.

Proposition 3.2. *A labeled trellis T is linear if and only if there exists a linear representation code for T or, equivalently, if it can be described by a representation matrix.*

Proof. ($\Rightarrow$) If T is a linear trellis, then its label code $\mathbb{S}(T)$ is a linear representation code for T, and a generator matrix for $\mathbb{S}(T)$ is a (complete) representation matrix for T. ($\Leftarrow$) Let G be a representation matrix for T, and let $\mathbb{S}$ be the linear representation code generated by G. By definition, cycles corresponding to the label sequences in $\mathbb{S}$ cover every edge in T. Therefore T is reduced, and the edge classes $E_0, E_1, \ldots, E_{n-1}$ of T can be identified with $\tau_0(\mathbb{S}), \tau_1(\mathbb{S}), \ldots, \tau_{n-1}(\mathbb{S})$, where the linear mappings $\tau_i(\cdot)$ are as defined in (6). Since $\mathbb{S}$ is a linear code, we conclude that every edge class in T has the structure of a linear space. This, in turn, implies that a linear combination of any two label sequences in T is also a label sequence in T. Hence $\mathbb{S}(T)$ is a linear code, and T is linear. ∎

An alternative proof of Proposition 3.2 may be obtained by showing that every representation matrix can be extended to a complete representation matrix for the same trellis. Although we will not pursue such a proof here, it can be verified that the labeling algorithms of the next section always produce a complete representation matrix as a by-product.

Of course, linear trellises have many other interesting properties. However, their proof relies on the factorization theorem (Theorem 6.2) of Section 6, and is therefore deferred until Section 7. Also, we do not discuss here the minimality of linear tail-biting trellises. We shall see in Section 6 that the minimal conventional trellis is linear. For a thorough treatment of minimality of tail-biting trellises, the reader is referred to [7].

4. How to determine whether a trellis is linear?

Given an unlabeled trellis T over $\mathbb{F}_q$, how can one decide whether T is linear? An obvious necessary condition is that the size of each vertex class in T is a power of q, namely $|V_i| = q^{s_i}$ for all i and for some integers $s_0, s_1, \ldots, s_{n-1}$. Given that this condition is satisfied, we could simply try all the possible labelings of the vertices of T and see whether one of them produces a linear trellis. Since the vertices of V_i can be labeled with q-ary strings of length s_i in $q^{s_i}!$ different ways, the complexity of this approach is about $O(q^{s_0}!\, q^{s_1}! \cdots q^{s_{n-1}}!)$.

In this section, we develop vertex labeling algorithms whose complexity is only polynomial in the size of the trellis. These algorithms either label all the vertices of T in such a way that the resulting labeled trellis is linear, or halt if no such labeling exists.

We first present an algorithm suitable for labeling *one-to-one* trellises. To describe the algorithm, we have to introduce some notation. To start with, we need a set of linearly independent vectors: thus we let

$$e_i \stackrel{\text{def}}{=} (0\cdots0\underbrace{10\cdots0}_{i})$$

denote a unit vector with the single nonzero entry in the i-th position. Next, we need notation and terminology pertinent to partially labeled trellises, namely trellises in which some of the vertices are labeled while others are not. Let Γ be a cycle in a partially labeled trellis of depth n, and let $L = (\nu_0, \varepsilon_0, \nu_1, \varepsilon_1, \ldots, \nu_{n-1}, \varepsilon_{n-1})$ be a sequence of (potential) edge and vertex labels. If for all $i = 0, 1, \ldots, n-1$, the following conditions are satisfied:

a. the edge $\Gamma \cap E_i$ is labeled with ε_i (in a one-to-one trellis, this makes Γ unique if it exists)

b. if there exists a vertex in V_i labeled with ν_i, then $\Gamma \cap V_i$ is labeled with ν_i ($\Gamma \cap V_i$ is this vertex); otherwise $\Gamma \cap V_i$ is unlabeled

we say that Γ *agrees* with L. Given a cycle Γ that agrees with L, we can label all the yet unlabeled vertices (if any) of Γ so that $\Gamma \cap V_i$ is labeled with ν_i for all i. We refer to this procedure as *labeling* Γ *according to* L. Finally, we extend the notation of (4) and (5) to codewords of $\mathbb{C}(T)$. If a trellis is one-to-one, then each codeword $c \in \mathbb{C}(T)$ corresponds to a unique cycle $\Gamma(c)$, so that $\nu_i(c)$ and $\varepsilon_i(c)$ are well-defined.

A necessary condition for a trellis T to be linear is that its edge-label code $\mathbb{C}(T)$ is linear. The general idea of our labeling algorithm is to iteratively label subtrellises of T that represent linear subcodes of $\mathbb{C}(T)$, until a full vertex labeling of T is achieved. Thus we choose an arbitrary set of generators $x_1, x_2, \ldots, x_k$ for $\mathbb{C}(T)$, and define

$$\mathbb{C}_j \stackrel{\text{def}}{=} \langle x_1, x_2, \ldots, x_j \rangle \qquad \text{for } j = 1, 2, \ldots, k$$

where $\langle \cdot, \cdot \rangle$ denotes the linear span. We set $\mathbb{C}_0 = \{\mathbf{0}\}$ by convention, where $\mathbf{0}$ is the all-zero vector. The following algorithm may be used to label any one-to-one trellis T, either conventional or tail-biting. We describe the algorithm in the context of tail-biting trellises; for conventional trellises, simply replace cycles by paths.

Labeling Algorithm A

Input: One-to-one reduced trellis T, a set of generators $x_1, x_2, \ldots, x_k$ for $\mathbb{C}(T)$.

Initialization Step: Find $\Gamma(\mathbf{0})$, the unique cycle in T corresponding to $\mathbf{0} \in \mathbb{C}(T)$. Label all the vertices of $\Gamma(\mathbf{0})$ by $\mathbf{0}$. Set $j := 0$.

Iteration Step: If $j = k$, exit (*the labeling is complete, and the trellis is linear*). Otherwise, set $j := j + 1$.

Innovation Step: Find the unique cycle $\Gamma(x_j)$ and label all the yet unlabeled vertices (if any) of $\Gamma(x_j)$ with e_j.

Completion Step: Label all the cycles in T that correspond to codewords of $\mathbb{C}_j = \langle \mathbb{C}_{j-1}, x_j \rangle$. That is, for all $c \in \mathbb{C}_{j-1}$ and all $\beta \in \mathbb{F}_q^*$ form the following sequence of edge and vertex labels:

$$\begin{aligned} L(c, x_j; \beta) \;=\; & (\nu_0(c) + \beta\nu_0(x_j), \varepsilon_0(c) + \beta\varepsilon_0(x_j), \;\ldots\;, \\ & \nu_{n-1}(c) + \beta\nu_{n-1}(x_j), \varepsilon_{n-1}(c) + \beta\varepsilon_{n-1}(x_j)) \end{aligned}$$

Find the unique cycle Γ in T that agrees with $L(c, x_j; \beta)$. If it does not exist, halt (*the trellis T is nonlinear*). Otherwise, label the vertices of Γ according to $L(c, x_j; \beta)$ and repeat. Once done with all $c \in \mathbb{C}_{j-1}$ and $\beta \in \mathbb{F}_q^*$, go to the iteration step.

We now sketch a proof of why the algorithm works. First, we show that if the algorithm finds a vertex labeling, then T is indeed linear. Let T_j denote the labeled subtrellis of T induced by those vertices that have been labeled after j iterations of Algorithm A. The trellis $T_0 = \Gamma(\mathbf{0})$ is obviously linear. As an induction hypothesis, assume that T_{j-1} is linear. Then the fact that T_j is linear follows immediately from the definition of $L(c, x_j; \beta)$ in the completion step. The converse direction is somewhat more complicated. What we need to show is this: if a trellis T is linear, then the algorithm will indeed find a linear labeling of the vertices of T. To this end, we exhibit a representation matrix for T that conforms to the labeling procedure of Algorithm A. We start with an arbitrary generator matrix G for $\mathbb{S}(T)$. W.l.o.g., assume that the projection of G on the edge positions is the chosen set of generators $x_1, x_2, \ldots, x_k$; otherwise perform the appropriate elementary operations on the rows of G. Now suppose that the projection of G on the positions corresponding to the vertex space V_0 is $v_1, v_2, \ldots, v_k$. We replace $v_1, v_2, \ldots, v_k$ by $\xi(v_1), \xi(v_2), \ldots, \xi(v_k)$, where the mapping $\xi(\cdot)$ is defined, recursivley, as follows. If $v_1 = \mathbf{0}$ then $\xi(v_1) = \mathbf{0}$, otherwise $\xi(v_1) = e_1$. For $j = 2, 3, \ldots, k$, if $v_j = \beta_1 v_1 + \beta_2 v_2 + \cdots + \beta_{j-1} v_{j-1}$ for some $\beta_1, \beta_2, \ldots, \beta_{j-1} \in \mathbb{F}_q$ then $\xi(v_j) = \beta_1 \xi(v_1) + \beta_2 \xi(v_2) + \cdots + \beta_{j-1} \xi(v_{j-1})$, otherwise $\xi(v_j) = e_j$. It is easy to see that $\xi(\cdot)$ is an isomorphism, and therefore replacing $v_1, v_2, \ldots, v_k$ with $\xi(v_1), \xi(v_2), \ldots, \xi(v_k)$ preserves the linearity of the vertex labeling. We repeat this procedure, independently, for each of the vertex spaces $V_0, V_1, \ldots, V_{n-1}$. The resulting representation matrix G^* has the desired property: labeling the trellis with Algorithm A is precisely equivalent to reconstructing it cycle-by-cycle from G^*. Thus if the trellis T is linear, the algorithm will not halt.

Example 4.1. Let $\mathcal{H}_8$ denote the $(8,4,4)$ extended binary Hamming code generated by (11111111), (00001111), (00110011), and (01010101). A labeled linear tail-biting trellis T for $\mathcal{H}_8$ is depicted in Figure 1. Now, consider the following two matrices:

$$G_1 = \begin{bmatrix} \underline{0}\,1\,\underline{01}\,1\,\underline{01}\,1\,\underline{01}\,1\,\underline{0}\,1\,\underline{01}\,1\,\underline{01}\,1\,\underline{01}\,1 \\ \underline{1}\,0\,\underline{10}\,1\,\underline{10}\,1\,\underline{10}\,0\,\underline{1}\,0\,\underline{10}\,1\,\underline{10}\,1\,\underline{10}\,0 \\ \underline{0}\,0\,\underline{00}\,0\,\underline{00}\,1\,\underline{11}\,1\,\underline{1}\,0\,\underline{10}\,0\,\underline{01}\,1\,\underline{01}\,1 \end{bmatrix}$$

$$G_2 = \begin{bmatrix} \underline{0}\,1\,\underline{01}\,1\,\underline{01}\,1\,\underline{01}\,1\,\underline{0}\,1\,\underline{01}\,1\,\underline{01}\,1\,\underline{01}\,1 \\ \underline{1}\,0\,\underline{10}\,1\,\underline{10}\,1\,\underline{10}\,0\,\underline{1}\,0\,\underline{10}\,1\,\underline{10}\,1\,\underline{10}\,0 \\ \underline{0}\,0\,\underline{00}\,0\,\underline{00}\,1\,\underline{11}\,1\,\underline{1}\,0\,\underline{10}\,0\,\underline{01}\,1\,\underline{01}\,1 \\ \underline{0}\,0\,\underline{00}\,0\,\underline{00}\,0\,\underline{00}\,0\,\underline{0}\,1\,\underline{01}\,1\,\underline{01}\,1\,\underline{01}\,1 \end{bmatrix}$$

It can be verified by direct inspection of Figure 1 that both G_1 and G_2 are representation matrices for T. Notice that the corresponding representation codes satisfy $\mathbb{S}_1 \subset \mathbb{S}_2 = \mathbb{S}(T)$. In fact, $\mathbb{S}_2 = \mathbb{S}(T)$ can be obtained by starting with G_1 and appending to it the label sequence in T that corresponds to the "missing" generator for $\mathcal{H}_8$, namely (00001111).

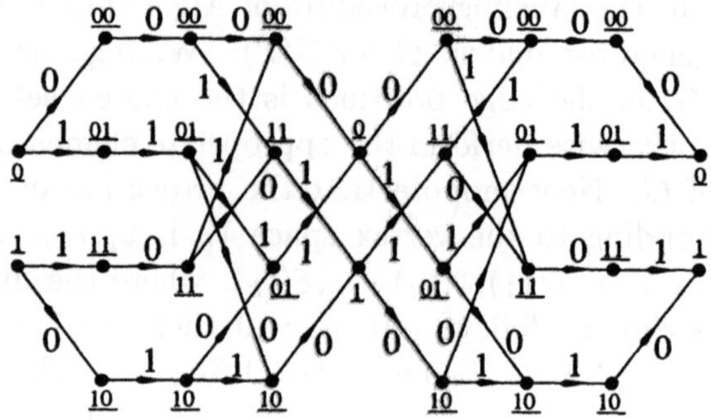

Figure 1. *A labeled tail-biting trellis for the* (8, 4, 4) *code*

The vertex labeling in Figure 1 was obtained using Algorithm A with the following set of generators: (11111111), (01100110), (00001111), and (00110011). Notice that after the first two iterations the entire trellis is already labeled. Thus the remaining two iterations serve for verification purposes only: the algorithm will not assign any new labels, but will verify that all the cycles in T agree with the already existing labeling. □

Algorithm A uses the edge-label code $\mathbb{C}(T)$ as a guide to labeling the vertices of T. This approach will not work if the trellis is not one-to-one. Indeed, let T be a linear trellis which is not one-to-one, and let T_0 denote the subtrellis of T consisting of those cycles that correspond to $\mathbf{0} \in \mathbb{C}(T)$. It is easy to see that T is a union of cosets of T_0 (more precisely, the label code $\mathbb{S}(T)$ is a union of cosets of $\mathbb{S}(T_0)$ under *any* linear labeling of the vertices of T). It follows that an algorithm that finds a linear labeling of the vertices of T should be also able to label the vertices of T_0 in a linear fashion — in fact, it is possible that $T = T_0$. However, the edge-label code of T_0 is the trivial code $\mathbb{C}(T_0) = \{\mathbf{0}\}$.

To deal with trellises that are not necessarily one-to-one, we abandon the edge-label code and work directly with the representation code of T that is being constructed, iteratively, during the algorithm. Modifying Algorithm A along these lines leads to labeling Algorithm B, presented below. The algorithm also verifies whether the trellis is reduced.

Labeling Algorithm B

Initialization Step: Find $\Gamma(\mathbf{0})$, an arbitrary cycle in T that corresponds to $\mathbf{0} \in \mathbb{C}(T)$, and label all the vertices of $\Gamma(\mathbf{0})$ by $\mathbf{0}$. Set $j := 0$ and $\mathbb{S}_0 := \{\mathbf{0}\}$.

Iteration Step: If the cycles corresponding to the label sequences in $\mathbb{S}_j$ cover all the edges of T, exit (*the trellis is linear*). Otherwise, set $j := j + 1$.

Innovation Step: Find an edge in T which does not belong to $\cup_{c \in \mathbb{S}_{j-1}} \Gamma(c)$ and an arbitrary cycle Γ_j which passes through this edge. If Γ_j does not exist, halt (*the trellis T is not reduced, and hence nonlinear*). Otherwise, label all the yet unlabeled vertices (if any) of Γ_j with e_j. Let y_j denote the resulting edge-vertex label sequence of Γ_j.

Completion Step: Set $\mathbb{S}_j := \langle \mathbb{S}_{j-1}, y_j \rangle$, and label all the cycles in T that correspond to codewords of $\mathbb{S}_j$. Specifically, for all $c \in \mathbb{S}_{j-1}$ and all $\beta \in \mathbb{F}_q^*$ form the following sequence of edge and vertex labels:

$$L(c, y_j; \beta) = (\nu_0(c)+\beta\nu_0(y_j), \varepsilon_0(c)+\beta\varepsilon_0(y_j), \ldots, \nu_{n-1}(c)+\beta\nu_{n-1}(y_j), \varepsilon_{n-1}(c)+\beta\varepsilon_{n-1}(y_j))$$

Find an arbitrary cycle Γ in T that agrees with $L(c, y_j; \beta)$. If it does not exist, halt (*the trellis is nonlinear*). Otherwise, label Γ according to $L(c, y_j; \beta) = c + \beta y_j$ and repeat. Once done with all $c \in \mathbb{S}_{j-1}$ and $\beta \in \mathbb{F}_q^*$, go to the iteration step.

To see that this algorithm works, observe that $\mathbb{S}_j$ is a linear representation code for the subtrellis T_j induced by those vertices that have been labeled after j iterations. Thus if the algorithm successfully terminates at the iteration step, then T is linear by Proposition 3.2. The converse direction is considerably more complicated, and we will not pursue a rigorous proof. The general idea is similar to the proof we sketched for Algorithm A just before Example 4.1. Starting with an arbitrary linear label code $\mathbb{S}(T)$, one needs to transform it into a linear representation code for T that conforms to the labeling procedure of Algorithm B.

Example 4.2. Consider the tail-biting trellises T_1, T_2, and T_3 depicted in Figure 2. It can be verified by inspection that all the three trellises are reduced. The edge-label alphabet for all the three trellises is $\mathcal{A} = \{0\}$. Since all the edge labels are zero, they are suppressed in Figure 2, and we will also suppress them in the discussion that follows.

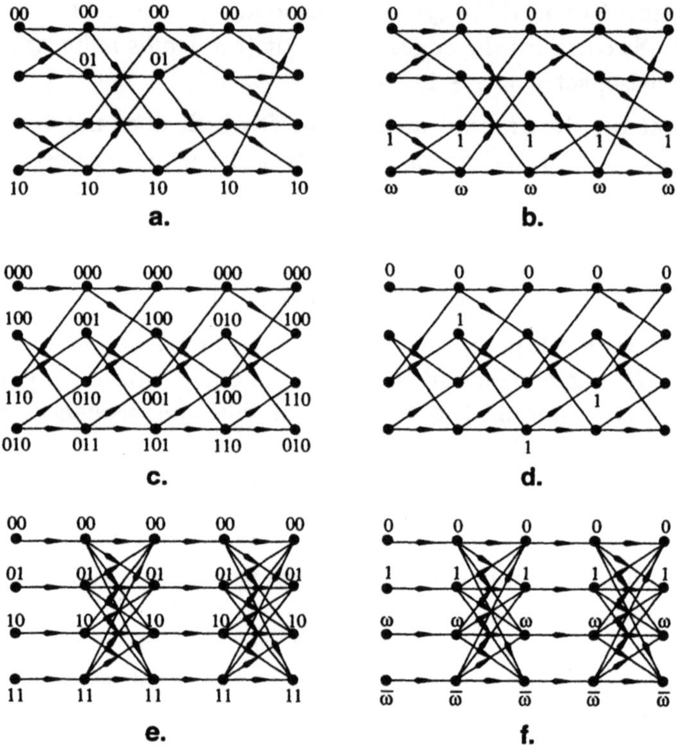

Figure 2. *Linear and nonlinear trellises over the alphabet* $\mathcal{A} = \{0\}$

In Figures 2a, 2c, and 2e, we apply Algorithm B to determine whether the trellises T_1, T_2, and T_3 are linear over $\mathbb{F}_2$. Observe that different vertex labelings may result from Algorithm B, depending upon the choice of specific cycles at the innovation step and at the completion step. Thus Figure 2 provides only typical examples. In Figure 2a, the algorithm halts at the completion step, since there is no cycle that agrees with $(10, 11, 11, 10)$. In Figure 2c, the algorithm successfully terminates after four iterations, with $y_1 = (000, 001, 001, 000)$, $y_2 = (010, 010, 000, 010)$, $y_3 = (100, 010, 100, 100)$, and $y_4 = (110, 000, 000, 010)$. In Figure 2e, the algorithm also exits successfully after four iterations. In Figures 2b, 2d, and 2f, we use Algorithm B to determine linearity of the same trellises over $\mathbb{F}_4$. The algorithm halts in Figure 2b, respectively Figure 2d, since there is no cycle that agrees with $(\bar{\omega}, \bar{\omega}, \bar{\omega}, \bar{\omega})$, respectively $(0, \omega, \omega, \omega)$. It terminates successfully after four iterations in Figure 2f.

The overall results are that T_1 is nonlinear, T_3 is linear over both $\mathbb{F}_2$ and $\mathbb{F}_4$, while T_2 is linear over $\mathbb{F}_2$ but not linear over $\mathbb{F}_4$. The same results would be reached regardless of the specific order in which the sixteen cycles in T_1, T_2, T_3 are labeled (try this out). □

An interesting conclusion from Example 4.2 is that trellis linearity is essentially a graph-theoretic property: it may be defined and thought of as a characteristic of unlabeled graphs. However, we are not aware of any prior work on linearity, or similar concepts, in graph theory.

Remark. For group trellises over an abelian group $\mathcal{G}$, Algorithm B works in essentially the same way. The only changes are as follows: $\langle \cdot, \cdot \rangle$ should be understood as the span over $\mathcal{G}$, and the dependence on $\beta \in \mathbb{F}_q^*$ should be removed from the completion step. That is, the label sequence in the completion step should be defined simply as $L(c, y_j) = c + y_j$.

5. The product construction

The notion of a trellis product and the corresponding product construction were first introduced by Kschischang and Sorokine [9] for conventional trellises. See also [8, 10, 13]. The product construction was extended to tail-biting trellises in [2], where it was used to construct minimal tail-biting trellises for several well-known codes.

Definition 5.1. *Let $T' = (V', E', \mathcal{A})$ and $T^* = (V^*, E^*, \mathcal{A})$ be two trellises of depth n, either conventional or tail-biting, over the alphabet $\mathcal{A}$, and assume that $\mathcal{A}$ is endowed with the addition operation. Then the product*

$T' \times T^*$ is the trellis $T = (V, E, \mathcal{A})$ whose vertex classes and edge classes are the Cartesian products, defined as follows:

$$V_i \stackrel{\text{def}}{=} \{ (v', v^*) \ : \ v' \in V_i' \ \text{ and } \ v^* \in V_i^* \}$$

$$E_i \stackrel{\text{def}}{=} \{ \big((v_1', v_1^*), a' + a^*, (v_2', v_2^*)\big) \ : \ (v_1', v_2', a') \in E_i', \ (v_1^*, v_2^*, a^*) \in E_i^* \}$$

Specifically, there is an edge $e \in E_i$ from a vertex $(v_1', v_1^*) \in V_{i-1}$ to a vertex $(v_2', v_2^*) \in V_i$ if and only if (v_1', a', v_2') is an edge in E_i' for some $a' \in \mathcal{A}$, and (v_1^*, a^*, v_2^*) is an edge in E_i^* for some $a^* \in \mathcal{A}$.

Kschischang-Sorokine [9] showed, in the context of conventional trellises, that if $\mathbb{C}_1 = \mathbb{C}(T')$ and $\mathbb{C}_2 = \mathbb{C}(T^*)$, then the product trellis $T' \times T^*$ represents the code

$$\mathbb{C}_1 + \mathbb{C}_2 \stackrel{\text{def}}{=} \{ c_1 + c_2 \ : \ c_1 \in \mathbb{C}_1 \ \text{ and } \ c_2 \in \mathbb{C}_2 \} \tag{7}$$

It is obvious that the trellis product operation is associative, and hence an expression of the form $T_1 \times T_2 \times \cdots \times T_k$ is well-defined. Now let $\mathbb{C}$ be a linear code of length n and dimension k over the field $\mathbb{F}_q$, and let $\{x_1, x_2, \ldots, x_k\}$ be a basis for $\mathbb{C}$. Each x_i generates a one-dimensional subcode of $\mathbb{C}$, which we denote by $\langle x_i \rangle$. Thus

$$\mathbb{C} = \langle x_1 \rangle + \langle x_2 \rangle + \cdots + \langle x_k \rangle \tag{8}$$

It follows from (7) and (8) that if $T_1, T_2, \ldots, T_k$ are trellises for $\langle x_1 \rangle$, $\langle x_2 \rangle, \ldots, \langle x_k \rangle$, respectively, then their product $T = T_1 \times T_2 \times \cdots \times T_k$ represents $\mathbb{C}$. This almost completes the description of the product construction. It remains to specify the trellises $T_1, T_2, \ldots, T_k$.

To do so, we need to introduce the notions of *span* and *elementary trellis*. Although simple, these concepts have turned out to be fundamental in trellis theory, and special care has to be taken to define them properly. Given a codeword $x \in \mathbb{C}$, the ***span*** of x, denoted $[x]$, is a semi-open interval $(i, j] \in \mathcal{I}$ such that the corresponding closed interval $[i, j]$ contains all the nonzero positions of x. If the span of a codeword x is $(i, j]$ we define the symbols $\triangleleft(x)$ and $\triangleright(x)$ as $\triangleleft(x) = i$ and $\triangleright(x) = j$. Although a semi-open interval cannot contain the entire time axis $\mathcal{I} = \{0, 1, \ldots, n-1\}$ by definition, we override this technicality by explicitly allowing $[x] = \mathcal{I}$, in addition to the above. The foregoing definition of span holds for both conventional and tail-biting trellises. The only difference is in the interpretation of an interval: cyclic intervals for tail-biting trellises and conventional intervals for conventional trellises.

Example 5.1. Let $x = (x_0, x_1, \ldots, x_4) = (01011)$. There are 8 possible spans for x in a tail-biting trellis, namely $(1,4]$, $(3,1]$, $(0,4]$, $(1,0]$, $(2,1]$, $(3,2]$, $(4,3]$, and $[0,4] = \mathcal{I}$. Of these, only $(1,4]$, $(0,4]$ and $[0,4]$ are possible spans for x in a conventional trellis.

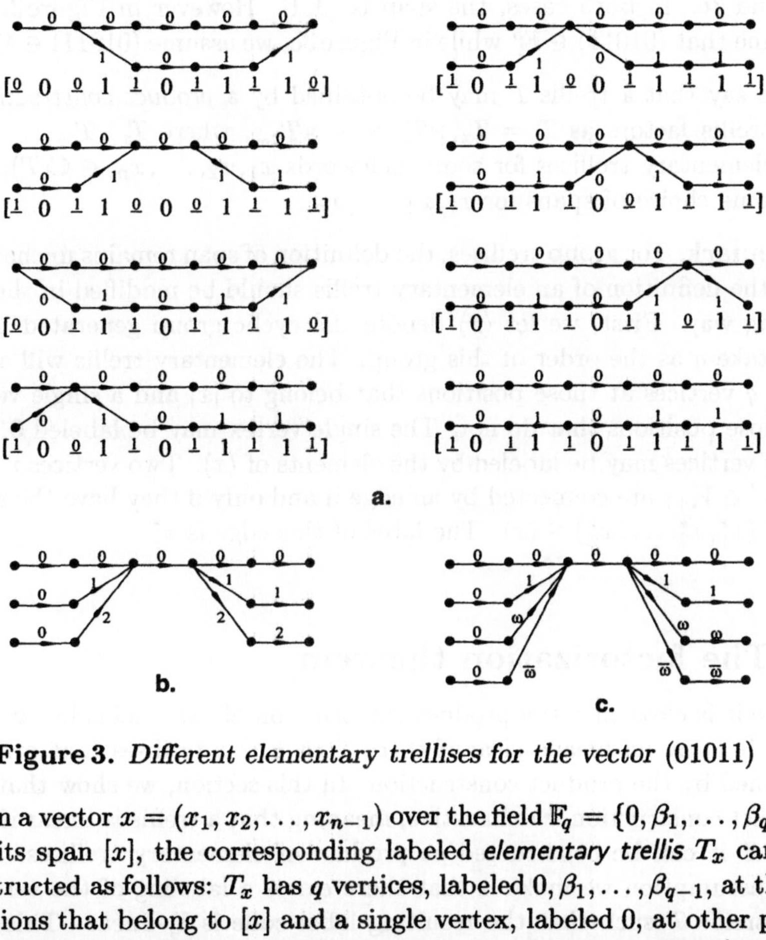

Figure 3. *Different elementary trellises for the vector* (01011)

Given a vector $x = (x_1, x_2, \ldots, x_{n-1})$ over the field $\mathbb{F}_q = \{0, \beta_1, \ldots, \beta_{q-1}\}$, and its span $[x]$, the corresponding labeled *elementary trellis* T_x can be constructed as follows: T_x has q vertices, labeled $0, \beta_1, \ldots, \beta_{q-1}$, at those positions that belong to $[x]$ and a single vertex, labeled 0, at other positions. There is an edge $e \in E_i$ from a vertex $v \in V_i$ to a vertex $v' \in V_{i+1}$ if and only if these two vertices have the same label $\beta \in \mathbb{F}_q$. The label of this edge is βx_i. It is easy to see that the elementary trellis T_x is always linear, contains exactly q cycles (or paths, in the case of conventional trellises), and represents the one-dimensional code $\langle x \rangle$ over $\mathbb{F}_q$.

For example, eight different elementary trellises for $x=(01011)\in\mathbb{F}_2^5$, corresponding to the different spans for x, are depicted in Figure 3a along with the corresponding representation matrices. Note that an elementary trellis T_x depends not only on x and $[x]$, but also on the ambient field $\mathbb{F}_q$. For example, two elementary trellises for (01011) are depicted in Figures 3b and 3c. In both cases, the span is $(3,1]$. However in Figure 3b, we assume that $(01011)\in\mathbb{F}_3^5$ while in Figure 3c, we assume $(01011)\in\mathbb{F}_4^5$. □

We say that a trellis T may be obtained by a *product construction* if the trellis factors as $T = T_{x_1}\times T_{x_2}\times\cdots\times T_{x_m}$, where $T_{x_1}, T_{x_2},\ldots,T_{x_m}$ are elementary trellises for some codewords $x_1,x_2,\ldots,x_m\in\mathbb{C}(T)$, and for some choice of spans for $x_1,x_2,\ldots,x_m$.

Remark. For group trellises, the definition of span remains unchanged and the definition of an elementary trellis should be modified in the following way. First, we let $\langle x\rangle$ denote the cyclic group generated by x, and take q as the order of this group. The elementary trellis will again have q vertices at those positions that belong to $[x]$ and a single vertex at those positions that do not. The single vertex may be labeled $\mathbf{0}$, and the q vertices may be labeled by the elements of $\langle x\rangle$. Two vertices $v\in V_i$ and $v'\in V_{i+1}$ are connected by an edge if and only if they have the same label $(x_1^*,x_2^*,\ldots,x_n^*)\in\langle x\rangle$. The label of this edge is x_i^*.

6. The factorization theorem

While it is clear that the product construction always yields linear trellises, it is by no means clear if there exist linear trellises that are not obtained by the product construction. In this section, we show that the product construction is exhaustive, meaning that a trellis is linear if *and only if* it can be constructed as product of elementary trellises. The only assumption we make is that there *exists* a labeling of the vertices of a trellis T such that the resulting label code is $\mathbb{F}_q$-linear. This will be enough to show that T, as an unlabeled graph, may be obtained as a product of unlabeled graphs that correspond to elementary trellises.

We start by considering a special form of a representation matrix. Note that a representation matrix G_x for an elementary trellis T_x consists of a single row. The projection of G_x on those positions that correspond to the edges of T_x is precisely x. Let $G_x|_{V_i}$ denote the projection of G_x on the position corresponding to the vertex class V_i. By definition, $G_x|_{V_i} = \mathbf{0}$ if and only if $i \notin [x]$. The value of $G_x|_{V_i}$ for $i \in [x]$ is arbitrary as long as it is nonzero, and we assume, w.l.o.g., that $G_x|_{V_i} = G_x|_{V_j}$ for all $i, j \in [x]$. Let us denote this common nonzero value by $G_x|_V$. Now, let T_x and T_y be elementary trellises with representation matrices G_x and G_y, respectively, and suppose that the vectors $v_x = G_x|_V$ and $v_y = G_y|_V$ are linearly independent. Then a representation matrix $G_{x,y}$ for the product $T_x \times T_y$ can be obtained by simply taking G_x, G_y as its two rows. To see this, let v' be a vertex in T_x labeled αv_x and let v^* be a vertex in T_y labeled βv_y, where α, β are elements of the ambient field. Referring to Definition 5.1, we can identify the vertex (v', v^*) in $T_x \times T_y$ with the label $\alpha v_x + \beta v_y$, provided that v_x, v_y are linearly independent. The incidence relations of Definition 5.1 are also preserved by the representation matrix $G_{x,y} = [G_x, G_y]^T$. A similar, though more elaborate, argument establishes the following lemma.

Lemma 6.1. *Let $T' = (V', E', \mathcal{A})$ and $T^* = (V^*, E^*, \mathcal{A})$ be two linear trellises with representation matrices G' and G^*, respectively. Then*

$$G = \begin{bmatrix} G' \\ G^* \end{bmatrix} \tag{9}$$

is a representation matrix for the product trellis $T = T' \times T^*$ if and only if $V_i' \cap V_i^* = \{\mathbf{0}\}$ for all $i = 0, 1, \ldots, n$, where the vertex classes V_i' and V_i^* are identified with their labels.

Lemma 6.1 extends straightforwardly to the product of an arbitrary number of elementary trellises. Let $\{x_1, x_2, \ldots, x_k\}$ be a basis for $\mathbb{C}$ and let $\{v_1, v_2, \ldots, v_k\}$ be an arbitrary set of k linearly independent vectors. A representation matrix for the trellis that is obtained as a product of the elementary trellises $T_{x_1}, T_{x_2}, \ldots, T_{x_k}$ is given by

$$G_{x_1, x_2, \ldots, x_k} = \begin{bmatrix} \underline{y_{1,0}} \; x_{1,0} \; \underline{y_{1,1}} \; x_{1,1} & \cdots & \underline{y_{1,n-1}} \; x_{1,n-1} \\ \underline{y_{2,0}} \; x_{2,0} \; \underline{y_{2,1}} \; x_{2,1} & \cdots & \underline{y_{2,n-1}} \; x_{2,n-1} \\ & \vdots & \\ \underline{y_{k,0}} \; x_{k,0} \; \underline{y_{k,1}} \; x_{k,1} & \cdots & \underline{y_{k,n-1}} \; x_{k,n-1} \end{bmatrix} \tag{10}$$

where $y_{i,j} = v_i$ if $j \in [x_i]$ and $y_{i,j} = \mathbf{0}$ otherwise, for all $i = 1, 2, \ldots, k$. Such matrices will be useful later on. We formalize the notion as follows.

Definition 6.1. *A $k \times n$ representation matrix G is said to be in product form if each row of G is a representation matrix for an elementary trellis, the nonzero vertex labels in each of the k rows of G are identical, and the set of these k vertex labels is linearly independent.*

It is easy to see that any representation matrix G in product form represents a trellis that is obtained by the product construction. Thus a trellis T may be obtained by the product construction if and only if there exists a representation matrix for T in product form.

Theorem 6.2. *Let T be a linear trellis, either tail-biting or conventional. Then there exists a representation matrix for T in product form. Equivalently, the trellis can be factored as $T = T_{x_1} \times T_{x_2} \times \ldots \times T_{x_k}$, where $T_{x_1}, T_{x_2}, \ldots, T_{x_k}$ are elementary trellises.*

Theorem 6.2 is of fundamental importance in the study of tail-biting trellises. In particular, this theorem implies that in constructing minimal linear tail-biting trellises for linear codes, we only need to investigate those trellises that are obtained by the product construction (cf. [7]).

We will, indeed, prove Theorem 6.2 in the context of tail-biting trellises. The claim for conventional trellises then easily follows. The somewhat technical proof consists of a long series of lemmas (Lemmas 6.4, 6.5, 6.6, 6.7, 6.8, and 6.9), presented in the remainder of this section.

First, however, we consider a degenerate special case. We say that a trellis T is *degenerate* if it consists of disconnected cycles or paths (such trellises were called SOI by McEliece [10]). If a degenerate trellis T is also linear over $\mathbb{F}_q$, then the number of cycles (paths) in T is q^k.

Lemma 6.3. *If T is a degenerate linear trellis, then it can be factored as $T = T_{x_1} \times T_{x_2} \times \ldots \times T_{x_k}$, where $T_{x_1}, T_{x_2}, \ldots, T_{x_k}$ are elementary trellises.*

Proof. Let $v_1, v_2, \ldots, v_k$ be a set of linearly independent vectors over $\mathbb{F}_q$. It is straightforward to verify that T can be represented by a matrix G of the form

$$G = \begin{bmatrix} \underline{v_1}\ x_{1,0}\ \underline{v_1}\ x_{1,1}\ \cdots\ \underline{v_1}\ x_{1,n-1} \\ \underline{v_2}\ x_{2,0}\ \underline{v_2}\ x_{2,1}\ \cdots\ \underline{v_2}\ x_{2,n-1} \\ \vdots \\ \underline{v_k}\ x_{k,0}\ \underline{v_k}\ x_{k,1}\ \cdots\ \underline{v_k}\ x_{k,n-1} \end{bmatrix}$$

Comparing this with (10), we see that G is already in product form. Hence, T factors as a product of k elementary trellises, each of span $\mathcal{I}$. ∎

For a non-degenerate trellis T, our first task is to find a cycle in T that "generates" an elementary trellis. The following lemma makes this possible.

Lemma 6.4. *If T is a non-degenerate linear tail-biting trellis of depth n, then there exists a label sequence $y \in \mathbb{S}(T)$ such that $\nu_i(y) = \mathbf{0}$ while $\nu_{i+1}(y) \neq \mathbf{0}$ for some $i \in \{0, 1, \ldots, n-1\}$.*

Proof. Since T is non-degenerate, there exist two distinct cycles in T that intersect in a vertex $v \in V_i$. Let y_1 and y_2 be the corresponding label sequences in $\mathbb{S}(T)$. Since $\mathbb{S}(T)$ is linear, $y_1 - y_2 = y$ is also a label sequence in $\mathbb{S}(T)$ and $\nu_i(y) = \mathbf{0}$. If $\nu_{i+1}(y) \neq \mathbf{0}$, we are done. Otherwise, set $i \leftarrow i+1$ and check again whether $\nu_{i+1}(y) \neq \mathbf{0}$. Eventually this condition will be satisfied, since we cannot have $\nu_i(y_1) = \nu_i(y_2)$ for all i. ∎

W.l.o.g., we assume that the position i exhibited in Lemma 6.4 is $i = 0$. The reader is invited to verify that this assumption has no effect on the proof other than simplifying notation. Let $\mathbb{S}_0(T)$ denote the set of all label sequences $y \in \mathbb{S}(T)$ such that $\nu_0(y) = \mathbf{0}$ and $\nu_1(y) \neq \mathbf{0}$. Lemma 6.4 proves that this set is not empty. We shall say that $y \in \mathbb{S}_0(T)$ *terminates* at time i and write $\triangleright(y) = i$ if $\nu_{i+1}(y) = \nu_{i+2}(y) = \cdots = \nu_{n-1}(y) = \mathbf{0}$ and $\nu_i(y) \neq \mathbf{0}$. All the label sequences in $\mathbb{S}_0(T)$ terminate at some time $i \in \{1, 2, \ldots, n-1\}$. Let $y^* \in \mathbb{S}_0(T)$ be such that

$$t \overset{\text{def}}{=} \triangleright(y^*) \leq \triangleright(y) \qquad \text{for all } y \in \mathbb{S}_0(T) \tag{11}$$

In other words, $\Gamma(y^*)$ is a cycle in T that diverges from the all-zero cycle $(\underline{\mathbf{0}}, 0, \underline{\mathbf{0}}, \ldots, \underline{\mathbf{0}}, 0, \underline{\mathbf{0}})$ at time $i = 0$ and merges back to it at the *earliest possible* time t. This is the cycle we are after.

Lemma 6.5. *Let $y^* = (\underline{\mathbf{0}}, x_1, \underline{v_1}, x_2, \underline{v_2}, \ldots, \underline{v_t}, x_{t+1}, \underline{\mathbf{0}}, 0, \ldots, \underline{\mathbf{0}}, 0, \underline{\mathbf{0}})$ for some $(x_1, x_2, \ldots, x_{t+1}, 0, \ldots, 0) \in \mathbb{C}(T)$. Then the vertices $v_1, v_2, \ldots, v_t$ are all nonzero.*

Proof. The vertex (label) v_1 is nonzero by the definition of $\mathbb{S}_0(T)$, and v_t is nonzero by the definition of $\triangleright(y^*)$ and (11). For $i = 2, 3, \ldots, t-1$, the vertex v_i is nonzero since otherwise $\mathbb{S}_0(T)$ would contain the label sequence $(\underline{\mathbf{0}}, x_1, \underline{v_1}, x_2, \underline{v_2}, \ldots, \underline{v_{i-1}}, x_i, \underline{\mathbf{0}}, 0, \ldots, \underline{\mathbf{0}}, 0, \underline{\mathbf{0}})$ that terminates before time t, in contradiction with the definition of t in (11). ∎

Given a labeled linear trellis T, we can always apply a nonsingular linear transformation to the vertex labels of a vertex class V_i, without changing the structure of the trellis. Such transformation amounts to a simple re-labeling of the vertices in V_i. Since the vertex labels

$v_1, v_2, \ldots, v_t$ are nonzero by Lemma 6.5, applying the appropriate linear transformations to each of the vertex classes $V_2, V_3, \ldots, V_t$, we can obtain a linear labeling of T such that

$$y^* = (\underline{0}, x_1, \underline{v_1}, x_2, \underline{v_1}, \ldots, \underline{v_1}, x_{t+1}, \underline{0}, \ldots, \underline{0}, 0, \underline{0}) \qquad (12)$$

Since $y^* \in \mathbb{S}(T)$ and $\mathbb{S}(T)$ is a linear code, there exists a generator matrix G for $\mathbb{S}(T)$ which contains (12) as its first row. Let T^* be the linear subtrellis of T defined by the representation matrix

$$G^* = \left[\, \underline{0}, x_1, \underline{v_1}, x_2, \underline{v_1}, \ldots, \underline{v_1}, x_{t+1}, \underline{0}, \ldots, \underline{0}, 0, \underline{0} \,\right] \qquad (13)$$

It is easy to see that T^* is an elementary trellis — it is the sought after elementary factor of T. Our next task is to show that T can be factored as $T = T^* \times T'$, where T' is again a linear trellis. Let G' denote the matrix obtained from G by deleting the first row (which is equal to y^*), and let T' be the linear subtrellis of T having G' as its representation matrix. We say that a set $\{y_1, y_2, \ldots, y_m\}$ of rows of G' is *inadmissible* if there exist $\alpha_1, \alpha_2, \ldots, \alpha_m \in \mathbb{F}_q$, not all of them zero, so that $y = \alpha_1 y_1 + \alpha_2 y_2 + \cdots + \alpha_m y_m$ is such that $\nu_i(y) = v_1$ for at least one $i \in \{1, 2, \ldots, t\}$. The following lemma justifies this definition.

Lemma 6.6. *The trellis T factors as $T = T^* \times T'$ if and only if the matrix G' does not contain inadmissible sets of rows.*

Proof. This follows directly from (13) and Lemma 6.1. Since $V_i^* = \langle v_1 \rangle$ for $i = 1, 2, \ldots, t$ and $V_i^* = \{\mathbf{0}\}$ otherwise, it can be readily seen that $V_i' \cap V_i^* = \{\mathbf{0}\}$ unless there exists a linear combination of rows of G' that produces the vertex label v_1 in position $i \in \{1, 2, \ldots, t\}$. ∎

To get rid of inadmissible sets of rows of G', we will perform elementary row operations on G as follows. We say that a row y of G' is *marked by zero* if $\nu_i(y) = \mathbf{0}$ for at least one $i \in \{1, 2, \ldots, t\}$. The objective of the elementary row operations is to make sure that every row in every inadmissible set of rows of G' is marked by zero. If this objective is achieved, we are done. Otherwise, let $\{y_1, y_2, \ldots, y_m\}$ be an inadmissible set of rows of G' and assume without loss of generality that y_1 is not marked by zero. We then replace the row y_1 in G' by the linear combination

$$y_1 \leftarrow \alpha_1 y_1 + \alpha_2 y_2 + \cdots + \alpha_m y_m - y^*$$

where $\alpha_1, \alpha_2, \ldots, \alpha_m \in \mathbb{F}_q$ are the coefficients of the linear combination of $y_1, y_2, \ldots, y_m$ that produces the vertex label $v_1 = \nu_i(y^*)$ for some

$i \in \{1, 2, \ldots, t\}$. Since every elementary row operation of this type increases the number of rows marked by zero in G' by one, the objective will be reached in at most $k-1$ steps, where k is the number of rows in G.

Let us refer to the procedure described in the foregoing paragraph as the *factorization algorithm*. It is obvious that when the factorization algorithm terminates, then G' either

(i) does not contain inadmissible sets of rows; or

(ii) contains inadmissible sets, with every row in every such set marked by zero.

The key to our proof is that the choice of $\Gamma(y^*)$ in (11) ensures that condition (ii) never occurs. Thus the factorization algorithm indeed eliminates all the inadmissible sets of rows of G'. To show this, we first need two lemmas that pertain to conventional linear trellises.

We shall say that a trellis T, either conventional or tail-biting, is *connected*, if it consists of a single connected component when viewed as an *undirected graph*. That is, a trellis $T = (V, E, \mathcal{A})$ is connected if for any two vertices $v, v' \in V$, there exists a path from v to v' in T that does not necessarily preserve the direction of the edges in E.

Lemma 6.7. *Let T be a linear trellis of depth n, and let G be a generator matrix for $\mathbb{S}(T)$. If every row y of G satisfies $\nu_i(y) = \mathbf{0}$ for at least one $i \in \{0, 1, \ldots, n\}$, then T is connected.*

Proof. The proof is by induction on the number k of rows of G. Clearly, the lemma is true if $k = 0$, in which case T consists of a single zero-labeled path. As an induction hypothesis, assume that the lemma is true whenever a generator matrix has $k-1$ rows, and let $T' = (V', E', \mathbb{F}_q)$ denote the connected subtrellis of $T = (V, E, \mathbb{F}_q)$ represented by the first $k-1$ rows of G. Let y be the last row of G, and suppose that $\nu_i(y) = \mathbf{0}$. But then $V_i = V_i'$, since T and T' are both linear. It follows that all the paths in T pass through a vertex (in the i-th vertex class) that is already in T'. Hence, since T' is connected, so is T. ∎

The next lemma is concerned with paths in a conventional trellis $T = (V, E, \mathcal{A})$. We will distinguish between *directed paths* that preserve the direction of edges in E and *undirected paths* that do not necessarily preserve the direction of edges in E. A directed path P of length η from a vertex $v_0 \in V_j$ to a vertex $v_\eta \in V_{j+\eta}$ can be uniquely specified by a sequence of vertices $P = (v_0, v_1, \ldots, v_{\eta-1}, v_\eta)$, where $v_i = V_{j+i}$ for all $i = 0, 1, \ldots, \eta$. An undirected path P' of length η from a vertex $v_0 \in V$ to a vertex $v_\eta \in V$ is also uniquely specified by the corresponding sequence of vertices $P' = (v_0, v_1, \ldots, v_{\eta-1}, v_\eta)$. Even though P' is not directed, the trellis structure of T implies that if $v_i \in V_j$ then v_{i+1} is either in V_{j-1} or in V_{j+1}, for all $i = 0, 1, \ldots, \eta - 1$.

Lemma 6.8. *Let $T = (V, E, \mathbb{F}_q)$ be a conventional labeled linear trellis of depth n, and let $P = (v_0, v_1, \ldots, v_n)$ be a path of length n in T from $v_0 \in V_0$ to $v_n \in V_n$. Suppose that there exists an <u>undirected</u> path in T from some vertex v^* on P to a vertex labeled $\mathbf{0}$. Then there also exists a <u>directed</u> path of length n in T from $v_0 \in V_0$ to the vertex $\mathbf{0} \in V_n$.*

Proof. Let $P' = (u_0, u_1, \ldots, u_\eta)$ be the undirected path in T from the vertex $u_0 = v^*$ on P to the vertex $u_\eta = \mathbf{0}$. Further let $u_0 \in V_r$ and $u_\eta \in V_s$. Then the trellis structure of T implies that either $u_1 \in V_{r+1}$ or $u_1 \in V_{r-1}$, and either $u_{\eta-1} \in V_{s+1}$ or $u_{\eta-1} \in V_{s-1}$. This leads to four different cases to consider. We henceforth assume that $u_1 \in V_{r+1}$ and $u_{\eta-1} \in V_{s+1}$. The other three cases can be proved by a similar argument.

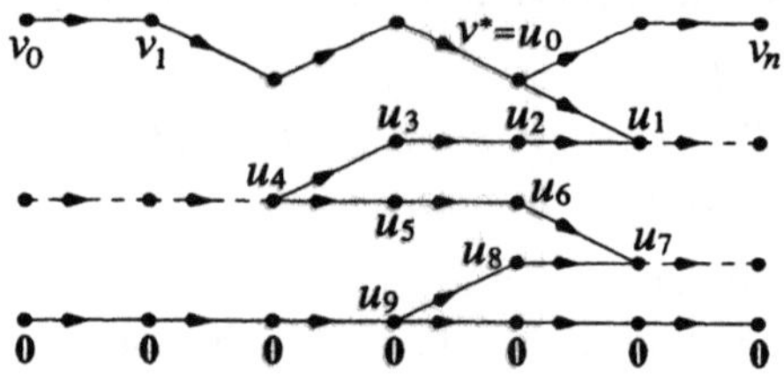

Figure 4. *Example of an undirected path from v^* to* $\mathbf{0}$

We say that a vertex u_i on $P' = (u_0, u_1, \ldots, u_\eta)$ is *pivotal* if the path P' reverses direction when it reaches u_i. Thus u_i on P' is pivotal if and only if the vertices u_{i-1} and u_{i+1} belong to the same vertex class in T. Now suppose that P' contains exactly m pivotal vertices, say $u_{i_1}, u_{i_2}, \ldots, u_{i_m}$. We can then parse P' into $m+1$ directed paths as follows

$$\underset{\longrightarrow}{(u_0, u_1, \ldots, u_{i_1})},\ \underset{\longleftarrow}{(u_{i_1}, u_{i_1+1}, \ldots, u_{i_2})},\ \cdots,\ \underset{\longrightarrow}{(u_{i_m}, u_{i_m+1}, \ldots, u_\eta)} \qquad (14)$$

where the direction alternates from $\longrightarrow$ to $\longleftarrow$ as shown. For example, the undirected path $(u_0, u_1, \ldots, u_9)$ in Figure 4 reverses direction three times and has $m = 3$ pivotal vertices, namely u_1, u_4, and u_7. This path is therefore parsed as follows:

$$\underset{\longrightarrow}{(u_0, u_1)},\ \underset{\longleftarrow}{(u_1, u_2, u_3, u_4)},\ \underset{\longrightarrow}{(u_4, u_5, u_6, u_7)},\ \underset{\longleftarrow}{(u_7, u_8, u_9)}$$

Note that our assumption that $u_1 \in V_{r+1}$ and $u_{\eta-1} \in V_{s+1}$ implies that the number m of pivotal vertices in P' must be odd. We say that the pivotal vertices $u_{i_1}, u_{i_3}, \ldots, u_{i_m}$ are odd and the pivotal vertices $u_{i_2}, u_{i_4}, \ldots, u_{i_{m-1}}$ are even. To each odd pivotal vertex u_{i_j} we associate a tail, denoted $\mathsf{tail}\{u_{i_j}\}$, which is a directed path in T from u_{i_j} to a vertex in V_n. To each even pivotal vertex u_{i_j} we associate a head, denoted $\mathsf{head}\{u_{i_j}\}$, which is a directed path in T from a vertex in V_0 to u_{i_j}. We can then extend the parsing of P' in (14) into $m+1$ directed paths from V_0 to V_n in T as follows:

$$\begin{array}{lc}
\big(v_0, v_1, \ldots, v^* = u_0, u_1, \ldots, u_{i_1-1}, \mathsf{tail}\{u_{i_1}\}\big) & + \\
\big(\mathsf{head}\{u_{i_2}\}, u_{i_2-1}, u_{i_2-2}, \ldots, u_{i_1+1}, \mathsf{tail}\{u_{i_1}\}\big) & - \\
\big(\mathsf{head}\{u_{i_2}\}, u_{i_2+1}, u_{i_2+2}, \ldots, u_{i_3-1}, \mathsf{tail}\{u_{i_3}\}\big) & + \\
\vdots & \\
\big(\mathsf{head}\{u_{i_{m-1}}\}, u_{i_{m-1}+1}, \ldots, u_{i_m-1}, \mathsf{tail}\{u_{i_m}\}\big) & + \\
\big(0, 0, \ldots, 0,\ u_\eta, u_{\eta-1}, \ldots, u_{i_m+1},\ \mathsf{tail}\{u_{i_m}\}\big) & -
\end{array}$$

where we have reversed the direction of every second path in the original parsing of P', as indicated by the $-$ signs above. For example, in Figure 4, the heads and tails of pivotal vertices are represented by dashed lines, and the corresponding $m+1 = 4$ directed paths in T are given by

$$\begin{array}{lc}
\big(v_0, v_1, \ldots, v^* = u_0, \mathsf{tail}\{u_1\}\big) & + \\
\big(\mathsf{head}\{u_4\}, u_3, u_2, \mathsf{tail}\{u_1\}\big) & - \\
\big(\mathsf{head}\{u_4\}, u_5, u_6, \mathsf{tail}\{u_7\}\big) & + \\
\big(0, 0, \ldots, 0, u_9, u_8, \mathsf{tail}\{u_7\}\big) & -
\end{array}$$

Since T is a linear trellis, any linear combination of the above $m+1$ directed paths of length n in T is another directed path of length n in T. Adding these $m+1$ paths with alternating signs as indicated, we observe that all the heads and tails of pivotal vertices cancel. Thus the first vertex of the resulting path in T is $v_0 \in V_0$ and the last vertex is $\mathbf{0} \in V_n$. ∎

Now recall that, according to Lemma 6.6, the original non-degenerate linear tail-biting trellis T factors as $T = T^* \times T'$, where T^* is the elementary trellis represented by (13), if and only if the matrix G' does not contain inadmissible sets of rows.

Lemma 6.9. *When the factorization algorithm terminates, the matrix G' does not contain inadmissible sets of rows.*

Proof. Assume to the contrary that $\{y_1, y_2, \ldots, y_m\}$ is an inadmissible set of rows of G'. Let G'' be the $m \times 2n$ submatrix of G' having $y_1, y_2, \ldots, y_m$ as its rows. Let $G''(y^*)$ be the $(m+1) \times 2n$ submatrix of G having $y^*, y_1, y_2, \ldots, y_m$ as its rows, where y^* is given by (12). Let T'' and $T''(y^*)$ be the subtrellises of T represented by G'' and $G''(y^*)$, respectively. We convert T'' and $T''(y^*)$ into conventional trellises by projecting them on the positions in the span of y^*. Specifically, let $T|_{[y^*]}$ be the subtrellis of T induced by the vertices in $V_1 \cup V_2 \cup \cdots \cup V_t$, where t is given by (11). Thus $T|_{[y^*]}$ consists of the vertex classes $V_1, V_2, \ldots, V_t$ and those edges of T whose endpoints are (both) in $V_1 \cup V_2 \cup \cdots \cup V_t$. Note that $T|_{[y^*]}$ is a conventional (not tail-biting) trellis, since $V_0 \not\subseteq V_1 \cup V_2 \cup \cdots \cup V_t$. We then consider the conventional trellises

$$T''|_{[y^*]} \stackrel{\text{def}}{=} T'' \cap T|_{[y^*]} \qquad \text{and} \qquad T''(y^*)|_{[y^*]} \stackrel{\text{def}}{=} T''(y^*) \cap T|_{[y^*]}$$

It is straightforward to verify that these trellises are represented by the matrices $G''|_{[y^*]}$ and $G''(y^*)|_{[y^*]}$ obtained from G'' and $G''(y^*)$, respectively, by retaining only the appropriate $2t-1$ columns.

When the factorization algorithm terminates, every row in every inadmissible set, in particular $\{y_1, y_2, \ldots, y_m\}$, is marked by zero. It follows that every row y in the matrix $G''|_{[y^*]}$ satisfies $\nu_i(y) = \mathbf{0}$ for at least one $i \in \{1, 2, \ldots, t\}$. Therefore, the trellis $T''|_{[y^*]}$ is connected by Lemma 6.7. By the definition of an inadmissible set, there exist $\alpha_1, \alpha_2, \ldots, \alpha_m \in \mathbb{F}_q$ so that $y = \alpha_1 y_1 + \alpha_2 y_2 + \cdots + \alpha_m y_m$ is such that $\nu_i(y) = v_1$ for some $i \in \{1, 2, \ldots, t\}$. Thus the cycles $\Gamma_T(y)$ and $\Gamma_T(y^*)$ intersect in the vertex $\nu_i(y) = \nu_i(y^*) = v_1$. It follows that in the trellis $T''(y^*)|_{[y^*]}$, the path corresponding to y^* passes through a vertex that also belongs to $T''|_{[y^*]}$, which is a subtrellis of $T''(y^*)|_{[y^*]}$. Since $T''|_{[y^*]}$ is linear and connected, there must be a path (possibly undirected) from this vertex $\nu_i(y) = \nu_i(y^*) = v_1$ to a vertex labeled $\mathbf{0}$. We can now apply Lemma 6.8 to $T''(y^*)|_{[y^*]}$ along with the path $(v_1, v_1, \ldots, v_1)$ that corresponds to y^*

in $T''(y^*)|_{[y^*]}$, to establish that there exists a directed path from the vertex $v_1 \in V_1$ to the vertex $\mathbf{0} \in V_t$. Continuing this path along the all-zero cycle in T, we exhibit a cycle in T whose label sequence is given by

$$(\underline{0}, x_1, \underline{v_1}, x_2', \underline{v_2'}, \ldots, \underline{v_{t-1}'}, x_t', \underline{0}, 0, \ldots, \underline{0}, 0, \underline{0}) \tag{15}$$

This label sequence belongs to $\mathbb{S}_0(T)$ and terminates at time $\le t-1$, which contradicts our choice of t and y^* in (11). Hence, the inadmissible set $\{y_1, y_2, \ldots, y_m\}$ does not exist. ∎

We are now in a position to complete the proof of the factorization theorem (Theorem 6.2). If the trellis T is degenerate, then it factors as $T = T_{x_1} \times T_{x_2} \times \ldots \times T_{x_k}$ directly by Lemma 6.3. Otherwise, we find a cycle $\Gamma_T(y^*)$ in T that diverges from the all-zero cycle at time $i = 0$, say, and merges back to the all-zero cycle at the earliest possible time. Such a cycle exists by Lemma 6.4, and up to a linear re-labeling of the vertices of T, its label sequence y^* is given by (12). We then find a generator matrix G for $\mathbb{S}(T)$ that contains (12) as its first row, and perform the factorization algorithm on the remaining rows of this matrix. In view of Lemma 6.6 and Lemma 6.9, the trellis T then factors as $T = T^* \times T'$, where T^* is the elementary trellis represented by the first row of G and T' is another linear trellis. Since T' is linear, we can now apply the same procedure to T', and proceed in this manner until T factors completely into a product of elementary trellises.

This completes the proof of Theorem 6.2 for the case of tail-biting trellises. Conventional trellises may be seen as tail-biting trellises with only one vertex in a vertex class V_i. Thus the foregoing proof applies in this case as well, and any linear conventional trellis can be also factored into a product of elementary trellises. We point out that this was already known for *minimal* conventional trellises [9, 10], but was not known for conventional trellises in general.

7. More properties of linear trellises

We conclude this chapter with a proof of several structural properties of linear tail-biting trellises that follow from the factorization theorem of the previous section. These properties are formulated in Theorems 7.2 through 7.4 below. First, however, we need to some notation.

Given two vertices u and v that belong to the same vertex class in a trellis T, one can *merge* these vertices by replacing them with a single vertex that inherits all the edges incident upon u and v. We say that the vertices u and v are *mergeable* if merging them does not alter the edge-label code of T. A trellis T is said to be *non-mergeable* if there are no mergeable vertices in T (cf. [8, 13, 14]). A trellis T, either conventional or tail-biting, is said to be *biproper* if the edges beginning at any vertex of T are labeled distinctly and the edges ending at any vertex of T are also labeled distinctly (for conventional trellises, we also require that $|V_0| = |V_n| = 1$). We begin with a simple lemma, whose proof is left as an exercise for the reader.

Lemma 7.1. *Suppose that $T = T' \times T^*$ is a linear trellis, either conventional or tail-biting. If either T' or T^* is mergeable, then so is T.*

This lemma, in conjunction with Theorem 6.2, is used in the proof of the following theorem, which in turn will be used to prove Theorem 7.3.

Theorem 7.2. *A non-mergeable linear tail-biting trellis T is biproper.*

Proof. Assume to the contrary that $T = (V, E, \mathcal{A})$ is not proper. Then there exist two distinct edges $e_1 = (v, a, v_1)$ and $e_2 = (v, a, v_2)$ in T that begin at the same vertex $v \in V$ and have the same label $a \in \mathcal{A}$. We assume w.l.o.g. that $v \in V_0$. Since T is linear, it must also contain the edge $e_1 - e_2 = (\mathbf{0}, 0, u)$, where $u = v_1 - v_2 \neq \mathbf{0}$. By Theorem 6.2, there exists a representation matrix G for T in product form, which generates a representation code $\mathbb{S}$. Let $y = (\underline{0}, 0, \underline{u}, \dots)$ be a codeword of $\mathbb{S}$ such that $\Gamma(y)$ contains the edge $e_1 - e_2$. Then y is a linear combination of the rows of G. Thus $y = \alpha_1 x_{i_1} + \alpha_2 x_{i_2} + \cdots + \alpha_m x_{i_m}$ for some nonzero $\alpha_1, \alpha_2, \dots, \alpha_m \in \mathcal{A}$ and some m rows $x_{i_1}, x_{i_2}, \dots, x_{i_m}$ of G. Since G is in product form, the vertex labels of different rows in G are linearly independent. It follows that $x_{i_1,0} = x_{i_2,0} = \cdots = x_{i_m,0} = \mathbf{0}$, otherwise the linear combination

$$\alpha_1 x_{i_1} + \alpha_2 x_{i_2} + \cdots + \alpha_m x_{i_m} \;=\; y \;=\; (\underline{0}, 0, \underline{u}, \dots) \tag{16}$$

would be nonzero in the first position. Now, suppose that $T' = (V', E', \mathcal{A})$ $= T_{x_{i_1}} \times T_{x_{i_2}} \times \cdots \times T_{x_{i_m}}$ is the product of the corresponding m elementary trellises. Since $x_{i_1,0} = \cdots = x_{i_m,0} = \mathbf{0}$, we have $V'_0 = \{\mathbf{0}\}$, and T' can be regarded as a conventional trellis. It is obvious that T' is linear, and therefore it contains the edge $(\mathbf{0},\mathbf{0},\mathbf{0}) \in E'_0$. But it also contains the edge $e_1 - e_2 = (\mathbf{0},\mathbf{0},u) \in E'_0$ by construction. Thus T' is not proper. It is known [13, 14] that a conventional trellis which is not proper must be mergeable. Hence T' is mergeable. Since T can be factored as $T = T' \times T^*$, Lemma 7.1 implies that T must also be mergeable. A similar argument shows that T must be mergeable if it is not co-proper. ∎

We have not defined minimal tail-biting trellises in this chapter. For a precise definition of minimality in this context, see [7]. However, it is clear that a minimal trellis T, either conventional or tail-biting, cannot be mergeable, since otherwise we could merge vertices in T and obtain a smaller trellis for the same code. This leads to the following result.

Theorem 7.3. *Every linear minimal trellis, either conventional or tail-biting, is biproper.*

Here is an interesting consequence from Theorem 7.3. Let $x_1, x_2, \ldots, x_k$ be a set of generators (a basis) for a linear code $\mathbb{C}$, let $[x_1], [x_2], \ldots, [x_k]$ be a choice of spans for these generators, and let $T = T_{x_1} \times T_{x_2} \times \cdots \times T_{x_k}$ be the corresponding trellis for $\mathbb{C}$.

Corollary 7.4. *If the trellis T is minimal, then no two spans in the set $[x_1], [x_2], \ldots, [x_k]$ start in the same position and no two spans in the set $[x_1], [x_2], \ldots, [x_k]$ end in the same position.*

Proof. If there are two spans that start at the same position, then T is not proper. If there are two spans that end at the same position, then T is not co-proper. Such a trellis cannot be minimal by Theorem 7.3. ∎

The condition described in Corollary 7.4 is known [9, 13] as the *minimal-span form.* Indeed, it is well known [9, 10] that a set of generators for the minimal conventional trellis must be in minimal-span form. However, to the best of our knowledge, this is the first time this property is established in the context of tail-biting trellises.

References

[1] S. Aji, G. Horn, R.J. McEliece, and M. Xu, "Iterative min-sum decoding of tail-biting codes," pp. 15–24 in *Proc. Allerton Conference Comm., Control, and Computing*, Monticello, IL, September 1998.

[2] A.R. Calderbank, G.D. Forney, Jr., and A. Vardy, "Minimal tail-biting trellises: the Golay code and more," *IEEE Trans. Inform. Theory*, **45**, pp. 1435–1455, July 1999.

[3] G.D. Forney, Jr., "Codes on graphs: normal realizations," *IEEE Trans. Inform. Theory*, **47**, pp. 520–548, February 2001.

[4] G.D. Forney, Jr. and M.D. Trott, "The dynamics of group codes: state spaces, trellis diagrams and canonical encoders," *IEEE Trans. Inform. Theory*, **39**, pp. 1491–1513, September 1993.

[5] A.B. Kiely, S. Dolinar, R.J. McEliece, L. Ekroot, and W. Lin, "Trellis decoding complexity of linear block codes," *IEEE Trans. Inform. Theory*, **42**, pp. 1687–1697, November 1996.

[6] R. Koetter and A. Vardy, "Construction of minimal tail-biting trellises," pp. 72–74 in *Proceedings IEEE International Workshop on Information Theory*, Killarney, Ireland, June 1998.

[7] R. Koetter and A. Vardy, "The structure of tail-biting trellises: minimality and basic principles," preprint, March 2002.

[8] F.R. Kschischang, "The trellis structure of maximal fixed-cost codes," *IEEE Trans. Inform. Theory*, vol. 42, pp. 1828–1838, 1996.

[9] F.R. Kschischang and V. Sorokine, "On the trellis structure of block codes," *IEEE Trans. Inform. Theory*, **41**, pp. 1924–1937, 1995.

[10] R.J. McEliece, "On the BCJR trellis for linear block codes," *IEEE Trans. Inform. Theory*, **42**, pp. 1072–1092, July 1996.

[11] Y. Shany and Y. Be'ery, "Linear tail-biting trellises, the square-root bound, and applications for Reed-Muller codes," *IEEE Trans. Inform. Theory*, **46**, pp. 1514–1523, 2000.

[12] P. Ståhl, J.B. Anderson, and R. Johannesson, "Optimal and near-optimal encoders for short and moderate-length tail-biting trellises," *IEEE Trans. Inform. Theory*, **45**, pp. 2562–2571, November 1999.

[13] A. Vardy, "Trellis structure of codes," pp. 1989–2118 in HANDBOOK OF CODING THEORY, V.S. Pless and W.C. Huffman, Elsevier, 1998.

[14] A. Vardy and F.R. Kschischang, "Proof of a conjecture of McEliece regarding the expansion index of the minimal trellis," *IEEE Trans. Inform. Theory*, **42**, pp. 2027–2034, November 1996.

[15] N. Wiberg, H.-A. Loeliger and R. Kötter, "Codes and iterative decoding on general graphs," *Euro. Trans. Telecom.*, **6**, pp. 513–526, 1995.

Coding Over Graphs

Anxiao (Andrew) Jiang   Jehoshua Bruck
Parallel and Distributed Systems Lab
California Institute of Technology
{jax,bruck}@paradise.caltech.edu

Abstract

This paper introduces the concept of *Coding Over Graphs*: given an arbitrary graph, assign information to vertices of the graph in such a way that the information can be retrieved at every vertex by exploring its close proximity. This framework is a generalization of RAID (Redundant Array of Inexpensive Disks). We study a specific *coding over graphs* scheme which has the property that as the number of failures increases, the information can be retrieved by exploring slightly larger proximities—therefore a graceful trade-off between reliability and performance is provided. An efficient algorithm for that scheme for the graphs being trees is presented. The algorithm can always find a 'perfect' *coding over trees* scheme.

Keywords: Channel coding, data allocation, file assignment problem, graph coloring, layered diversity coloring.

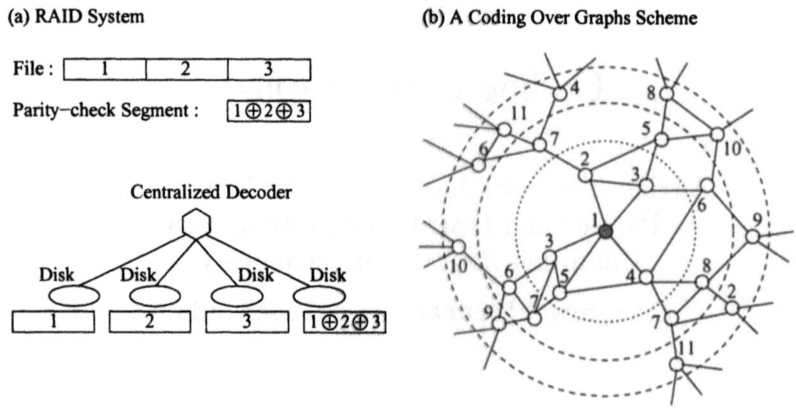

Figure 1: Examples of RAID and the Coding Over Graphs Scheme.

1 Introduction

RAID (Redundant Array of Inexpensive Disks) is a well-known data storage system that uses codeword layout to improve reliability [1]. In RAID, a file is split into n equally long segments. Then an extra parity-check segment is created which is the exclusive-OR of the n segments. Those $n+1$ segments are stored on $n+1$ disks. If any disk loses its data, a centralized decoder can use the data on the other n disks to recover the file. Fig. 1 (a) shows an example of RAID with $n = 3$.

The concept of RAID can be generalized for data storage on networks of arbitrary topologies—namely, to distributively store information on nodes of a network and use coding to improve reliability. And we can relax the constraint of having a centralized decoder—instead, every node should be able to recover the information by decoding only the data stored in a small proximity of the node. We call such a generalized scheme *Coding Over Graphs.*

In real networks, the radius of the proximity from which a vertex can get sufficient data for decoding usually determines the data-retrieving delay. When data stored on some nodes are lost, a node may need to explore a larger proximity to find enough data. It's very desirable that the radius of such proximities can increase gracefully as the data loss becomes more severe. Below we propose a specific scheme with such a desired property.

In the scheme, a file is first split into K equally long segments. Then an

MDS code is used to encode those K segments to get totally N segments. ($N \geq K$.) Because of the maximum-distance-seperable property of an MDS code, any K of those N segments can be used to recover the original file. Now given a graph (network), we store one segment on each vertex of the graph—different vertices can store copies of the same segment—in such a way that for every vertex, the larger a proximity it explores, the more different segments it can find. Such a scheme provides a graceful trade-off between fault-tolerance and delay.

Fig. 1 (b) illustrates the scheme with an example. In the figure, different numbers represent the different file segments. We assume that here $K = 5$. The shaded vertex can find 5 different segments within 1 hop, 8 different segments within 2 hops, and 11 different segments within 3 hops $\cdots\cdots$ Such a situation should also be true for every other vertex in the graph.

If we represent every file segment with a color, then the above scheme can be formulated as a graph coloring problem called the *Layered Diversity Coloring Problem* [2]. The problem is defined as follows:

Layered Diversity Coloring Problem: Given a graph $G(V, E)$ and N colors, how to assign one color to each vertex, (different vertices can have the same color), so that every vertex can find at least K_1 different colors within m_1 hops, at least K_2 different colors within m_2 hops, $\ldots\ldots$, at least K_p different colors within m_p hops? (Here parameters N, K_1, $K_2, \ldots\ldots, K_p, m_1, m_2, \ldots\ldots, m_p$ and p are all positive integers, and $N \geq K_1 > K_2 > \cdots > K_p$, $m_1 > m_2 > \cdots > m_p$.)

Parameter N is the number of colors that can be used for the layered diversity coloring. It influences the existance of the coloring as well as the difficulty in constructing the coloring. If a layered diversity coloring uses $N = K_1$ colors, then we call it a *perfect* layered diversity coloring.

We'll study the layered diversity coloring problem for the graph being a tree in Section II. In Section III, we'll conclude this paper.

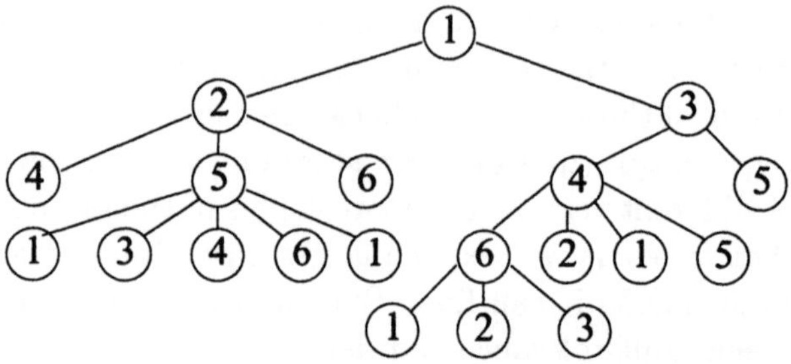

Figure 2: An example of layered diversity coloring on a tree, with parameters $p = 3$, $N = 6$, $K_1 = 6$, $K_2 = 5$, $K_3 = 4$, $m_1 = 4$, $m_2 = 3$ and $m_3 = 2$.

2 Layered Diversity Coloring For Trees

Trees are an important family of graphs. Besides being widely used as network structures, trees also often act as spanning graphs in graph algorithms.

In [2] it's proven that trees have a very unique property for layered diversity coloring, as the following theorem says.

Theorem 2.1 *There exists a layered diversity coloring on a tree $G(V, E)$ if and only if for any vertex $v \in V$ and for $1 \le i \le p$, there are at least K_i vertices within m_i hops from v (including v itself).*

Note that Theorem 2.1 doesn't mention the parameter N. Therefore any tree that can have a layered diversity coloring can also have a *perfect* layered diversity coloring (because we can simply let N be equal to K_1). Clearly such a property doesn't hold for general graphs.

Example 1: Fig. 2 shows an example of layered diversity coloring on a tree. It's easy to verify that every vertex can find at least $K_1 = 6$ different colors within $m_1 = 4$ hops, at least $K_2 = 5$ different colors within $m_2 = 3$ hops, and at least $K_3 = 4$ different colors within $m_3 = 2$ hops. And clearly the layered diversity coloring is *perfect.*
□

In this section we'll introduce an efficient algorithm for constructing the layered diversity coloring on trees. Before we present the algorithm, we first define some basic terms: (1) We use $G(V, E)$ to denote a tree, and there is one vertex in the tree called its *root.* (2) For any two vertices $u \in V$ and $v \in V$, the *distance between u and v*, $d(u,v)$, is defined to be the number of edges in the path connecting u and v. (3) For every vertex $v \in V$, we say v is at level $L(v)$, where the level number $L(v)$ is defined to be the distance between v and the root of $G(V,E)$. (4) For any two vertices $u \in V$ and $v \in V$, we say '*u is an ancestor of v, and v is a descendant of u*' if and only if $L(v) - L(u) = d(u,v) > 0$; we say '*u is a quasi-ancestor of v, and v is a quasi-descendant of u*' if and only if $L(v) - L(u) = d(u,v)$.

Next we define two operations that will appear in the algorithm. The first operation, called '*to find knots that are quasi-descendants of v with parameters K_i and m_i*', is to find a set A of vertices that satisfy the following properties (here each vertex in A is called a *knot*; A can be empty):

- For any vertex $u \in A$, the following conditions hold: (1) u is a quasi-descendant of v; (2) u has at least one quasi-descendant that cannot find K_i different colors within m_i hops, and u has at least one uncolored quasi-descendant. If we use w_1 to denote a quasi-descendant of u at the smallest level that cannot find K_i different colors within m_i hops, and use w_2 to denote an uncolored quasi-descendant of u at the smallest level, then $L(w_1)+L(w_2)-2L(u) \leq m_i$. (3) u does not have an ancestor for which the previous two conditions both hold.
- If for a vertex u, the three conditions above hold, then $u \in A$.

The second operation, called '*to color quasi-descendants of v with parameters K_i and m_i*', is a process described as follows:

- Let w denote a quasi-descendant of v at the smallest level that cannot find K_i different colors within m_i hops, and define $C = \{$ the N different colors $\} - \{$ colors within m_i hops from $w\}$. Use colors in C to color the uncolored quasi-descendants of v, with each color used at most once. While coloring, always color vertices at a smaller level before coloring vertices at a greater level. Keep coloring until all colors in C are used or until there is no uncolored quasi-descendant of v left.

Now we present the layered diversity coloring algorithm for trees:

Algorithm: Layered Diversity Coloring on Tree $G(V, E)$

Prerequisite: $\forall\ 1 \leq i \leq p$ and $\forall\ v \in V$, there are at least K_i vertices within m_i hops from v (including v itself).

Algorithm:

for $1 \leq i \leq p$ do

{ Let $R = \emptyset$. Find knots that are quasi-descendants of the root of $G(V, E)$ with parameters K_i and m_i, and insert those knots into R.

while $R \neq \emptyset$ do

{ Remove a vertex v from R. Color quasi-descendants of v with parameters K_i and m_i. Then find knots that are quasi-descendants of v with parameters K_i and m_i, and insert those knots into R.

}

}

Arbitrarily color the remaining uncolored vertices. The algorithm ends.

□

Below we use an example to illustrate how the algorithm colors a tree.

Example 2: Use the above algorithm to get a layered diversity coloring on the tree $G(V, E)$ shown in Fig. 3 (a). Here the parameters are $p = 3$, $N = 6$, $K_1 = 6$, $K_2 = 5$, $K_3 = 4$, $m_1 = 4$, $m_2 = 3$ and $m_3 = 2$. The vertex v_1 is the root of the tree.

The process of the algorithm's coloring the tree is as follows:

- $i = 1$. Find the knots that are quasi-descendants of v_1 with parameters K_1 and m_1. Then $R = \{v_1\}$.

  Remove v_1 from R. Color quasi-descendants of v_1 with parameters K_1 and m_1. (See Fig. 3 (b) for the coloring.) Find knots that are quasi-descendants of v_1 with parameters K_1 and m_1. Then $R = \{v_3\}$.

  Remove v_3 from R. Color quasi-descendants of v_3 with parameters K_1 and m_1. (See Fig. 3 (c) for the coloring.) Find knots that

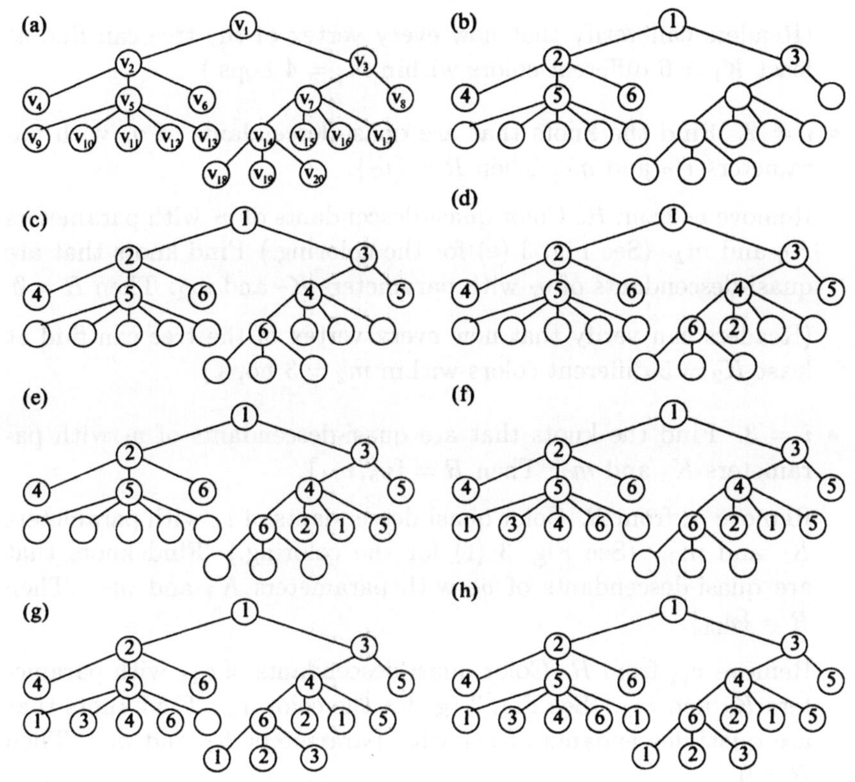

Figure 3: An example of using the layered diversity coloring algorithm to color a tree.

are quasi-descendants of v_3 with parameters K_1 and m_1. Then $R = \{v_7\}$.

Remove v_7 from R. Color quasi-descendants of v_7 with parameters K_1 and m_1. (See Fig. 3 (d) for the coloring.) Find knots that are quasi-descendants of v_7 with parameters K_1 and m_1. Then $R = \emptyset$.

(Readers can verify that now every vertex of the tree can find at least $K_1 = 6$ different colors within $m_1 = 4$ hops.)

- $i = 2$. Find the knots that are quasi-descendants of v_1 with parameters K_2 and m_2. Then $R = \{v_7\}$.

  Remove v_7 from R. Color quasi-descendants of v_7 with parameters K_2 and m_2. (See Fig. 3 (e) for the coloring.) Find knots that are quasi-descendants of v_7 with parameters K_2 and m_2. Then $R = \emptyset$.

  (Readers can verify that now every vertex of the tree can find at least $K_2 = 5$ different colors within $m_2 = 3$ hops.)

- $i = 3$. Find the knots that are quasi-descendants of v_1 with parameters K_3 and m_3. Then $R = \{v_5, v_{14}\}$.

  Remove v_5 from R. Color quasi-descendants of v_5 with parameters K_3 and m_3. (See Fig. 3 (f) for the coloring.) Find knots that are quasi-descendants of v_5 with parameters K_3 and m_3. Then $R = \{v_{14}\}$.

  Remove v_{14} from R. Color quasi-descendants of v_{14} with parameters K_3 and m_3. (See Fig. 3 (g) for the coloring.) Find knots that are quasi-descendants of v_{14} with parameters K_3 and m_3. Then $R = \emptyset$.

  (Readers can verify that now every vertex of the tree can find at least $K_3 = 4$ different colors within $m_3 = 2$ hops.)

- Arbitrarily color the uncolored vertex v_{13}. (See Fig. 3 (h) for the coloring.) The algorithm ends.

The coloring shown in Fig. 3 (h) is a layered diversity coloring. Readers might have noticed that it's actually the same coloring as in Example 1. □

The coloring algorithm has p iterations. It's proven in [2] that after the i-th iteration ($1 \leq i \leq p$), every vertex in the tree can find at least K_i different colors within m_i hops. So when the algorithm ends, the coloring on the tree is a layered diversity coloring. It's worthwhile to point out that during the i-th iteration, the algorithm doesn't need to know any parameter K_j or m_j where $j \neq i$. However, it's very interesting that the coloring done in different iterations can 'cooperate' with each other very well to get the desired result.

In [2], details on how to carry out the coloring algorithm efficiently are described. It is proved that the algorithm has time complexity $O(|V|(K_1 + \sum_{i=1}^{p} m_i))$. And by replacing N with K_1, we can use the algorithm to find perfect layered diversity coloring on trees.

3 Conclusions

In this paper we introduced the concept of *Coding Over Graphs*, and studied a specific coding-over-graphs scheme which can be formulated as the *Layered Diversity Coloring Problem.* A coloring algorithm for trees is presented. The algorithm can always be used to get perfect layered diversity coloring on trees.

Many different coding-over-graphs schemes can be derived, corresponding to different constraints or optimization criteria [3]. And as a last note, besides RAID, Coding Over Graphs can also be seen as generalizing the classical *file assignment problem* (FAP) [4] by introducing coding.

References

[1] D. A. Patterson, G. A. Gibson, and R. Katz. A case for redundant arrays of inexpensive disks. In *Proc. SIGMOD Int. Conf. Data Management*, pages 109–116, 1988.

[2] A. Jiang and J. Bruck. DNAS: Dispersed network attached storage for reliability and performance. Technical Report ETR 043, Parallel and Distributed Systems Lab, California Institute of Technology, http://www.paradise.caltech.edu/papers/etr043.pdf, 2002.

[3] M. Naor and R. M. Roth. Optimal file sharing in distributed networks. *SIAM J. Comput.*, 24(1):158–183, 1995.

[4] L. W. Dowdy and D. V. Foster. Comparative models for the file assignment problem. *Computing Surveys*, 14(2):287–313, 1982.

On Approaching the Capacity of Finite-State Intersymbol Interference Channels

Joseph B. Soriaga, Henry D. Pfister, and Paul H. Siegel
Signal Transmission and Recording (STAR) Group
University of California, San Diego
La Jolla, CA 92093-0407
{jsoriaga,hpfister,psiegel}@ucsd.edu

Abstract

In this paper, we examine several methods of coding which approach the capacity of finite-state intersymbol interference (ISI) channels. We begin by showing the effectiveness of multilevel coding (MLC) on such channels, and then apply MLC to the binary-input, discrete-time Gaussian channel with ISI. In cases where the input spectrum must be shaped, we discuss a method for combining MLC with spectral shaping codes, and provide an example using the biphase code. Finally, we characterize the absolute minimum value of signal-to-noise ratio (per input bit) required for reliable communication on these channels, and describe a simple coding scheme which achieves this threshold.

Key Words: Channel capacity, intersymbol interference, multilevel coding, low-rate Shannon limit, finite-state channel.

1 Introduction

In this paper, we consider the discrete-time Gaussian channel with intersymbol interference, and refer to this as a partial-response (PR) channel. These channels are often used to model the magnetic recording channel with partial-response equalization. Recently, computing the capacity of PR channels has received much attention. Arnold and Loeliger [1], Pfister, Soriaga, and Siegel [2], and Kavčić [3] have developed efficient methods to estimate information rates with Markov sources, and have optimized the sources to tighten lower bounds on capacity. Vontobel and Arnold [4] have also presented improved upper bounds.

Here we focus on coding techniques which achieve reliable communication while approaching these rates. We begin by examining multilevel coding (MLC), with multistage decoding, on finite-state intersymbol interference (ISI) channels defined by,

$$Pr(y_k|x_k, \ldots, x_{k-\nu}) \tag{1}$$

where $x_k, x_{k-1}, \ldots, x_{k-\nu}$ are drawn from a finite alphabet and $q_k = (x_{k-1}, \ldots, x_{k-\nu})$ defines the state of the channel. The component codes are designed and decoded under the assumption that channel inputs are i.i.d. and that the subchannel for each level is memoryless. This assumption will be discussed later in more detail.

We show that by increasing the number of component codes, m, one can achieve the mutual information rate of (1) for i.i.d., equiprobable inputs, also known as the Symmetric Information Rate (SIR) [5]. We estimate MLC information rates (via Monte Carlo simulation) on PR channels, and optimize *low-density parity-check* (LDPC) component codes to approach these rates. The LDPC optimization is not discussed in detail, but some properties are stated which allow the application of *density evolution* [6]. At low rates, this system is not sufficient to approach capacity. We provide a simple example of concatenating modulation codes with MLC as a method to more closely approach capacity.

As the code rate approaches zero, we consider a simpler coding system which achieves the absolute minimum E_b/N_0 required for reliable communication on a binary-input PR channel. In some cases, this minimum is identical to that achieved without the binary-input constraint, and other cases it is strictly larger.

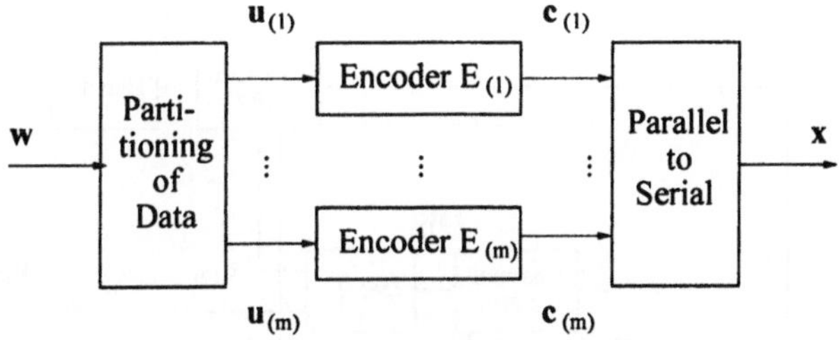

Figure 1: Multilevel encoder.

This work represents an extension of our previous results, reported in [2]. Alternative methods for optimizing LDPC codes for use on ISI channels are presented in [7],[8].

2 Multilevel Coding

Henceforth, uppercase represents random variables, and lower case their realizations. A vector is denoted $\mathbf{v} = (v_1 \dots v_N)$, a subvector $\mathbf{v}_i^j = (v_i \dots v_j)$, and a labeled vector, $\mathbf{v}_{(i)} = (v_{(i),1} \cdots v_{(i),N})$.

In MLC, as described in [9], the encoder partitions the input block into m subblocks, and separately encodes each of them. For convenience, we assume all subcodes have blocklength N. Each m-tuple of coded symbols is then mapped onto a symbol in a modulation alphabet for transmission over the channel.

Here, we employ MLC to combat ISI caused by the channel memory. The m-tuples of symbols from the multiple code levels are distributed across time, via interleaving, rather than mapped to a modulation symbol. This scheme is illustrated in Figure 1, with the block interleaving

$$\mathbf{X} = (C_{(1),1} \cdots C_{(m),1} C_{(1),2} \cdots C_{(m),2} \ \cdots \ C_{(m),N}).$$

The receiver consists of an m-level, multistage decoder (MSD), depicted in Figure 2. At each successive stage, channel symbols are re-detected, using decisions provided from previous stages as *a priori* information. To simplify the analysis, we use a "window-APP" module, as defined below.

Definition 2.1. *A "window-*a posteriori probability*" (window-APP) detector calculates the probability $Pr(x_k|\mathbf{y}_{k-w}^{k+w})$, for all values of x_k. When*

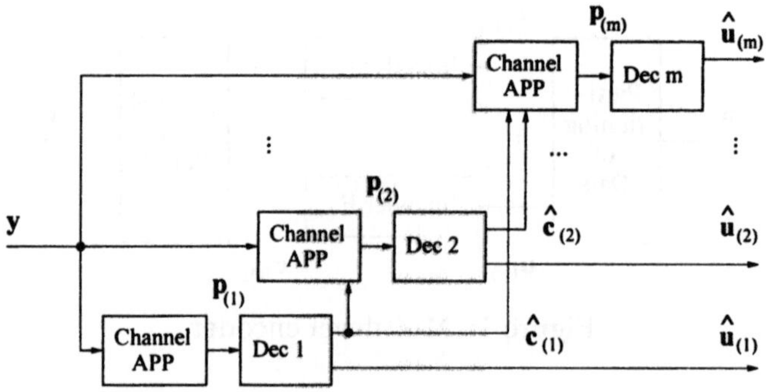

Figure 2: Multistage decoder.

given some set of previous decisions $\{\hat{x}_t : t \in T\}$*, the detector calculates* $Pr(x_k|\mathbf{y}_{k-w}^{k+w}, X_t = \hat{x}_t, t \in T)$.

Let $P_{(i),k}$ be the window-APP detector output for $C_{(i),k}$ given past decisions $\hat{\mathbf{c}}_{(1)}, \ldots, \hat{\mathbf{c}}_{(i-1)}$ (as shown in Figure 2). In general, independence among the components of $\mathbf{C}_{(i)}$ does not imply independence among the components of $\mathbf{P}_{(i)}$. Using random codes and a decoder which assumes each $P_{(i),k}$ depends only on $C_{(i),k}$, one can show that all rates less than $I(C_{(i)}; P_{(i)})$ are achievable. This is because the $P_{(i),k}$ are identically distributed for $w < km \leq N - w$, so we can neglect edge effects and drop the subscript k.

This system, with m levels, can achieve the rate $R_{av,m} = \frac{1}{m}\sum_{i=1}^{m} R_i$, where

$$R_i = I(C_{(i)}; P_{(i)}|\hat{\mathbf{C}}_{(l)} = \mathbf{C}_{(l)}, l = 1, \ldots, i-1). \tag{2}$$

Notice that this expression for the rate assumes perfect decisions are fed to each stage. This assumption is justified by upper bounding the error probability of MSD with

$$Pr(E_{MSD}) < Pr(E_1) + \ldots + Pr(E_m),$$

where the event E_i occurs when stage i fails and all previous stages were successful [9].

3 Achieving the Symmetric Information Rate

We now consider the maximum average rate that can be achieved by the MLC scheme incorporating codes optimized for memoryless channels.

Definition 3.1. *The Symmetric Information Rate (SIR), denoted I_s, is the mutual information rate*

$$\lim_{n\to\infty} \frac{1}{n} I(\mathbf{X}_1^n; \mathbf{Y}_1^n)$$

when the input is both i.i.d. and equiprobable.

Lemma 3.1. *$R_{av,m} \to I_s$ as $m \to \infty$.*

Proof. (Sketch) Using the chain rule, one can express I_s as the sum $\frac{1}{m}\sum_{i=1}^{m} \tilde{R}_i$, where

$$\tilde{R}_i = \lim_{N\to\infty} \frac{1}{N} I(\mathbf{C}_{(i)}; \mathbf{Y}_1^{mN} | \mathbf{C}_{(1)} \dots \mathbf{C}_{(i-1)}). \tag{3}$$

For all $i > \nu$, perfect state information is provided periodically (with period m) throughout the block $\mathbf{y}_1^{mN}$. Consequently, (3) converges to the mutual information,

$$\begin{aligned} \tilde{R}_i &= I(C_{(i),k}; \mathbf{Y}_{(k-1)m+i}^{km+i} | Q_{(k-1)m+i}, Q_{km+i}) \\ &= I(C_{(i),k}; P_{(i),k} | \hat{\mathbf{C}}_{(l)} = \mathbf{C}_{(l)}, l = 1, \dots, i-1), \end{aligned}$$

where the second equality requires that window parameter $w \geq m$. Finally, dropping the subscript k we have $\tilde{R}_i = R_i$ from (2). Thus, the discrepancy between I_s and $R_{av,m}$ vanishes asymptotically as m increases. □

As an example, we examine the PR channel known as the dicode channel in AWGN, which is characterized by

$$y_k = \frac{1}{\sqrt{2}}(x_k - x_{k-1}) + n_k$$

where $x_k \in \{\pm 1\}$, and n_k is white Gaussian noise with mean zero and variance σ^2. In this case, our APP detector returns log-APP ratios

$$L_{(i),k} = \log \frac{P(C_{(i),k} = 1 | \mathbf{Y}_{(k-1)m+i-w}^{(k-1)m+i+w})}{P(C_{(i),k} = -1 | \mathbf{Y}_{(k-1)m+i-w}^{(k-1)m+i+w})}.$$

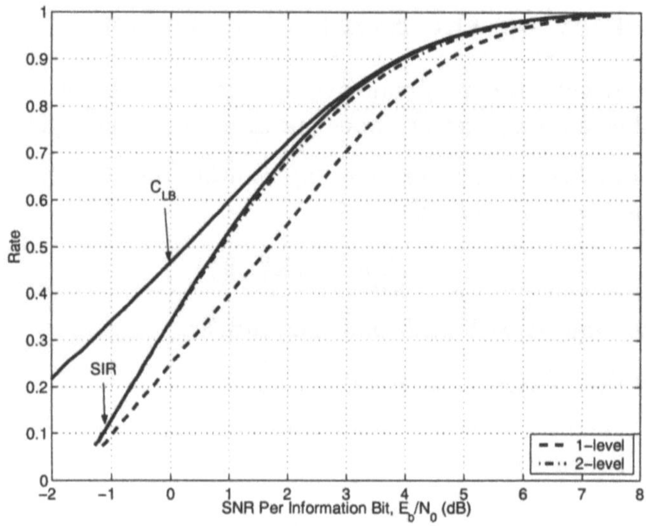

Figure 3: MLC Achievable Rates on the Dicode Channel.

Calculating the optimal component code rates requires determining the conditional densities, $f(L_{(i),k} = l|C_{(i),k} = c) = f_{(i)}(l|c)$, for the log-APP ratios. In general, these cannot be expressed in closed form. Therefore, we approximate them by computing a histogram of outputs from a simulated APP detector at each stage of decoding.

In Figure 3, we show achievable rates for $m = 1, 2$ plotted against the SIR and a lower bound on capacity, C_{LB} [2]. For $m = 2$, the gap from the SIR is less than 0.2 dB, and at high rates this is sufficient for approaching the capacity (bound). On the so-called EPR4 channel with AWGN, with input-output relation given by

$$y_k = \frac{1}{2}(x_k + x_{k-1} - x_{k-2} - x_{k-3}) + n_k,$$

a similar gap of roughly 0.2 dB was achieved with only $m = 3$ levels [2].

4 Component Code Design

We now address the problem of code design for ISI channels at high-rates, moderate rates, and asymptotically low rates. In the previous section, we showed that MLC with memoryless codes can achieve the SIR at all rates. At high rates, this is sufficient because the SIR is close to capacity, while other, possibly hybrid, techniques will be required to approach capacity at lower rates.

4.1 Code Design for High Rates

For binary-input PR channels, we can prove the following symmetry properties of the component channels and the corresponding probability density functions of the log-APP ratios. This allows us to apply density evolution [6] to optimize component LDPC codes for the MLC/MSD scheme described above.

Proposition 4.1. *When the inputs are i.i.d. and Bernoulli one-half, the component channels are block symmetric, i.e.,*

$$P(\mathbf{y}_{k-w}^{k+w}|x_k) = P(-\mathbf{y}_{k-w}^{k+w}| - x_k).$$

Corollary 4.1. *The marginals of the log-APP ratio probability density functions exhibit symmetry and exponential symmetry, i.e.,*

(i) $f_{(i)}(l|c) = f_{(i)}(-l| - c)$

(ii) $f_{(i)}(l|1) = e^l f_{(i)}(l| - 1)$.

Using these results, we can optimize the LDPC code ensemble degree-distribution pairs, represented by the polynomials $\lambda(x)$ and $\rho(x)$, to maximize the rate (following [6]), given the approximate densities computed earlier. Figure 4 shows simulated performance results for a 2-level coding scheme with component code rates $R_1 = 0.63$ and $R_2 = 0.77$, which makes the overall rate $R_{av} = 0.7$. The component LDPC codes reflect optimized degree distribution pairs. MLC system block lengths $N = 2 \times 10^4$ and $N = 2 \times 10^5$ were evaluated. The figure also shows the SIR estimate, the 2-level MLC threshold, and the density evolution threshold for optimized ensembles.

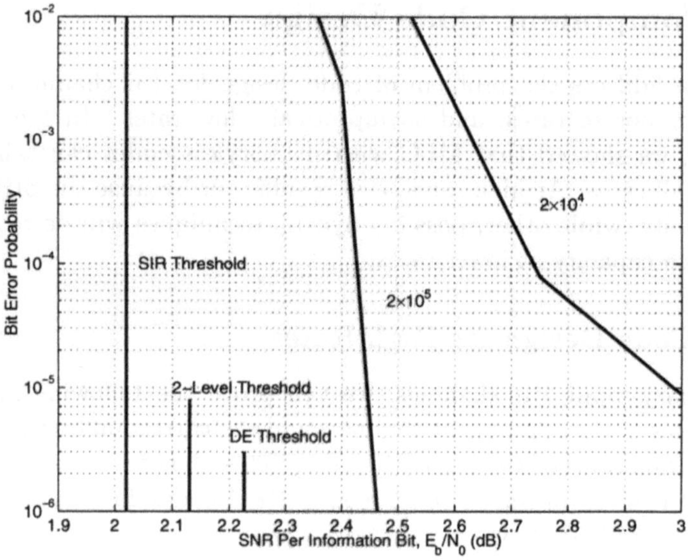

Figure 4: Optimized LDPC codes in a 2-level system at rate 0.7.

4.2 Code Design for Moderate Rates

As can be seen in Figure 3, the SIR does not approximate the capacity lower bound very well below $E_b/N_0 \approx 2$ dB. This is because spectral shaping of the input is needed to approach capacity at low SNR. Since the frequency response of the dicode channel has a peak at $\omega = \pi$, we would expect that the optimal source would generate frequent transitions in symbol values. One simple approach is to concatenate the MLC with the rate 1/2 biphase mapping, defined by $x_i \rightarrow (x_i, -x_i)$. In the decoder, the APP detectors are modified to operate on the trellis representing the biphase-coded dicode channel. In Figure 5, we plot the achievable rates for an $m = 1$ level system, and for the limit as $m \rightarrow \infty$. We see that just one level realizes practically all of the gain and that the achievable rate comes close to the capacity bound, even though the mapping incurs initial rate penalty of 1/2. Other modulation codes or simple mappings can be used in this manner; for instance, in [8] an encoder is designed to closely match the information rate of the optimal Markov source.

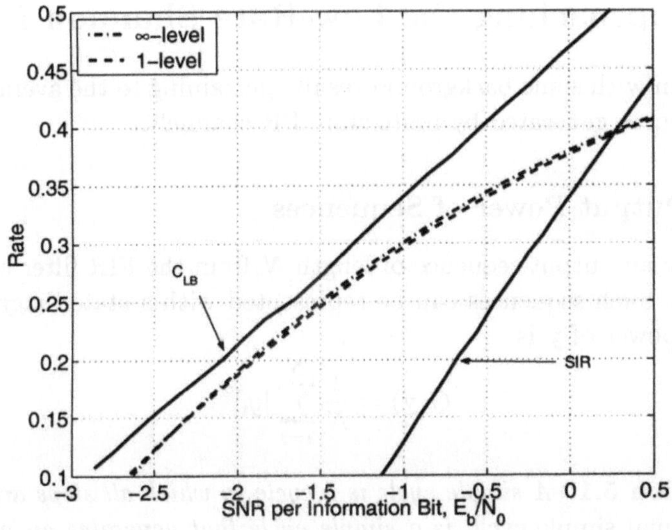

Figure 5: Achievable rates for MLC followed by a biphase mapper.

4.3 Low-Rate Shannon Limit

As the rate approaches zero on a binary-input PR channel, the minimum required SNR per bit approaches a limit

$$\left(\frac{E_b}{N_0}\right)^* = \lim_{\sigma\to\infty} \frac{1}{2\sigma^2 C_{PR}(1/\sigma)}$$

where C_{PR} is the capacity of the binary-input channel represented as a function of the noise variance σ^2. Specifically, E_b/N_0 is the ratio of input power (prior to the PR filter) to noise power, normalized by rate.

To the best of our knowledge, this quantity has not been explicitly determined. In the next section, we characterize this limit for finite-length impulse-response channels and describe a coding scheme that approaches this limit.

5 Approaching the Low-Rate Shannon Limit

We begin with some background results pertaining to the average power of sequences generated by finite-state PR channels.

5.1 Output Power of Sequences

Let $\mathbf{y}$ be an output sequence of length N from the FIR filter of the PR channel. Such sequences can be represented with a state diagram. The output power of $\mathbf{y}$ is

$$G(\mathbf{y}) = \frac{1}{N}\sum_{i=1}^{N} |y_i|^2 .$$

Definition 5.1. *A simple cycle is a cycle in which all steps are unique. An* optimal simple cycle *is a simple cycle that generates an output sequence with power G_{opt}, which is the maximum power over all output sequences generated by simple cycles.*

Proposition 5.1. *For any finite length cycle in the state diagram, the corresponding output $\mathbf{y}_c$ satisfies $G(\mathbf{y}_c) \leq G_{opt}$.*

Proof. (Sketch) Every cycle is composed of simple cycles. Since the linear combination of their output powers (each one weighted by the fraction of total length) gives the total power, this total cannot exceed G_{opt}. □

Corollary 5.1. *For any $\epsilon > 0$, there exists an n_0 such that, for any $\mathbf{y}$ with length $N > n_0$.*

$$G(\mathbf{y}) < G_{opt} + \epsilon. \tag{4}$$

Proof. (Sketch) Any path through the graph can be closed on itself in at most ν steps. Therefore, there exists a cycle with an output sequence $\mathbf{y}'$, where $\mathbf{y}'$ is the concatenation of $\mathbf{y}$ and $\mathbf{t}$ (of length at most ν). The proposition then bounds the output power of this concatenation according to

$$NG(\mathbf{y}) + \nu G(\mathbf{t}) \leq G_{opt}(N + \nu),$$

from which we can derive (4). □

5.2 Coding Near the Low-Rate Shannon Limit

The following two results characterize the E_b/N_0 threshold for reliable communication at asymptotically low rates. The proof of the first theorem demonstrates a family of codes that approach the limit.

Theorem 5.1. *For a binary-input (±1) PR channel, let G_{opt} be the output power of an optimal simple cycle in the state diagram. Reliable communication can be achieved for any value $E_b/N_0 > \ln 2/G_{opt}$.*

Proof. (Sketch) Let $\mathbf{y}_{opt}$ be any sequence which achieves G_{opt}, and has some length L.

Consider a modulation scheme for the *Gaussian channel* where we only send one of two sequences: $\mathbf{y}_{opt}$ and $-\mathbf{y}_{opt}$. We can then achieve any $R < \frac{1}{L}C_b(\sqrt{G_{opt}}/\sigma)$, where C_b is the binary-input Gaussian channel capacity.

Note that the PR channel can be transformed into this system. Specifically, before sending the inputs which generate $\pm\mathbf{y}_{opt}$, we send a prefix (of length ν) to put the channel into the appropriate state. In this case, we incur a rate penalty of $1/(L+\nu)$, and thus we can achieve any

$$\frac{E_b}{N_0} > \lim_{\sigma\to\infty} \frac{(L+\nu)}{2\sigma^2 C_b(\sqrt{LG_{opt}}/\sigma)} = \frac{(L+\nu)\ln 2}{LG_{opt}}. \tag{5}$$

(See [10, problem 4.14] for the evaluation of this limit). We can bring this limit arbitrarily close to $\ln 2/G_{opt}$ by modulating with M repetitions of $\mathbf{y}_{opt}$, letting $M \to \infty$. □

Theorem 5.2. *Reliable communication can be achieved only if $E_b/N_0 > \ln 2/G_{opt}$.*

Proof. (Sketch) This is a consequence of Corollary 5.1, since the gain realized from the filter is limited to G_{opt}, asymptotically. Equivalently, we have a Gaussian channel for which the power is constrained to G_{opt}. Therefore, reliable communication implies that

$$E_b/N_0 > \left[\sigma^2 \log_2(1 + G_{opt}/\sigma^2)\right]^{-1},$$

which is merely an application of the converse coding theorem on the Gaussian channel [10]. This expression also monotonically decreases to $\ln 2/G_{opt}$ as $\sigma \to \infty$. □

Channel	$h(D)$	G_{max}	G_{opt}	$\left(\frac{E_b}{N_0}\right)^*$ (dB)	Gap (dB)
PR4	$(1-D^2)/\sqrt{2}$	2	2	-4.69	0
EPR4	$(1-D)(1+D)^2/2$	64/27	2	-4.69	0.75
E^2PR4	$(1-D)(1+D)^3/\sqrt{10}$	27/10	12/5	-5.39	0.51

Table 1: Low-rate Shannon limit for some PR channels.

We remark that, with Gaussian inputs, water-filling arguments show that we can achieve reliable communication at E_b/N_0 as low as $\ln 2/G_{max}$, where G_{max} is the square of the peak magnitude of the channel frequency response. However, for some channels, this value cannot be achieved with binary signaling. See Table 1 in the next section for examples.

We also note that these results can be easily extended to the case of any finite alphabet of the form $\{-a_n \ldots - a_1, a_1 \ldots a_n\}$, $a_i \in \mathbb{R}$.

5.3 Numerical Results

In Section 5.2, we characterized the low-rate Shannon limit for a binary-input PR channel. Determining this value numerically involves finding the optimal simple cycle for a given channel graph. For several PR channels of interest, this can be accomplished by application of dynamic programming techniques.

Table 1 shows results for three commonly used PR channel models. The normalized impulse response is denoted $h(D)$, and the values of G_{max} and G_{opt} are shown. The minimum E_b/N_0 corresponding to the low-rate Shannon limit for binary-inputs is then given, along with the gap to the Gaussian-input capacity.

6 Conclusion

In this paper we examined several methods for approaching the capacity of partial-response (PR) channels. We began by showing that multi-level coding (MLC) and multistage decoding (MSD) could theoretically achieve the symmetric information rate (SIR) of finite-state intersymbol-interference (ISI) channels. We then explored the application of this scheme to PR channels. In the high rate region, we found that the MLC/MSD architecture could be used to approach capacity (bounds). We showed that at lower rates, the MLC/MSD scheme was not adequate, and we illustrated an improved hybrid coding method involving the concatenation of the MLC encoder with spectral shaping codes. Finally, we determined the minimum signal-to-noise ratio per bit required for reliable communication at asymptotically low rates, and demonstrated a simple coding technique that achieves it.

References

[1] D. Arnold and H. Loeliger, "On the information rate of binary-input channels with memory," in *Proceedings IEEE International Conference on Communications*, (Helsinki, Finland), pp. 2692–2695, June 2001.

[2] H. D. Pfister, J. B. Soriaga, and P. H. Siegel, "On the achievable information rates of finite state ISI channels," in *Proceedings IEEE Global Telecommunications Conference*, (San Antonio, US), IEEE, November 2001.

[3] A. Kavčić, "On the capacity of markov sources over noisy channels," in *Proceedings IEEE Global Telecommunications Conference*, (San Antonio, US), IEEE, November 2001.

[4] P. Vontobel and D. M. Arnold, "An upper bound on the capacity of channels with memory and constraint input," in *Proceedings of the IEEE Information Theory Workshop*, (Cairns, Australia), pp. 147–149, IEEE, September 2001.

[5] W. Hirt, *Capacity and Information Rates of Discrete-Time Channels with Memory*. PhD thesis, E.T.H., Zurich, Switzerland, 1988.

[6] T. J. Richardson, M. A. Shokrollahi, and R. L. Urbanke, "Design of capacity-approaching irregular low-density parity-check codes," *IEEE Transactions on Information Theory*, vol. 27, pp. 619–637, January 2001.

[7] A. Kavčić, "Optimized LDPC codes for partial response channels." Submitted to ISIT, 2002.

[8] A. Kavčić, "Matched information rate codes for binary ISI channels." Submitted to ISIT, 2002.

[9] H. Imai and S. Hirakawa, "A new multilevel coding method using error-correcting codes," *IEEE Transactions on Information Theory*, vol. 23, pp. 371–377, May 1977.

[10] R. J. McEliece, *The Theory of Information and Coding.* Reading, Massachusetts: Addison-Wesley, 1977.

MIX
Papier aus verantwortungsvollen Quellen
Paper from responsible sources
FSC® C105338

If you have any concerns about our products,
you can contact us on
ProductSafety@springernature.com

In case Publisher is established outside the EU,
the EU authorized representative is:
Springer Nature Customer Service Center GmbH
Europaplatz 3, 69115 Heidelberg, Germany

Printed by Libri Plureos GmbH
in Hamburg, Germany